A Textbook of Arthropod Anatomy

A TEXTBOOK OF

ARTHROPOD ANATOMY

BY R. E. SNODGRASS

Collaborator, Bureau of Entomology and Plant
Quarantine, U.S. Department of Agriculture

Comstock Publishing Associates

A DIVISION OF CORNELL UNIVERSITY PRESS

ITHACA, NEW YORK, 1952

Open access edition funded by the National Endowment for the Humanities/Andrew W. Mellon Foundation Humanities Open Book Program.

First paperback printing 2019

ISBN 978-1-5017-4079-4 (pbk.: alk. paper)
ISBN 978-1-5017-4080-0 (pdf)
ISBN 978-1-5017-4081-7 (epub/mobi)

Librarians: A CIP catalog record for this book is available from the Library of Congress

PREFACE

THE arthropods are a group of related invertebrates; arthropodists, for the most part, are a group of unrelated vertebrates. Each specialist, whether an entomologist, a myriapodist, a carcinologist, or an arachnologist, works in his own particular field and gives little thought to the work of specialists in other fields. As a result, the relationships of the various kinds of arthropods to one another are by some ignored, while others propose theories of arthropod phylogeny based on an insufficient knowledge of the anatomy of the arthropods in general. Since the insects are conceded to be at the top of the arthropod line of evolution, entomologists in particular have been concerned with the ancestry of insects. Some have sought to derive the insects from myriapods, others from symphylans, others from crustaceans, while some would carry the insects back to the trilobites, or even in a direct line of descent to the annelid worms. Clearly, all these claims of insect ancestry cannot be true. The writer, therefore, himself for many years an entomologist, has attempted to evaluate the various theories of insect origin by browsing around in the fields of other specialists. The final result has been the disconcerting conclusion that the facts of arthropod structure are not consistent with any proposed theory of arthropod interrelationships. The investigation, however, has added much to the writer's own information about the comparative anatomy of the arthropods, and this information is set forth in the following chapters in the hope of making a general knowledge of the arthropods more readily available to students who expect to be specialists in one arthropod group or another. Just as a cone sits best on its base, so specialization should taper upward from a broad foundation.

R. E. SNODGRASS

U.S. National Museum, Washington, D.C.

CONTENTS

A Textbook of Arthropod Anatomy

INTRODUCTION

IN THEIR fundamental organization the arthropods show that they are segmented animals related to the segmented annelid worms and the Onychophora. Though the body of the onychophoran is not segmented in the adult stage, it is fully segmented during embryonic development, and in many respects the structure of the mature animal retains clear evidence of its primitive metamerism. The Annelida, the Onychophora, and the Arthropoda, therefore, may rightly be classed together as members of the superphylum Annulata (Articulata of Cuvier). The interrelationships of the three groups, however, is a matter on which there is no specific evidence, though plenty of theoretical opinion.

The arthropods, the onychophorans, and the polychaetes among the annelids have segmental locomotor appendages. The appendages of the aquatic polychaetes are bilobed, chaeta-bearing flaps along the sides of the body, known as *parapodia,* which normally serve for swimming, but can be used for progression on solid surfaces when the worms are taken out of the water. The polychaete parapodium, therefore, has been much exploited as the prototype of the arthropod leg. A parapodium, however, is a lateral appendage having no resemblance to the jointed limbs of the arthropods, and moreover, the Polychaeta are a specialized group of annelids, so highly individualized, in fact, that it is hardly to be supposed they have ever produced anything else than polychaete worms. Furthermore, since the antiquity of the polychaetes is not known, it is quite possible that the arthropod progenitors antedated them.

The locomotor appendages of the terrestrial Onychophora are short

1

legs in the position on the body of arthropod legs; they are not truly jointed, but they are transversely ringed, and some of the distal rings are individually musculated, so that the leg might be said to have an incipient segmentation. In its embryonic origin the onychophoran leg and the leg rudiment of an arthropod are both mere lobelike outgrowths of the body wall. The arthropod leg, therefore, would seem to be much more probably related in its origin to the onychophoran leg than to the annelid parapodium, and neither the structure nor the development of the parapodium gives any reason for believing that the onychophoran leg originated from a parapodium. In other respects also the Arthropoda have characters that are more onychophoran than annelidan. The excretory organs of the Onychophora are direct derivations of the segmental coelomic sacs, and the coelomic excretory organs of the arthropods have essentially the same structure. The onychophoran reproductive organs are composite coelomic sacs with a single pair of coelomic exit ducts, and furnish the basic pattern of structure for the reproductive system of the arthropods. On the whole, therefore, the arthropods appear to be much more closely related to the Onychophora than to the Polychaeta. The annelid connections of the two groups, then, must be with some primitive member of the Annelida, far more generalized than the modern polychaetes or their immediate ancestors.

The oldest known arthropods are the fossil trilobites, but certainly a trilobite has little resemblance to any wormlike animal, either an onychophoran or an annelid. Its appendages are fully segmented ambulatory legs; the body is flattened, the integument shell-like, the body segments are grouped in well-defined tagmata, in the first of which the component segments are highly integrated, and the animal has filamentous antennae and compound eyes. To believe that a trilobite is a direct descendant from a polychaete worm requires much faith in a theory, and it is not any easier to visualize the descent of a trilobite from an onychophoran. In short, the only logical concept of annulate interrelationships is that, from some primitive, segmented wormlike animal, there evolved, on the one hand, a branch culminating in the parapodia-bearing polychaetes and, on the other, a lobe-legged form (lobopod) with coelomic excretory and reproductive organs discharging through coelomaducts, which soon split into the progenitors of the Onychophora and the Arthropoda, the one developing a sclerotized integument and jointed

2

legs, the other remaining soft-skinned and wormlike, and sufficiently accommodated for locomotion by its primitive leg stumps.

Modern Onychophora are all terrestrial animals, but there are fossils of onychophoralike forms from Cambrian and Pre-Cambrian times that must have been aquatic. A wormlike animal with legs would be much better fitted for crawling out of the water and becoming a land animal than would a parapodia-bearing polychaete, but probably no animals were able to live on land before the Devonian, and the trilobites were fully developed at the beginning of the Cambrian. Hence, a terrestrial onychophoran was not the ancestor of the arthropods. Whenever an aquatic onychophoran did take to the land, it acquired tracheal organs of a simple kind for the respiration of air. The terrestrial Onychophora, therefore, might be said to have invented the mode of breathing by means of ingrowths of the integument; the idea was adopted by later land arthropods, but was applied in different ways, so that the presence of tracheae is no evidence of close relationship between different tracheate forms. The pretrilobite history of the arthropods probably never will be known, since it has been erased from the records of Pre-Cambrian time, but there can be little doubt that arthropod forms were already well differentiated before the Cambrian, since the trilobites ended their career in the Permian as trilobites, and among modern arthropods only the horseshoe crab has any resemblance to them.

The arthropods are so named because they have jointed legs, a feature clearly not distinctive of them, but the segmentation of the legs taken in combination with segmentation of the body might serve as a definition for any arthropod that preserves its ancestral form. Body segmentation is fundamentally muscle segmentation (i.e., mesoderm segmentation), with correlated segmentation of the nervous system, giving a more efficient mechanism for bodily movement than that possessed by an unsegmented animal; legs are adjuncts adding further locomotor efficiency. Body segmentation is practical in a soft-skinned animal, but elongate flexible legs would be of little use as locomotor organs. The arthropod type of limb, therefore, must have been developed in an animal with a hard integument, since the jointing of the appendages into individually musculated segments thus became possible and followed as a mechanical necessity. Since the segmentation of the legs is essentially the same in all the arthropods, from trilobites to insects, the form and segmentation of the limb must have been established in some early ancestor of the group.

The principal evolutionary changes in the body of the arthropods have been a grouping of the segments into different body regions, or *tagmata,* accompanied often by a fusion of the grouped segments, or of some of them. Thus there have been produced the characteristic features of the several arthropod classes, as the trilobites, the arachnids, the crustaceans, the myriapods, and the insects. It is to their numerous jointed limbs, however, that the arthropods owe the major part of their evolution and the multitude of activities of which they are capable. The large number of legs were at first all ambulatory in function, but it was soon found that they were not all needed for locomotion. Hence the great structural and functional diversification of the limbs in modern arthropods; no other animals carry on their bodies such an assortment of tools. The possession of tools implies the ability to use them, and their use involves a high degree of development in the nervous system, which, in the arthropods, is expressed in a marvelous development of instinctive intelligence.

Taxonomically the arthropods are divisible into eleven well-defined classes, namely, the Trilobita, the Xiphosurida, the Eurypterida, the Pycnogonida, the Arachnida, the Crustacea, the Chilopoda, the Diplopoda, the Pauropoda, the Symphyla, and the Hexapoda, representatives of which will be the subjects of the following chapters. In a larger way the classes are separated into three major groups: the Trilobita, the Chelicerata, and the Mandibulata. The Trilobita have filamentous antennules, but their distinguishing feature is the uniformity in structure of the other appendages, which are all ambulatory legs. The Chelicerata lack antennules, and take their name from the pincerlike structure of the first pair of ventral appendages, known as *chelicerae;* they include the Xiphosurida, Eurypterida, Pycnogonida, and Arachnida. The Mandibulata have antennules, but their important feeding organs are the jawlike second pair of ventral appendages, the *mandibles.* The first pair are suppressed except in Crustacea. The mandibulate arthropods are the Crustacea, Chilopoda, Diplopoda, Pauropoda, Symphyla, and Hexapoda. Again, the Xiphosurida and Eurypterida are generally classed together as Merostomata or, together with the Trilobita, Pycnogonida, and Arachnida, as Arachnomorpha. The chilopods, diplopods, pauropods, and symphylans were formerly combined in the Myriapoda, but modern zoologists do not generally recognize the myriapods as a natural group, though the chilopods are often termed the Myriapoda opisthogoneata, and the other three the Myriapoda progoneata, because of the different position of the genital opening.

Finally there are two small groups of animals, the minute Tardigrada and the parasitic Pentastomida, that are often classed with the arthropods, or thought to be somehow related to them, because they have short stumps of legs. However, since the taxonomic affinities of these creatures are very uncertain, they will not be included in the present text.

THE TRILOBITA

THE trilobites (fig. 1) are arthropods of particular interest because of their great antiquity. They appear on the geological scene at the very beginning of the Paleozoic already fully developed as trilobites; they flourished during the Cambrian and Silurian periods, and continued in diminishing numbers through the Carboniferous into the Permian. Several thousand species have been described, referred to numerous genera and many families in four recognized orders. Such highly organized and diversified animals, therefore, must have had a long evolutionary history in Pre-Cambrian times, though the rocks of this period have so far furnished no evidence of their existence. The trilobites may truly be said to be the oldest of known arthropods, and in some respects they are the most generalized of known arthropods, but if the arthropods have been developed from a segmented, wormlike progenitor provided with jointed legs, there is a vast gap between the trilobites and their vermiform ancestors. A trilobite is in no sense a primitive arthropod, and, notwithstanding all the claims that have been made in favor of a trilobite ancestry for the other arthropods, it is not probable that any other group of arthropods was derived directly from the trilobites. Any specialized form of animal produces only more specialized and diversified forms of its own kind.

Very soon after the beginning of the Cambrian the trilobites are accompanied in the geologic record by representatives of the Crustacea, the Eurypterida, and the Xiphosurida, but also there are various fossil arthropods found in the Cambrian rocks that cannot accurately be classified in these groups, but which appear to be related to them.

6

Hence, it is to be supposed that the arthropods present in Cambrian and Ordovician periods represent lines of descent from common progenitors that lived far back in the immeasurable period before the Cambrian, and which, therefore, have little chance of ever being known. It is curious, however, that in the later Paleozoic rocks arachnids, myriapods, and insects appear as fully formed animals

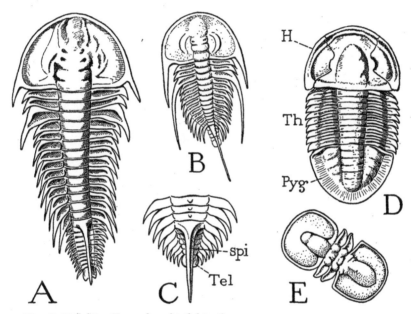

Fig. 1. Trilobita. Examples of trilobite forms.

A, *Olenellus vermontanus* Hall (from Walcott, 1910). B, *Olenellus gilberti* Meek (from Walcott, 1910). C, *Schmidtiellus mickwitzi* Schmidt (from Walcott, 1910). D, *Asaphiscus wheeleri* Meek (from Walcott, 1916). E, *Peronopsis montis* (Matthew) (from Walcott, 1908).

H, head, or prosoma; *Pyg*, pygidium; *spi*, spine; *Tel*, telson; *Th*, thorax.

of their kinds, while there is nothing to show from what they came during the earlier Devonian, Ordovician, and Cambrian times, when trilobites, eurypterids, xiphosurids, and crustaceans were well preserved. Fossilization, however, depends largely on the presence of a hard skeleton, so that it is likely that animals first appear as fossils only when they have acquired a sufficiently resistant integument, and even then they must meet with favorable external conditions for fossilization, while, finally, fossils themselves may be utterly destroyed by subsequent metamorphosis of the containing rocks.

The form and structure of the trilobites (fig. 2) show clearly that in these ancient animals the fundamental arthropod organization was already fully developed and had attained a specific type of specialization. Only in the lack of differentiation in the postoral appendages are the trilobites generalized, but their specialization is of a relatively simple kind that fitted them for life on the ocean bottom in shallow water along the shore, where most of them lived probably in the manner of the modern *Limulus,* though some species are thought to have been pelagic, or even deepwater inhabitants. Lacking jaws or grasping organs of any kind other than the legs, the trilobites could not have obtained active prey by raptatory methods, and it has been thought that probably they were mud feeders, but animals with long filamentous antennae can hardly be supposed to have made a practice of burrowing in mud or sand. In some species the projecting mesal ends of the coxae (fig. 2 B) are armed with spines, which fact suggests that the coxal lobes had some use in the obtaining of food; Raymond (1920) says, "The primary function of these spiny lobes of the coxa was doubtless the gathering and preparation of food, and carrying it to the mouth by passing it forward from one to the next." On the other hand, since the coxal lobes do not meet in the middle line, and the spines are not well developed in all species, Störmer (1944) thinks it unlikely that they functioned as jaws. In any case, whatever may have been the food of the trilobites, and however they obtained it, the great numbers of the animals would indicate that they had an ample food supply within their reach. Certainly worms of various kinds as well as other soft-bodied creatures living in the mud or sand along the ocean shores were abundant in both Cambrian and Pre-Cambrian times. It would seem, in fact, that a trilobite should be quite fit to live under modern conditions, and paleontologists have no positive evidence to account for their early extinction.

General Structure of a Trilobite

Since few perfectly preserved specimens of trilobites are known, it is not possible to give a full description of the trilobite structure in any one species, but inasmuch as the details of structure have been carefully studied in different species, we can reconstruct diagrammatically, as given in figure 2, the general form and make-up of a typical representative of the group.

8

The body of a simple trilobite of usual form is oval and dorso-ventrally flattened (figs. 1 D, 2 A); it is divided transversely into three parts known as the *head* (H), the *thorax* (*Th*), and the *pygidium* (*Pyg*). The thorax and pygidium together, however, may be said to constitute the *body* as distinguished from the *head*, but,

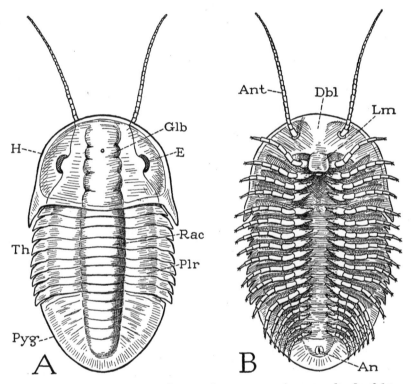

Fig. 2. Trilobita. Diagrams showing the structure of a generalized trilobite. A, dorsal. B, ventral.

An, anus; *Ant*, antenna; *Dbl*, doublure; *E*, compound eye; *Glb*, glabella; *H*, head or prosoma; *Lm*, labrum or hypostome; *Plr*, pleura (not the pleuron of other arthropods); *Pyg*, pygidium; *Rac*, rachis or axis; *Th*, thorax.

since the so-called head bears the first four pairs of legs, the terms *prosoma* and *opisthosoma* are preferable names for the two principal parts of the animal. The prosomatic head is unsegmented in the adult, though it may show evidence of coalesced segments; the thorax is completely segmented; and the pygidium is clearly composed of united segments. In the earlier trilobites the pygidial region is fully segmented (fig. 1 A, B, C). Extending lengthwise usually through

9

the three parts is a rounded, median elevation, which is flanked by wide, flattened or decurved lateral areas. On the head the median elevation is known as the *glabella* (fig. 2 A, *Glb*), the lateral parts as the *cheeks*, or *genae;* on the body the median elevation is termed the *axis,* or *rachis* (*Rac*), the lateral parts the *pleurae* (*Plr*). These terms and others to be introduced later are those current in trilobite taxonomy, and the parts they connote have no necessary relation to those similarly named in other arthropod groups. The name *trilobite* is derived from the apparent *lengthwise* division of the animal into three parts.

Though the trilobite as seen from above (fig. 2 A) looks like no other animal with which we are familiar, the undersurface (B) at once shows that the trilobite is an arthropod; on each side is a long row of jointed *legs,* and from the anterior part of the head there projects forward a pair of slender, filamentous *antennae* (*Ant*). Behind the bases of the antennae and extended over the mouth is a median lobe (*Lm*), such as all arthropods have and which is commonly known as the *labrum,* though students of trilobites have generally called it the *hypostome.* The mouth presumably is covered by the labrum, and behind it is a small metastomal lobe. Of the legs, the first four pairs pertain to the head, the others to the thorax and pygidium, there being a pair for each body segment except the last. It will be noted that the legs are attached to the body at each side of a narrow median space, and that the pleural areas of the dorsum (fig. 4 A, *D*) are inflected on the undersurface to the bases of the legs. The inflected ventral surfaces laterad of the leg bases constitute the *doublure* (*Dbl*), which includes also the inflected undersurface of the head (figs. 2 B, 3 K).

A cross section through the thorax of a trilobite (fig. 4 A) clearly shows that the principal body cavity of the animal is in the median part, or rachis (*Rac*), which is strongly convex on the back, and that the so-called pleurae are merely flat hollow extensions of the body segments over the appendages. Since various other flattened arthropods have a similar structure, the "three-lobed" character is not distinctive of the trilobites. The special feature of the trilobites is the uniform, leglike structure of all the appendages except the antennae, in which respect the trilobites are more generalized than any modern arthropods, since in all of the latter at least some of the appendages are modified and specialized for purposes other than that of locomo-

10

tion; the trilobites apparently did not even have appendages that served specifically as jaws. Details of the leg structure will be described in a later section.

The Head, or Prosoma

The head section, or prosoma, of a typical adult trilobite (fig. 2 A, H) is usually somewhat semicircular in outline, with its rounded anterior and lateral margins produced posteriorly in a pair of large *genal spines*. Between the bases of the spines the head is directly attached to the body by its transverse posterior margin, without the intervention of a neck. The dorsal surface of the head, as already noted, generally presents a median, elevated glabellar area (*Glb*) and broad, lateral genal areas, but in many species the head is covered by a perfectly smooth, rounded, shieldlike plate. The features of the head given in the following description, and illustrated diagrammatically in figure 3 L, have been made out from a study of many specimens of more generalized trilobites; the beginning student, however, is likely to see little trace of them in ordinary museum examples, owing partly to the imperfection or corrosion of the specimens, but also to the fact that the characters themselves were suppressed in the later evolution of the trilobites.

The grooves that separate the glabella from the genae are known as the *dorsal furrows* (fig. 3 L, *df*). A subdivision of the glabella into five successive parts, of which the first is the *frontal lobe* (*frl*), may be indicated by lateral notches in the dorsal furrows, or by imperfect transverse grooves in the glabellar surface. Each genal area is divided lengthwise by a *facial sulcus*, or "suture," before and behind the eye (*afs*, *pfs*), which separates it into a median part called the *fixed cheek*, or *fixigene* (*Fg*), and a lateral part distinguished as the *free cheek*, or *libragene* (*Lg*). The free cheeks are produced posteriorly into the genal spines (*gspi*). An *ocular ridge* (*er*) goes from each side of the frontal lobe to the eye. The entire median part of the head shield between the facial sulci, including the glabella and the fixed cheeks, is termed the *cranidium* (*Crn*). A *marginal furrow* (*mf*) surrounds the head inside a narrow *border area* (*b*). Most trilobites have a pair of large, lateral *compound eyes* (*E*) on the mesal parts of the free cheeks, and above each eye a protective *palpebral lobe* (*pbl*) on the edge of the fixed cheek. On the anterior part of the glabella in some species there is present a small tubercle,

11

which has been regarded as a median eye, but its ocular nature is questionable. A complete terminology for all parts of the trilobite is given by Howell, Frederickson, Lochman, Raasch, and Rasetti (1947).

Inasmuch as the trilobite head carries the first four pairs of legs, it must include at least four primitive postoral somites, and the glabellar grooves of the adult evidently represent the primary inter-segmental lines of the head segments. The best understanding of the adult head has been derived from studies, such as those of Raw (1925), Warburg (1925), Lalicker (1935), Störmer (1942, 1944), and others, on specimens of very young trilobites in successive stages of development. All writers are in close agreement as to the visible facts, though they differ somewhat in their interpretations.

The youngest-known developmental stage of a trilobite is a minute oval thing, from half a millimeter to a millimeter and a half in length, and is termed the *protaspis* (fig. 3 A, E). The major part of the protaspis represents only the head of the adult trilobite. On the back of the youngest specimens a median glabellar elevation is already differentiated from wide lateral genal areas and very soon becomes divided transversely into five primary subdivisions, of which the last four (*I–IV*) represent the somites of the four pairs of legs carried by the adult head; faint intersegmental lines may be seen extending laterally in the genal areas (E). The first glabellar subdivision of the protaspis includes the area of the frontal lobe of the adult (J, *frl*) and its lateral extensions curving posteriorly around the anterior margin, which eventually become the free cheeks bearing the compound eyes. This anterior, or acronal, section of the larva may be termed the *acron* (A, *Acr*), though the term has been used with various other applications. Its underfolded anterior part forms the doublure of the head, on which the antennae and labrum of the adult are situated (fig. 2 B).

The preoral antennae of the trilobites are clearly the first antennae, or antennules, of other arthropods. The four postoral pairs of legs on the head, then, should reasonably be supposed to correspond with the first four postantennular appendages of other arthropods, which primarily arise behind the mouth, these appendages being the chelicerae, the pedipalps, and the first two pairs of legs in the Chelicerata, or the second antennae, the mandibles, and the two pairs of maxillae in the Mandibulata. If the acron of the trilobite

12

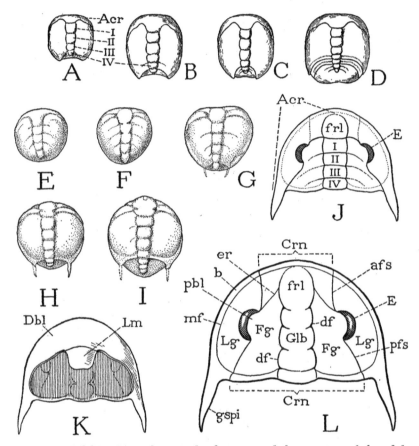

Fig. 3. Trilobita. Postembryonic development and the structure of the adult head, or prosoma.

A–D, developmental stages of *Liostracus linnarssoni* Brogger (from Warburg, 1925). E–I, developmental stages of *Olenus gibbosus* (Wahlenberg) (from Störmer, 1942). J, diagram of structure of mature trilobite head, dorsal. K, diagram of undersurface of head. L, diagram of dorsal surface of head, with parts named according to Howell, Frederickson, Lochman, Raasch, and Rasetti (1947).

Acr, acron, anterior unsegmented part of larva; *afs,* anterior facial sulcus; *b,* border area; *Crn,* cranidium; *df,* dorsal furrow; *E,* compound eye; *er,* ocular ridge; *Fg,* fixed cheek or fixigene; *frl,* frontal lobe; *Glb,* glabella; *gspi,* genal spine; *Lg,* free cheek or libragene; *mf,* marginal furrow; *pbl,* palpebral lobe; *pfs,* posterior facial sulcus; *I–IV,* primary postacronal segments of larva.

13

larva is a single segment, the trilobite head is, therefore, composed of five primary somites, but the frontal lobe is sometimes divided into two parts. According to Störmer (1942), in the fourth protaspis stage of *Olenus gibbosus* (fig. 3 H) there appears in the sides of the frontal lobe a pair of pits, which eventually run together in a transverse groove, so that the frontal lobe becomes divided secondarily in development into two segments, which Störmer designates an antennal (i.e., antennular) segment and a preantennal segment. Henriksen (1926) and other writers, however, have contended that the posterior division of the frontal lobe is the segment of a suppressed pair of appendages representing the second antennae of Crustacea, and that the four persistent appendages are to be identified with the mandibles, the two pairs of maxillae, and the first maxillipeds. This contention that the trilobites have lost a pair of appendages, of which there is no concrete evidence, seems to be purely presumptive. "Such conclusions," Störmer asserts, "have apparently been too much influenced by the current opinion of a crustacean nature of the trilobites." In other words, a lost "second antenna" has been arbitrarily injected in order to make the trilobites conform with their supposed crustacean descendants. However, the trilobite, as it is, conforms with either the Chelicerata or the Crustacea on the assumption that its first legs represent the chelicerae of the former and the second antennae of the latter.

From the development of the protaspis it may be deduced that the pattern of the segmental composition of the adult trilobite head must be approximately that shown diagrammatically at J of figure 3. The oculoantennal part of the head, bearing the eyes dorsally and the antennae ventrally, which for convenience is here called the acron, has extended posteriorly on the sides from the frontal lobe, forming the regions of the free cheeks ending in the genal spines. The median part of the head, including the postfrontal part of the glabella and the fixed cheeks, is the region of the four postoral leg-bearing somites, but the intersegmental lines become obliterated in the adult except for remnants of them on the glabella.

The Body, or Opisthosoma

Though the five primary segments of the young protaspis all go into the formation of the adult head, there is a very small region behind the last head segment that is destined to generate the body.

14

In the developing protaspis a stage is soon reached when a new segment appears behind the last head segment (fig. 3 B, G), and this segment is followed by others successively formed between it and the extending end of the body (C, D, H, I). These segments are the beginning of the series of body segments. The pygidial and thoracic segments, therefore, are *secondary somites* generated in the usual teloblastic manner from a subterminal *zone of growth,* just as are the secondary somites of the polychaetes and of those modern arthropods that have an anamorphic postembryonic development.

The Thorax— The thoracic segments of the adult trilobite are usually all of similar size and shape, but as between different species they are highly variable in number, since there may be from two to forty or more, though the usual number is perhaps between seven or eight and fifteen. The segments undoubtedly were movable on each other and connected by infolded membranous conjunctivae; probably they were hinged by articulations between the tergal plates, as are the abdominal segments of the crayfish, so as to allow up and down movements, though evidently there could have been little movement in a transverse direction. Many trilobites are found rolled ventrally upon themselves in the manner of certain modern isopods. The elevated median parts of the thoracic terga are distinctly demarked from the long, flattened or decurved pleural lobes on the sides, but, as already noted in cross section (fig. 4 A), the pleural lobes are merely lateral expansions of the body extending out beyond the leg bases. A cross section of the trilobite thorax resembles a cross section of *Limulus* (fig. 7 A), and in each animal the tergal plate, or carapace, must be interpreted as including not only the dorsal integument, but also the doublure (*Dbl*) on the undersurface to the bases of the legs. The projecting ends of the pleural lobes of the trilobites generally form a row of *pleural spines,* which may be uniform in size and shape or varied; in some forms particular spines are greatly elongate (fig. 1 B), in others all may be long and slender. The parts called "pleura" by students of trilobites are not to be identified strictly with the parts so termed in other arthropods. The true sternal area of the trilobite is the narrow ventral space between the leg bases and appears to have been a relatively soft integument.

The Pygidium— The pygidium is composed of a number of segments that are not fully differentiated as they are generated in development, but since in some of the earlier trilobites the body is

15

completely segmented (fig. 1 A), it is probable that the pygidium has been formed in the course of evolution by a secondary union of the posterior body segments. The typical pygidium (figs. 1 D, 2 A, *Pyg*) is a large, smoothly rounded plate equal to the head in size, or sometimes larger than the head. It contains an extension of the rachis, and the pleural areas are marked by indistinct intersegmental lines. The pygidial appendages become successively smaller than those of the thorax (fig. 2 B). The anus (*An*) is situated ventrally in the apical segment of the pygidium, which, therefore, is to be regarded as the telson. In forms in which the pygidial region is segmented, one or more long median spines may project from the dorsum of the anterior pygidial segments (fig. 1 A, B, C). In one genus at least, the terminal segment bears a pair of long, multiarticulate tail filaments. It has been suggested that the pygidium was used as a swimming organ in the manner of the abdomen and "tail fan" of a crayfish.

The Appendages

The appendages of the trilobites are now fairly well known in a number of species; they include the *antennae* and the *legs*. In all the arthropods the antennae, or antennules, appear to be organs of a nature different from that of the other appendages, since in their normal development they never have the form or structure of ambulatory limbs, while, as the trilobites themselves attest, the other appendages, regardless of their form in modern arthropods, undoubtedly were all primarily walking legs. If the antennae were ever evolved from leglike appendages, they must have completed their transformation long before the time of the oldest-known trilobites.

The Antennae— The long, slender, tapering antenna of the trilobite is divided into a large number of short rings, and in appearance resembles the antenna of an orthopteroid insect, or the flagellum of a crustacean antenna, in which the rings are mere annulations without muscles, and therefore are not true segments, as are the musculated subdivisions in the antennae of the myriapods and the entognathous apterygote insects. The antennae of the adult trilobite arise from the doublure of the head anterior to the mouth (fig. 2 B, *Ant*), and are evidently the homologues of the antennules of Crustacea, though they are not branched, and of the antennae of the myriapods and insects.

16

The Legs— The legs of the trilobites are typical, uniramous, segmented arthropod limbs, with long, pinnate epipodites arising from their bases (fig. 4 A). Since each leg has seven clearly marked segments, the segmentation was formerly thought to correspond with that of a crustacean leg, and the epipodite was regarded as an

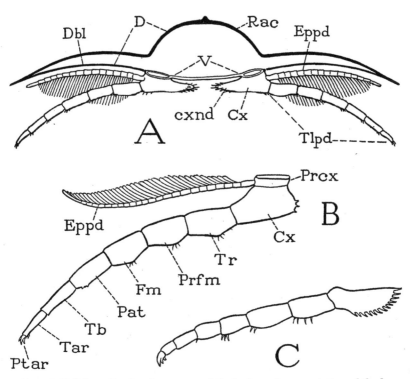

Fig. 4. Trilobita. Sectional structure of the body and segmentation of the legs.
A, diagrammatic cross section of a trilobite, showing attachment of the legs. B, a trilobite leg, with interpretation of segments according to Störmer (1944). C, leg of *Neolenus serratus* (Rominger) (from Raymond, 1920).
Cx, coxa; *cxnd,* coxal endite; *D,* dorsum; *Dbl,* doublure of dorsum; *Eppd,* epipodite; *Fm,* femur; *Pat,* patella; *Prcx,* precoxa; *Prfm,* prefemur; *Ptar,* pretarsus; *Rac,* rachis; *Tb,* tibia; *Tar,* tarsus; *Tlpd,* telopodite; *Tr,* trochanter; *V,* venter.

exopodite, three small claws on the end of the segment being treated as mere apical spines. More recently, Störmer (1944) has shown that there is at the base of the leg a small ring by which the appendage is attached to the body, and this proximal ring Störmer regards as the true basal segment of the limb, which he calls a "precoxa," or "subcoxa" (B, *Prcx*), the large segment following being the coxa. A

17

petrified specimen, however, can hardly be supposed to give conclusive evidence that a basal ring of the appendage is a true segment in the sense of having been an independently movable section of the limb when the animal was alive, and it seems very improbable that the body muscles that moved the leg as a whole should have been attached on such a relatively small piece. In modern arthropods the basal muscles of an appendage are always attached on the coxa, unless the coxa itself has become immovable, but the base of the coxa is often set off as a proximal ring, or *basicoxite*, by an external groove that forms internally a strengthening ridge, or shelf, for the muscle attachments, and the basal ring of the coxa might superficially resemble a small segment. That the "precoxa" of the trilobite leg is a real limb segment, therefore, is open to question.

The rest of the leg beyond the subcoxal ring is divided into seven cylindrical sections (fig. 4 B), and there is no reason for supposing that these parts were not movable on each other, though, lacking any knowledge of the leg musculature, we cannot be so sure they were all true segments. The seventh apparent segment bears the three apical spines or claws, and Störmer, with good reason, has identified the claws as constituting the true end segment of the limb, since a similar three-clawed endpiece of the legs in other arthropods is always found to be a reduced, clawlike, apical segment provided with its own muscles and armed with a pair of lateral claws. A figure of an appendage of *Neolenus serratus* (fig. 4 C) given by Raymond (1920) leaves little doubt of the segmental nature of the apical part of the limb, though Raymond himself describes the claws as "little spines" on the end of the "dactylopodite." The median claw in modern arthropods, however, is the true dactylopodite, of which the lateral claws are secondary outgrowths.

Discounting the segmental nature of the precoxa, then, the trilobite leg appears to be an eight-segmented limb. If so, the segments are to be identified as Störmer has named them (fig. 4 B): coxa (coxopodite), trochanter (basipodite), prefemur (ischiopodite), femur (meropodite), patella, tibia (carpopodite), tarsus (propodite), and pretarsus (dactylopodite), the names in parentheses being those used in carcinology. An eight-segmented limb depends on the coincidental occurrence of a patella and of two segments in the trochanteral region; it is not common in modern arthropods, but is present in the Pycnogonida and in some of the legs of the Solpugida among the arachnids. The patella is a segment characteristic of the legs of Arachnida and Pycnogonida and is not found in those of mandibulate arthropods. A tibia and tarsus normally follow the patella, or the femur if there is no patella; but it must be noted that the tarsus is prone to subdivision, particularly into two parts, which simulate segments, but are not interconnected by muscles. With regard to the trilobite leg, therefore, it might be questioned if the apparent tibia and tarsus are not two tarsal subsegments and the supposed patella the true tibia. There must, in other words, be some uncertainty as to the identity of limb segments where the musculature cannot be known. Yet the trilobite leg appears to be an eight-segmented appendage, not counting the subcoxal ring, and, if so, it is a truly generalized arthropod limb in that it contains *all* the segments present in the legs of any modern arthropod, but it particularly resembles the arachnid leg in the possession of a patella.

The long pinnate or plumose lateral branches arising from the bases of the trilobite legs (fig. 4 A, B, *Eppd*) are of particular interest. The earlier students of trilobite limbs regarded these branches as "exopodites," and from this interpretation they drew the conclusion that the trilobite leg is a biramous appendage, and therefore closely relates the trilobites to the Crustacea. Störmer (1933), however, showed that the supposed exopodite arises, as he then thought, on the coxa and not on the basipodite (first trochanter) as does the crustacean exopodite, from which fact it obviously is an epipodite and not an exopodite; later (1939) he assigned it to the precoxal ring (fig. 4 B). Furthermore, Störmer gives good reasons for regarding the organs in question as gills. The epipodites are covered by the pleural lobes of the body segments, and, if they are not branchiae, the trilobites have no evident organs of respiration. In any case, the idea that the crustacean biramous type of limb represents the primitive and fundamental structure of arthropod appendages is not supported by the trilobites. The crustacean exopodite is always borne by the basipodite, and is a special feature of the Crustacea, though not present on all their appendages.

⇥ II ⇤

LIMULUS

LIMULUS, and its living relatives generally assigned to two other genera, are the animals known as "horseshoe crabs" or "king crabs." They, together with various fossil forms, constitute the class Xiphosurida of the subphylum Chelicerata. *Limulus polyphemus* Latr., also known under the generic name of *Xiphosura*, lives along the Atlantic coast of North and Central America; other species are found in the waters of Asia, Japan, and the East Indies. The xiphosurids for the most part inhabit fairly shallow water over sandy bottoms along the ocean shores, but the Asiatic species, *Carcinoscorpius rotundicauda* Latr., is said to live in brackish estuaries, or even in rivers where the water is practically fresh (see Annandale, 1909). The horseshoe crab is an animal that adds romance to zoology; it is a relic of past ages that now stands alone in the midst of creatures of later origins and of more modern types of structure. The ancestors of the horseshoe crabs were companions of the trilobites.

General External Structure

The body of *Limulus* (fig. 5 A) is distinctly divided into an anterior *prosoma* and a posterior *opisthosoma*, the second part being freely movable up and down on the first; the opisthosoma carries a long, strong, independently movable *tail spine*. The animal gets its common name from the shape of the prosoma, which is covered by a convex dorsal shield, or *carapace*, in outline suggestive of a horseshoe. The opisthosoma is hexagonal, with its broad base attached to the prosoma on a strong transverse hinge between the projecting posterior angles of the prosomatic carapace. The lateral

20

margins of the opisthosoma are indented by six notches on each side, in which are seated slender movable spines; the tail spine arises from a deep indentation of the posterior margin. The body of the adult animal is entirely unsegmented.

The dorsal surface of both the prosoma and the opisthosoma is elevated along the middle (fig. 5 A), and on the broad lateral areas

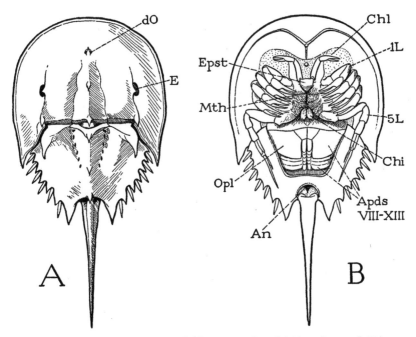

Fig. 5. Xiphosurida. *Limulus polyphemus* L., dorsal (A) and ventral (B).
An, anus; *Apds*, opisthosomatic appendages; *Chi*, chilarium; *Chl*, chelicera; *dO*, simple dorsal eye; *E*, compound lateral eye; *Epst*, epistome; *L*, leg; *Mth*, mouth; *Opl*, operculum.

of the prosomatic shield is a pair of compound eyes (*E*) at the sides of longitudinal ridges. In a general way, therefore, in its dorsal aspect, *Limulus* has a rather striking resemblance to a trilobite, except for the absence of segmentation in the opisthosoma and the presence of the tail spine. Six small pits along each side of the median elevation of the opisthosoma of *Limulus,* however, probably mark the lines of union between primitive segments of the opisthosoma corresponding with the lateral spines. In *Limulus* there is a pair of very small simple eyes (*dO*) situated on the sides of a median spine on the anterior

21

part of the prosoma, but in some trilobites a sense organ of some kind is present in the same position.

On the undersurface of *Limulus* (fig. 5 B) the appendages are most in evidence, but the xiphosurids in common with the arachnids have no antennae. On the prosoma are six pairs of jointed, leglike limbs, and on the opisthosoma six pairs of broad, flat lobes projecting posteriorly and underlapping each other, so that the posterior five are almost covered by the first. In the differentiated form and function of the appendages, and in the lack of antennae, the xiphosurids differ conspicuously from the trilobites. The prosomatic limbs include an anterior pair of small appendages termed the *chelicerae* (*Chl*), which are directed forward, and five pairs of *legs* (*L*) extended laterally. Behind and between the bases of the last legs are two small spiny lobes (*Chi*) known as the *chilaria*, which represent a seventh pair of prosomatic appendages. The legs are attached by their dorsoventrally elongated basal segments on the sides of a deep axial part of the body (fig. 7 A), and the margins of the prosomatic carapace are reflected ventrally to the leg bases as a broad *doublure* (*Dbl*). In cross section, therefore, *Limulus* again resembles a trilobite (fig. 4 A); the body of the animal in each case is a cylinder containing the principal viscera, with wide flat expansions of the dorsum extending over the limbs.

If the legs are spread apart (fig. 6 A), between their bases is exposed a median triangular area of the ventral integument, which is the sternal region of the prosoma. Anteriorly at its apex is the *mouth* (*Mth*), which has a central position beneath the prosoma (fig. 5 B). If the prosomatic appendages are removed from a specimen (fig. 7 F), it will be seen that the bases of the chelicerae and legs are radially arranged around the median sternal area, the chelicerae being directly in front of the mouth and the first two pairs of legs diverging forward from its sides; the chilaria (*Chi*) block the posterior end of the space between the legs. With the appendages removed, the doublure (*Dbl*) is fully exposed. Its peripheral part forms a wide horseshoe-shaped area, which has the hard, horny texture of the dorsal surface of the carapace and slopes steeply upward (A). The inner part immediately surrounding the bases of the appendages takes a horizontal position and becomes membranous (F), but it contains on each side a series of five small Y-shaped sclerites (*Pl*), to the stems of which the leg bases are specifically articulated (fig.

22

6 C). These sclerites, the "coxal pivots" of Benham (1885), clearly are to be regarded as *pleurites,* since, in that they support the appendages, they correspond with the plates called *pleura* in most other arthropods. On the membranous part of the doublure of *Limulus* before the bases of the chelicerae is a median papilla (fig. 7 F, *vO*), the supposed "ventral eye." Supporting the chelicerae is

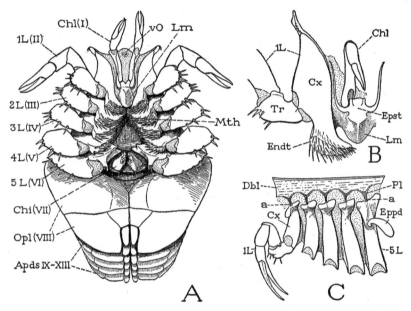

Fig. 6. Xiphosurida. *Limulus polyphemus* L.

A, the appendages in place on the body, ventral. B, epistome, labrum, right chelicera, and base of right first leg, anterior. C, coxae and first leg of left side, with pleural articulations.

a, pleural articulations of coxae; *Apds,* opisthosomatic appendages; *Chi,* chilarium; *Chl,* chelicera; *Cx,* coxa; *Dbl,* doublure; *Endt,* coxal endite; *Eppd,* epipodite; *Epst,* epistome; *L,* leg; *Lm,* labrum; *Mth,* mouth; *Opl,* operculum; *Pl,* pleurites; *Tr,* trochanter; *vO,* ventral sense organ; *I–XIII,* body segments beginning with cheliceral segment.

a small epistomal plate (fig. 6 B, *Epst*), which bears the *labrum* (*Lm*) projecting beneath the mouth (fig. 7 F).

On the opisthosoma the closely overlapping appendages (fig. 5 B, *ApdsVIII–XIII*) are set in a depression of the undersurface, which is surrounded laterally and posteriorly by a broad, flat doublure with a hard, polished surface. The appendages of the first pair are united to form an *operculum* (*Opl*) covering the other five, which

23

bear leaflike gills on their posterior surfaces. On the upper (posterior) side of the operculum are the openings of the genital ducts, situated in each sex on a pair of small papillae (fig. 12 B, *gp*). At the base of the tail spine is the anus (fig. 5 B, *An*). The appendages will be more fully described in a later section.

The prosoma and the opisthosoma are connected by a large median dorsal muscle, the *arthrotergal muscle* of Benham (1885), that crosses the hinge line between the two parts of the body and serves to flex the opisthosoma on the prosoma. In the ventral part of the body, long muscles arising in the prosoma are distributed by branches to the ventral wall of the opisthosoma and are evidently levators of the opisthosoma; a few ventral fibers from the prosoma, however, go to anterior dorsal apodemes of the opisthosoma and are therefore flexors, or depressors, of the opisthosoma.

The presence of seven pairs of appendages on the prosoma of *Limulus* should indicate that the xiphosurid prosoma contains at least seven segments. Studies on development show that all the appendages are postoral in the embryo (fig. 8 A), the definitive preoral position of the chelicerae and the first two pairs of legs being a secondary result of the posterior displacement of the mouth, by which the ventral parts of the anterior segments in the adult appear to be lapped forward around the sides of the mouth and the labrum (fig. 7 F). *Limulus* differs from a trilobite in having seven instead of four postoral segments in the prosoma. In appearance, however, there are only six pairs of appendages, since the chilaria have lost all semblance to legs, and in this respect the Xiphosurida resemble the Arachnida, in which the seventh segment is reduced or suppressed, and there are only six pairs of prosomatic limbs.

The area of the prosoma formed by the postoral segments cannot account for the whole of the prosomatic part of the animal, since there is a large preoral region in front of the cheliceral segment. Iwanoff (1933) has shown in a figure of a *Limulus* embryo (fig. 8 A) the presence of a procephalic lobe (*Prc*), which is a prominent anterior part of all arthropod embryos, and which appears to correspond with the acronal lobe of a young trilobite (fig. 3 A, *Acr*). Iwanoff, however, does not seem to attribute much of the adult structure of *Limulus* to this lobe, but a comparison of a larval stage of *Limulus* (fig. 8 B) with a trilobite would suggest that, as in the trilobites, the procephalic lobe has expanded laterally and posteriorly

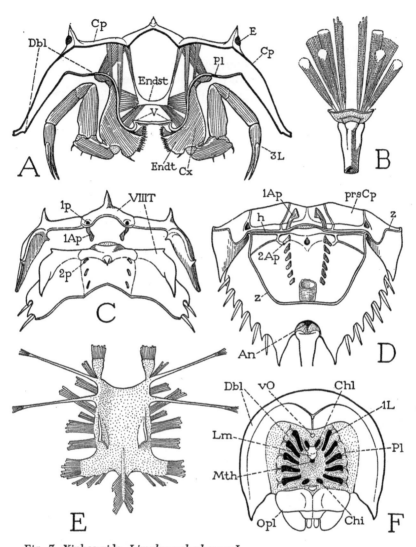

Fig. 7. Xiphosurida. *Limulus polyphemus* L.

A, section of prosoma behind bases of third legs. B, base of tail spine, with muscles. C, posterior part of prosomatic carapace, turned upward and separated along hinge line from anterior part of opisthosoma, showing division of eighth tergum (*VIIIT*) between the two parts of the body. D, inner surface of opisthosomatic carapace, and posterior part of prosomatic carapace, joined along hinge line (*h*). E, prosomatic endosternum of specimen 47 mm. long. F, ventral surface of prosoma with prosomatic appendages removed, showing position of their bases relative to the mouth.

1Ap, 2Ap, first and second tergal apodemes; *Cp,* carapace; *E,* compound eye; *Endst,* endosternum; *h,* dorsal hinge between prosoma and opisthosoma; *1p, 2p,* external pits of first and second tergal apodemes; *prsCp,* prosomatic carapace; *T,* tergum; *V,* venter; *z,* line of removal of ventral wall of opisthosoma. Other lettering as on figure 6.

around the postoral segments to form the lateral parts of the prosomatic carapace, on which the compound eyes are developed. Otherwise, it is difficult to account for the likeness of the prosomatic shield
of *Limulus* to the dorsal surface of the head of a trilobite. The forward radiation of the anterior postoral segments around the sides of
the mouth indicated by Iwanoff (1933, fig. 62) probably applies
only to the ventral surface of the animal. A division of the procephalic
area into an ocular and an antennal segment has not been observed in
the xiphosurid.

Segmentation of the opisthosoma in the adult of *Limulus* is indicated not only by the presence of the appendages, but also by the
double series of pits on the dorsal surface (fig. 5 A). The pits are
the points of ingrowth of apodemal processes on the inner surface
of the back (fig. 7 D), and, as shown by Benham (1885), the muscle
relation to the apodemes leaves no doubt that the apodemes arise
on the primitive intersegmental lines. The first pair of the opisthosomatic apodemes (*2Ap*), however, lies well behind the hinge
(*h*) between the two parts of the body, showing that the hinge is
not a true intersegmental line. Moreover, there is an anteriormost
pair of larger apodemes on the prosoma close to the posterior margin
of the carapace (C, D, *1Ap*). Therefore, the space between these
prosomatic apodemes and the first pair on the opisthosoma must
represent the dorsum of a segment that has been partitioned between the prosoma and the opisthosoma, and which contains the
intersomatic hinge. There is nothing unusual in this condition; it
occurs in other arthropods and is merely an anatomical adjustment
to the necessary mechanism of movement of one part of the body on
another. Since the longitudinal muscles are primarily intrasegmen-
tal, the opisthosoma of the xiphosurid can become movable on the
prosoma by its own muscles only by having the anterior part of its
first tergum fused with the prosomatic carapace and a line of flexibility developed across its middle.

The tergal area of the eighth body segment (fig. 7 C, *VIIIT*) is
discernible as a small arc on the posterior border of the prosoma
marked by the pits (*1p*) of the first dorsal apodemes, and as a
wider transverse area on the opisthosoma before the second apodemal pits (*2p*), ending laterally in large, free, pointed lobes. The
limits of the following segments can be judged only by the apodemal
pits and their apparent relation to the marginal spines. The posterior

26

part of the opisthosoma of *Limulus* must include a segment corresponding with the telson of other arthropods, which contains the anus. The anus of *Limulus* lies ventrally just before the base of the tail spine (fig. 7 D, *An*). It is questionable, therefore, if the spine itself represents the telson; it appears rather to be an appendage of the telson. The spine is not present in the embryo (fig. 8 A), but grows out from the end of the body during larval development (C, D). In the adult, the tail spine is an independently movable organ provided with a strong musculature (fig. 7 B) consisting of six long dorsal muscles and two large ventral muscles. The fibers of the median pair of dorsal muscles are distributed along the dorsal apodemes of the opisthosoma, those of the two lateral pairs arise on the opisthosomatic carapace; the ventral muscles divide each into a dorsal branch and a ventral branch. The origins of the spine muscles are thus far in front of the area of a presumed apical segment; their insertions, moreover, are not on the spine itself, but on a tough integument that connects the spine with the body. The spine is hollow to its tip, lined with an epithelium traversed by nerves and blood vessels.

In the decapod crustaceans and in the pterygote insects there is a more or less elaborately developed *endoskeleton* consisting of arms, ridges, or plates that grow in from the cuticle of the body wall and serve to strengthen the exoskeleton or to give attachment to muscles. The principal structures of this kind in *Limulus* are the dorsal apodemes of the carapace and the tendons of muscles. In the center of the prosoma, however, is a horizontal plate suspended by muscles from the carapace, lying between the alimentary canal and the ventral nerve cord, which, though it is not a skeletal derivative, takes the place of a ventral endoskeleton, since all the ventral prosomatic muscles are attached on it. The plate, therefore, is generally termed the *endosternum;* Lankester (1884, 1885) and Benham (1885) call it the "entochondrite," or "plastron." It differs somewhat in shape and consistency with the age of the animal. In a small specimen two inches long the endosternum is a thick membrane. In shape it is an oblong rectangle (fig. 7 E) with the anterior corners produced into a pair of spatulate anterior lobes, the posterior angles drawn out into broad posterolateral lobes; laterally on the upper surface is a pair of soft, triangular dorsal lobes, and from the posterior margin a taillike extension gives attachment to numerous muscle fibers. From the anterior end on each side two long, slender suspensory arms diverge outward and upward and are attached by muscles on the carapace. Numerous muscles depart from the margins of the plate and its lobes, and others arise directly on the dorsal and ventral surfaces. The endosternal muscles have been fully described and illustrated by Benham (1885). According to Lankester (1884), the endosternum, or "entochondrite," of the prosoma of an adult *Limulus* has a texture resembling that of hyaline vertebrate cartilage, though it lacks the essential constituents of cartilage. A similar endosternal plate is present in most Arachnida, and it is probable that the ligament connecting the ventral adductor muscles of the gnathal appendages in lower Crustacea and in Diplopoda is a structure of the same nature.

In its microscopic structure, Lankester (1884) says, the adult prosomatic endosternum "is a firm homogenous, or sparsely fibrillated matrix in which are imbedded nucleated cells generally arranged in rows of three, six, or even eight parallel with the adjacent lines of fibrillation." In a larval specimen 2 cm. in length, the endosternum is a thin membrane with a faintly fibrillated structure. Where muscles are attached on it, the fibrillae become more distinct and appear to run out directly into the fibrillae of the striated parts of the muscle fibers. The fibrillar continuity is plainly to be seen in the suspensory arms. In a specimen 5 mm. in length, the "plate" is a delicate membrane resting close upon the nerve mass beneath it, but it is not essentially different from that of older specimens. Where muscles are attached, the striated fibers break up into bundles of fine fibrillae that are soon lost in the tissue of the endosternal membrane. The endosternum is a nonchitinous tissue.

In the opisthosoma of *Limulus* the lateral systems of ventral muscles are segmentally connected by transverse ligamentlike bands, which are termed the "mesosomatic entochondrites" by Benham (1885) because they appear to be a tissue of the same kind as that of the prosomatic endosternum. These connectives, however, lie *below* the nerve cord, though they are free from the epidermis of the body wall beneath them.

The origin and nature of endosternal structures of arthropods have not been satisfactorily explained. Lankester (1885) suggests that the prosomatic endosternum of *Limulus* is derived from the subepidermal connective tissue of the sternal surface, but he does not explain how it brings with it the muscles, which presumably at first were attached on the cuticle of the body wall, and, to account for the definitive supraneural position of the plate, he has to assume that the endosternum of the prosoma was formed when the nerve

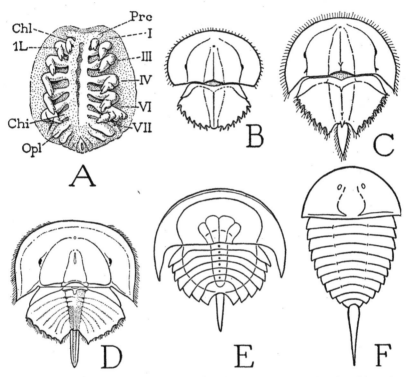

Fig. 8. Xiphosurida. Embryonic and larval stages and early fossil xiphosurids. A, *Tachypleus gigas* (Müller) (*Limulus moluccanus* Latr.), embryo with nine pairs of appendages (from Iwanoff, 1933). B, *Limulus polyphemus* L., larva 4 mm. long. C, same, larva 5 mm. long. D, same as A, second larva just before moulting, opisthosomatic segments indicated by mesoderm bands (from Iwanoff, 1933). E, *Prestwichianella rotundata*, a Carboniferous fossil xiphosurid (from Störmer, 1944, after Woodward). F, *Aglaspis aetoni*, a Cambrian fossil xiphosurid (from Störmer, 1944, after Raasch).

Chi, chilarium; *Chl*, chelicera; *Opl*, operculum; *Prc*, cephalic lobe; *I–VII*, prosomatic somites.

cords had a lateral position. Schimkewitsch (1895) has described the endosternum of Arachnida as a transformed muscle tissue. It may be noted that in some insects there is in the abdomen a series of transverse muscles over the nerve cord.

29

The Eyes

The functional eyes of *Limulus* include the lateral *compound eyes* (fig. 5 A, E), and the small *dorsal ocelli* (*dO*). The ventral organ lying in front of the chelicerae (fig. 6 A, *vO*) has been regarded by some investigators as a pair of degenerate ventral ocelli and by others as an olfactory organ.

The Compound Eyes— Each compound eye is covered by a thick, convex cornea having an entirely smooth surface, but on looking into it one sees in its deeper part a regular, honeycomblike pattern of minute, six-sided facets. This appearance is not due to any division of the cornea; an examination of the inner surface shows that the latter is projected into numerous small, peglike processes so arranged that the spaces between them have a hexagonal shape. Each corneal process (fig. 9 A, *Ln*) serves as a lens for a group of sensory cells (*Om*) lying beneath it, which is ensheathed in long cells of the epidermis (B). Each group of cells associated with a lens of the common cornea, therefore, is a light-receptive unit of the eye corresponding with an *ommatidium* of the compound eye of higher arthropods. The sensory cells constitute the *retina,* or *retinula,* of the ommatidium, and are closely arranged in a radial manner around a central axis (*C, rSCl*), so that, as Demoll (1914) says, the ommatidial retina resembles a peeled orange. The number of radial cells in a single ommatidium varies from 10 to 15; their inner ends are separated by an axial space, in which, according to Demoll, is a long, slender process (*axp*) from an *excentric retinal cell* (B, *ecSCl*) lying outside the bases of the other cells. The narrow inner margins of the radial cells and the axial process of the excentric cell are composed of a relatively clear cytoplasm with fine striations at the surface (B, C, *sb*). In the compound eyes of most arthropods the striated borders of the convergent retinal cells unite to form an axial, rodlike structure called a *rhabdom,* and are therefore termed *rhabdomeres.* The function of the rhabdom or the rhabdomeres is the diffraction of light from the lens into the light-sensitive parts of the retinal cells. Proximally the retinal cells are continued into nerve processes (B, *Nv*) that go to the brain. The corneagenous epidermal cells of the eye (B, *CgCls*) that surround the lens and ensheath the retina are shown by Demoll to converge between the lens and the retina, but they do not form here any specific dioptric body such as the crystalline cone of most compound eyes. The lateral eyes of *Limulus* are thus seen to be true compound eyes, but to have a simpler structure than those of higher arthropods.

The Dorsal Eyes— The small, median dorsal eyes of *Limulus*, as

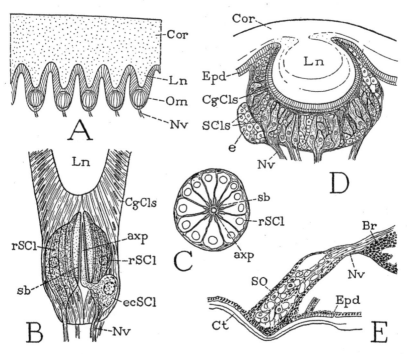

Fig. 9. Xiphosurida. Eyes of *Limulus*.

A, diagrammatic section of the cornea and five ommatidia of compound eye (from Lankester and Bourne, 1883). B, somewhat diagrammatic axial section of an ommatidium of compound eye (from Demoll, 1914). C, cross section of an ommatidium of compound eye (outline from Demoll, 1914). D, axial section of a dorsal ocellus (from Demoll, 1914). E, sagittal section of a "ventral eye" and nerve from brain (from Demoll, 1914).

axp, axial process of eccentric sensory cell (*ecSCl*); *Br*, brain; *CgCls*, corneagenous cells; *Cor*, cornea; *Ct*, cuticle; *e*, mass of undifferentiated cells; *ecSCl*, eccentric sensory cell; *Epd*, epidermis; *Ln*, lens; *Nv*, nerve; *Om*, ommatidium; *rSCl*, radial sensory cell; *sb*, striated border; *SCls*, sensory cells; *SO*, ventral sense organ.

described by Demoll (1914), are typical arthropod ocelli. The cornea of each of these eyes (fig. 9 D, *Cor*) is produced inwardly as a large spherical lens (*Ln*), which is enclosed in a layer of corneagenous cells (*CgCls*) continuous at the periphery of the eye with the surrounding epidermis (*Epd*). Beneath the corneagenous layer is the relatively thick retina, composed of the sensory cells (*SCls*), scattered interstitial cells, and a binding network of connecting tissue that forms also a limiting membrane around the base of the eye, penetrated by the nerve processes (*Nv*) of the sensory cells. At one side of the eye is a mass of undifferentiated cells (*e*) which Demoll regards as a degenerate eye, and he draws the inference that *Limulus* once had a group of four dorsal ocelli.

The "Ventral Eyes"— The supposed ventral eyes of *Limulus* (fig. 6 A, *vO*) are described by Demoll (1914) as consisting of two masses of irregularly arranged cells (fig. 9 E) abutting on the epidermis beneath convexities of the cuticle. Each organ consists for the most part of large cells associated in groups, with connective tissue between them. The large cells are produced into nerve fibers that go to the brain (*Br*), and where two of the cells adjoin each other their opposed surfaces, Demoll says, have rhabdomere borders, though they do not form rhabdoms. These cellular bodies certainly must be, or have been, sensory organs of some kind, and probably the idea that they are degenerate eyes is as good as any other; their structure does not suggest an olfactory function.

The Appendages

We may now give special attention to the appendages, including the seven pairs on the prosoma and the six on the episthosoma. A full account of the embryonic development of the appendages in *Limulus moluccanus* is given by Iwanoff (1933), who shows that all the appendicular organs, regardless of their differences in the adult stage, begin in the same way, as a pair of folds of the ventral body wall.

The Chelicerae— Since the Xiphosurida lack antennae, the first of the prosomatic appendages are the chelicerae (fig. 5 B, *Chl*), which, as already noted, lie directly in front of the mouth in the adult state, but take this position secondarily from a primary postoral position in the embryo (fig. 8 A). The chelicerae presumably represent the first postoral appendages of the trilobites. In the adult of *Limulus* the chelicerae are articulated on a small plate (fig. 6 B, *Epst*) that evidently is the epistome, since it supports the labrum (*Lm*) and is produced on each side into a slender arm curving forward along the mesal border of the coxa (*Cx*) of the first leg. Each

32

chelicera is three-segmented (fig. 10 D), but the terminal segment is a slender claw, or movable finger, opposed to a similar fixed process of the second segment, so that the two together form a pair of pincers. According to Benham (1885), the chelicerae are movable by muscles inserted on their bases that arise on both the carapace and the endosternum.

The Legs— The five pairs of legs (fig. 5 B, *L*) resemble the chelicerae in that their distal segments form pincers, but they have a greater number of segments, and in the adult male the chelae of the first legs assume an altered form (fig. 10 F). The basal segments, or *coxae*, of the legs are greatly elongate dorsoventrally and are attached on the membranous lateral walls of the deep median part of the prosoma beneath the lateral extensions of the carapace (fig. 7 A, *Cx*); their only firm articulations are with the small Y-shaped pleural sclerites in the membranous inner part of the doublure (fig. 6 C, *Pl*). The lower ends of the coxae of the first four pairs of legs are produced into large spine-covered mesal lobes, or *endites* (fig. 10 H, *Endt*), the spines of which curve forward toward the mouth (fig. 6 A). Above the endites of the second, third, and fourth legs is a small accessory endite arising from the membrane at the base of the coxa (fig. 10 G, H).

Each leg is movable anteriorly and posteriorly on the long base of its coxa (fig. 10 C, *a–b*), but also has a transverse movement because of the flexible, membranous nature of the body wall on which it is attached. As shown by Benham (1885), nine muscles are inserted on the basal rim of the coxa (fig. 10 C), five of which (*25, 26, 27, 28, 29*) are dorsal muscles arising on the carapace, and four (*32m, 32n, 33, 34*) are ventral muscles with their origins on the endosternum. Two of the ventral muscles (*32m, 32n*) branch from a common origin. The anterior and posterior dorsal muscles evidently are promoters and remoters, that is, they rotate the coxa on its dorsoventral axis (*a–b*); the ventral muscles must have an adductor function because of the dorsal suspension of the coxa on the pleural sclerite. Since there is no evident muscular mechanism to oppose the ventral adductors, abduction of the limb apparently results from the elasticity of its basal connections.

The legs of the first four pairs have each only six segments; the somewhat longer last legs, however, have seven segments, and therefore represent a more nearly complete appendage, for which reason

33

they should be studied first. The coxa of a fifth leg (fig. 10 A, *Cx*) resembles the coxae of the other legs except for the absence of spines on the endite lobe. A short section of its upper end is set off by a groove marking the base of a strong internal ridge (B, *R*), and bears a spatulate *epipodite* (A, *Eppd*). The next segment is the *trochanter* (*Tr*), which turns on the coxa in a vertical plane, and supports the longer third segment, which is the *femur* (*Fm*). At the end of the femur is the "knee" bend of the leg (*k*), beyond which are four segments. The segment adjoining the femur very evidently corresponds with the segment of an arachnid leg called the *patella* (fig. 19 A, *Pat*), and the following segments of the *Limulus* leg should then be the *tibia* (*Tb*), the *tarsus* (*Tar*), and the *pretarsus* (*Ptar*). A different idea concerning the identity of these segments, however, has been given by Hansen (1930), who regards the faint circular groove near the base of the segment beyond the knee as the division between patella and tibia, so that he has to invent the name "cotibia" for the next segment. The musculature of the leg, as will presently be seen, gives no support to this interpretation. The basal groove is present also on the preceding three pairs of legs (G, H), but not on the first legs (F). The segmentation of the hind leg of *Limulus* corresponds with the apparent segmentation of a trilobite leg (fig. 4 B), except for the lack of a second trochanter, or prefemur. The patellotibial joint has a characteristic structure (fig. 11 C), there being two dorsal lobes on the tibia, between which a median, tonguelike process curves down from the end of the patella. The end of the tibia of the last leg (fig. 10 A) bears four bladelike appendages that conceal the cylindrical tarsus. The pretarsus (*Ptar*) is a long, tapering, spinelike segment, which, with a shorter process projecting from the end of the tarsus, forms a slender chela.

In studying the other legs, the question arises as to what segment has been eliminated to leave only six segments; the answer is furnished by the musculature. The intrinsic musculature of the hind leg is sufficiently shown at A of figure 11, though some of the muscles of the inner, or posterior, side of the leg are not seen in the anterior view. Since there is no question as to the identity of the first three segments in all the legs, we may note first the muscles that move the part of the leg beyond the knee. In the ventral membrane of the knee joint is a small V-shaped sclerite. The arms of the V give attachment

34

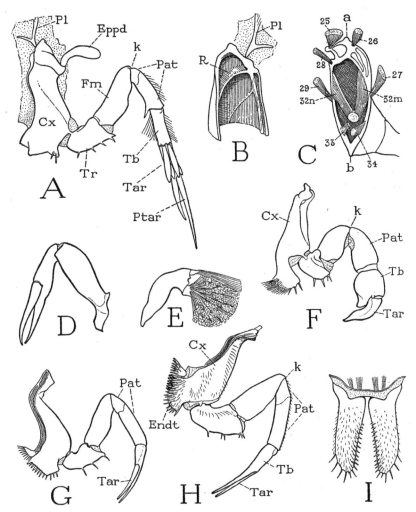

Fig. 10. Xiphosurida. *Limulus polyphemus* L. The prosomatic appendages. A, left last leg, anterior. B, inner surface of anterior wall of upper end of coxa. C, base of leg with attached muscles (from Benham, 1885). D, left chelicera, lateral. E, movable finger and muscles of chela of first leg of adult male. F, first left leg (pedipalp) of adult male, anterior. G, right third leg, posterior. H, left fourth leg, anterior. I, chilaria, posterior.

a, dorsal (pleural) articulation of coxa; *b,* ventral articulation of coxa; *Cx,* coxa; *Endt,* coxal endite; *Eppd,* epipodite; *Fm,* femur; *k,* knee bend of leg; *Pat,* patella; *Pl,* pleurite; *Ptar,* pretarsus; *R,* ridge of coxal wall; *Tar,* tarsus; *Tb,* tibia; *Tr,* trochanter.

to a pair of broad muscles (8) arising dorsally in the femur; the apex is produced into a long, thick, rodlike apodeme, or tendon (t), that extends through the femur and gives attachment to a thick conical muscle (10) in the trochanter. These three muscles are present in all the legs and serve to flex the distal part of the limb on the femur. The patella contains six muscles inserted on the base of the tibia; two are large dorsal muscles (12), two are fan-shaped lateral muscles (14), and two are long slender ventral muscles (16). Each ventral muscle has two branches, one branch (16a) arising in the distal end of the femur, the other (not seen in the figure) in the base of the patella. The tarsal muscles in the tibia are a dorsal levator of the tarsus (18) and a two-branched ventral depressor (19). The pretarsus has a simple levator (20) and depressor (21) arising in the tarsus.

If now we compare the musculature of one of the other legs with that of a hind leg, it is found that the first segment beyond the knee has the same flexor mechanism as this segment in the hind leg, and that it contains the same muscles (fig. 11 B). This segment, therefore, is the patella in all the legs (fig. 10 A, F, G, H, *Pat*). Between it and the next segment (fig. 11 B, *Tb*) there is no suggestion of a reduced or eliminated segment, and the nature of the musculature from the patella leaves no doubt that this segment is the tibia. Moreover, the structure of the joint between these two segments is exactly that of the patellotibial joint of the hind leg (C). Finally, the musculature in the tibia of the anterior legs is that within the tibia of the hind leg inserted on the tarsus. We must conclude, therefore, that in all the legs but the last, the terminal segment forming the movable finger of the chela is the *tarsus* (fig. 10 G, H, *Tar*), and that the opposed finger is a process of the tibia. Hansen (1930) made the same deduction without a study of the musculature. The chela of the anterior leg becomes much modified in the adult male (fig. 10 F). The tibia, or "hand" of the chela (*Tb*), is short but greatly swollen, and the long clawlike tarsus, or movable finger (*Tar*), is thick and bent, while the immovable process of the tibia is reduced to a short thumb. The globular tibia contains a small dorsal muscle of the tarsal claw (E) and two great masses of closing fibers inserted on a thick knob of the tarsal base. This seemingly deformed chela of the male, therefore, evidently has great strength.

The Chilaria— The appendages of the seventh postoral body seg-

36

ment, known as the chilaria, lie close to the posterior margin of the venter of the prosoma (fig. 6 A, *Chi*), between and behind the coxal endites of the last pair of legs. Each chilarium (fig. 10 I) is a

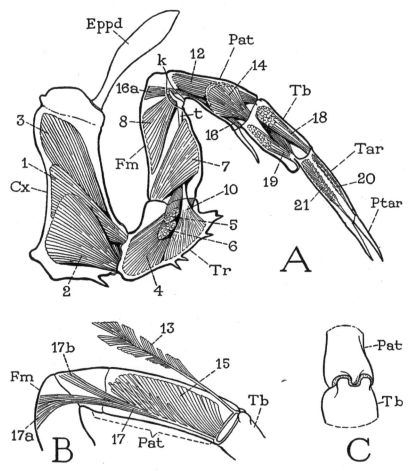

Fig. 11. Xiphosurida. *Limulus polyphemus* L. Structure and musculature of a leg.

A, left last leg and intrinsic musculature, anterior. B, patellar segment, and muscles of posterior half. C, patellotibial joint, dorsal, structure characteristic of this joint in all the legs.

Lettering as on figure 10; muscles explained in text.

small, simple, elongate, flattened lobe, armed with spines on its distal end and along its mesal margin. The two appendages arise close together and project downward, so that they close the posterior end of the space between the leg bases leading forward to the mouth (fig. 6 A). From their position the chilaria have acquired their name, *cheilarion* being Greek for a "little lip." According to Iwanoff (1933), the chilaria correspond with only the coxal endites of the preceding appendages. They have a very weak basal musculature.

The Opisthosomatic Appendages— The six appendages of the opisthosoma are broad, flat, closely appressed, horizontal plates lying in the ventral concavity of the opisthosoma (fig. 5 B, *ApdsVIII–XIII*), where they underlap from before backward. The first plate is a large operculum (*Opl*) mostly covering the others. Each plate is provided with a pair of promotor muscles, and a pair of remotor muscles arising on the dorsum of the body wall (fig. 12 A, B), both being inserted within the cavity of the appendage, the smaller promotor (*pmcl*) on the anterior lamella, the larger remotor (*rmcl*) on the posterior lamella. There is little reason to suppose that these opisthosomatic appendages of *Limulus* are not true segmental limbs, but it is difficult to identify their parts with those of a prosomatic leg.

The operculum hangs from the membranous ventral wall of the body between the prosoma and the opisthosoma, but, as we have seen, the tergum of the first opisthosomatic segment is partitioned between the opisthosoma and the prosoma. Accordingly, the promotor muscles of the operculum arise on the posterior apodemes of the prosomatic carapace, and the remotors on the anterior margin of the opisthosoma. The operculum differs from the following appendages in that its lateral halves are more closely united along the mid-line. On the anterior surface (fig. 12 A) two median series of sclerites appear to represent the shafts of the component appendages, and the broad lateral expansions can be interpreted as basal exites, probably coxal epipodites. A transverse line of flexibility (*1–1*) through the distal part of the operculum cuts off the apical parts of the exites as a pair of free lateral lobes and separates the two distal segments from the long basal segments of the median shafts, the apical segments of which form a pair of small, free median lobes. The posterior surface of the operculum (B), however, gives little evidence of limb structure, except for the apical lobes. At about the middle of the basal part are two small genital papillae (*gp*), present alike in both sexes, which contain the outlets of the

38

genital ducts (*Dct*). Since the operculum of *Limulus* pertains to the eighth postoral body segment, it is of interest to note that the genital opening in Arachnida also is on the eighth segment.

In the next five appendages of the opisthosoma (fig. 12 C, D) the broad lateral parts are separated by a wide, median membranous area, beyond which the component appendages are represented by a pair of three-segmented median lobes, with a slender tongue of

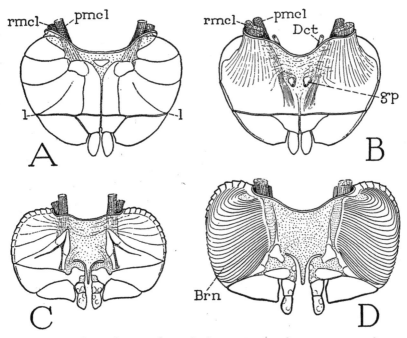

Fig. 12. Xiphosurida. *Limulus polyphemus* L. Opisthosomatic appendages. A, operculum, anterior. B, same, posterior. C, united second pair of appendages, anterior. D, same, more enlarged, posterior.

Brn, gills (branchiae); *Dct*, genital duct; *gp*, genital papilla, with genital orifice; *l*, line of flexion; *pmcl*, promotor muscle; *rmcl*, remotor muscle.

the basal membrane projecting between them. On the posterior surfaces of the appendages, turned upward, are the gills (D, *Brn*). The gills consist of two great lateral masses of thin, closely superposed branchial lamellae, at least 80 of them in a single gill mass of an adult specimen. Each gill is an extremely thin semilunar plate with a rigidly thickened margin fringed with delicate setae. The lamellae are attached transversely by their bases on the supporting appendage and slope distally on the latter.

Relationships of the Xiphosurida

A comparison of *Limulus* with a trilobite can hardly fail to give the impression that the two are related animals, though the principal likeness is in the form and structure of the prosoma, with its wide lateral parts inflected ventrally before the mouth and around the sides of the appendages. On the other hand, in the lack of antennae, the absence of segmentation in the opisthosoma, and the differentiation of the appendages, *Limulus* is far more specialized than a trilobite. The xiphosurids, however, have an ancient lineage; *Limulus* itself in almost its present form goes back to Jurassic times in the Mesozoic Era. Its ancestors in the Carboniferous have distinctly segmented abdomens (fig. 8 E), as do still earlier Paleozoic representatives that lived in the Cambrian (F). The two-part bodies and the tail spine of these ancient forms leave no doubt that they are xiphosurids. The union of segments in the opisthosoma, therefore, is only a specialization of the comparatively recent *Limulus*.

Associated in the Cambrian with the early xiphosurids and the trilobites were many other arthropods whose known remains show such a mixture of characters that paleontologists cannot agree among themselves just what they are, whether trilobites, xiphosurids, eurypterids, or crustaceans, which fact is probably good evidence that the animals themselves represent early intermediate forms between these several groups. *Limulus,* however, has survived all its ancient relatives and still flourishes in modern times, so it is not surprising that it has become the most specialized member of its tribe. Störmer (1944) has interestingly described and pictorially illustrated the evolutionary history of the xiphosurids, showing examples of the different forms found at successive paleontological levels. It is not necessary to suppose, however, that the Xiphosurida have been derived directly from trilobites; it is a safer assumption that the two groups of animals had a common ancestry.

III

THE EURYPTERIDA

THE eurypterids were aquatic animals that lived during the Paleozoic era from the Cambrian into the Permian, but they attained their greatest development in the Silurian and Devonian periods. They are known also as the Gigantostraca because some of them attained the great length of six or eight feet, though most species are from six to 15 inches in length. The remains of the earlier forms are found in marine deposits along ancient shores or in lagoons, but later in the Devonian and Carboniferous periods the animals became fresh-water inhabitants.

The body of a typical eurypterid (fig. 13 A) is elongate, broad in front, tapering posteriorly, and usually ends in a large tail spine. As in the trilobites and the xiphosurids, the body is divided into an unsegmented prosoma and an opisthosoma, but the opisthosoma is always fully segmented. The prosoma carries six pairs of appendages (B) corresponding with the chelicerae and legs of *Limulus*, and therefore contains at least six segments. The opisthosoma in dorsal view (A) presents 12 tergal plates, showing that the entire body of the animal is composed of 18 postoral segments, not counting the tail spine. On the undersurface (B), however, there is evidence of only 11 segments in the opisthosoma (*VIII–XVIII*), suggesting that the sternal part of the seventh body segment is either obliterated or united with the prosoma. The eurypterid differs from the xiphosurid in that the tergum of the seventh segment is free from the prosoma (A, *VII*).

The prosoma is covered dorsally by a carapace of variable shape, either quadrate (fig. 13 D), rounded (figs. 13 A, J; 14 B, C), or

41

triangular (fig. 14 A), but always of simpler structure than that of the xiphosurids. It bears a pair of lateral compound eyes (fig. 13 A, *E*) apparently similar in structure to those of *Limulus*, though in some species having an anterior position (fig. 14 A, C). Medially on the carapace is a pair of simple dorsal ocelli (fig. 13 A, *dO*). On the ventral surface of the prosoma (fig. 13 B) the limb bases occupy a greater area than in *Limulus*, because the prosoma is much less expanded and the doublure is therefore relatively narrow. The first of the appendages are a pair of three-segmented chelicerae, which are usually very small (B, G, *Chl*), but in one genus they are long arms (fig. 14 C) with large, strongly toothed chelae (E). The following appendages, or legs, are generally of simple form (fig. 13 G), increasing in size from before backward, and also in the number of segments, but in some species they are of diversified form and size (D, E). The usually very short first legs (G) have seven segments, counting the terminal spine as the apical segment, or pretarsus. The next two legs are successively a little longer and have eight segments each. The still longer fourth legs have nine apparent segments, as do also the much longer and broader last legs, which in most species have a paddlelike shape suggestive of their having been used as swimming organs.

It is difficult to account for the presence of nine segments in the last two pairs of eurypterid legs. The basal segment of these legs (fig. 13 H), as of the other legs, is undoubtedly the coxa (Cx), and the apical segment the pretarsus (*Ptar*). As in the case of the trilobite leg, however, it is impossible to be sure of the identity of the intervening segments without knowing the musculature. The penultimate and antepenultimate "segments" might be subdivisions of the tarsus, as they are in many other arthropods, including the scorpions, but in the eurypterid they appear to be true, independently movable segments. Suspicion, therefore, falls on the two small rings at the base of the trochanteral region of the leg, which may be supposed to be secondary subdivisions of the first trochanter (*1Tr*), since they are not present in the first three legs. In the lack of positive evidence, then, the other segments in the last two pairs of legs may be interpreted as the second trochanter, or prefemur (*2Tr*), the femur (*Fm*), the patella (*Pat*), the tibia (*Tb*), the tarsus (*Tar*), and the pretarsus (*Ptar*). In species having slender hind legs (fig. 14 B, D) the segmentation is the same as in the paddle-shaped legs. Between

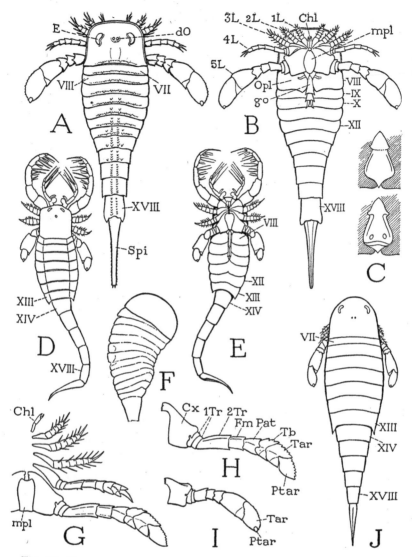

Fig. 13. Eurypterida. (A, B, F–I, outlines from Clarke and Ruedemann, 1912; C, from Störmer, 1936; D, E, J, from Störmer, 1934.)

A, *Eurypterus remipes* Dekay, dorsal. B, diagram of ventral surface of a eurypterid. C, *Pterygotus rhenaniae* Jaekel, opercular appendage of the broad type (female), ventral and dorsal. D, *Mixopterus kiaeri* Störmer, dorsal. E, same, ventral. F, *Strabops thatcheri* Beecher, upper Carboniferous merostome. G, *Dolichopterus macrochirus* Hall, left appendages and "metastome," ventral. H, same, hind leg. I, *Eurypterus remipes* Dekay, hind leg. J, *Hughmilleria norvegica* (Kiaer), dorsal.

Roman numerals indicate body segments beginning with cheliceral segment. *Chl,* chelicera; *Cx,* coxa; *dO,* dorsal ocellus; *E,* compound eye; *Fm,* femur; *go,* genital organ; *L,* leg; *mpl,* metastomal plate; *Opl,* operculum; *Pat,* patella; *Ptar,* pretarsus; *Spi,* tail spine; *Tar,* tarsus; *Tb,* tibia; *1Tr,* first trochanter; *2Tr,* second trochanter.

the coxae of the last legs is a large oval or elongate plate called the *metastome* (fig. 13 B, G, *mpl*). It has the appearance of being a sternal plate, perhaps the sternum of the seventh body segment pushed forward between the bases of the legs of the sixth segment, but some writers regard it as representing the united appendages of the seventh segment corresponding with the similarly placed chilaria of *Limulus*. Whatever the plate may be, it is not to be confused with the so-called "metastome" of Crustacea, which is an entirely different structure.

The mouth of the eurypterid lies between the radiating bases of the prosomatic appendages and thus has a central position on the underside of the prosoma, as in the Xiphosurida. In a depression between the mouth and the metastomal plate is a lobe termed the *endostoma*. Some investigators claim that there is to be seen on the doublure in front of the chelicerae a small median organ, which is thought to represent the "ventral eye" of *Limulus*.

Of the 11 segmental divisions of the undersurface of the opisthosoma (fig. 13 B), the first corresponds with the eighth tergum of the dorsum (A). The venter of segment VIII is covered by a pair of plates (B, *Opl*) separated medially by a deep cleft from which projects a median appendage (*go*) of different form in the two sexes. By comparison with *Limulus* the two lateral plates would appear to represent the two halves of the operculum of the eighth segment, and the median organ the united telopodites of the opercular appendages. The two types of structure presented by the median organ of the eurypterids are designated A and B. Though both are highly variable, type A is generally a long slender structure (B, *go*), segmented in its distal part, while type B is short and broad, and not so distinctly segmented (C). It has usually been assumed that type A pertains to the female, and type B to the male, but Störmer (1936), from studies on *Pterygotus rhenaniae* von Overath, concludes that the reverse is more probably the truth. On the dorsal side of the base of the more slender organ he finds evidence of a median aperture, which he takes to be the opening of the male genital outlet duct; on the corresponding part of the broad organ (C), however, are two oval apertures that Störmer assumes to be the outlets of paired oviducts of the female.

The ventral surfaces of the next four segments of the opisthosoma are covered by broad sternal plates (fig. 13 B, *IX–XII*), which are said to be separate from the corresponding tergal plates. The part of the opisthosoma composed of the first six segments, segments VII to XII inclusive, therefore, is termed the *mesosoma,* and the following part, composed of segments XIII to XVIII, in which the

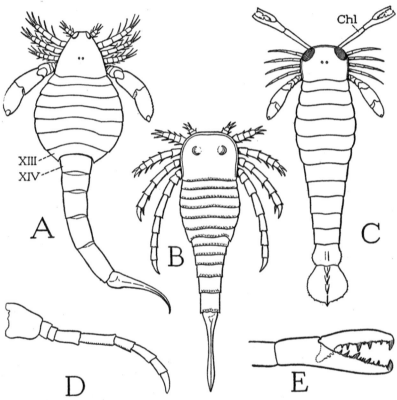

Fig. 14. Eurypterida. (A–D, outlines from Clarke and Ruedemann, 1912; E, from Störmer, 1936.)

A, *Eusarcus scorpionis* Grote and Pitt. B, *Stylonurus longicaudata* Clk. & Rued. C, *Pterygotus buffaloensis* Pohlman. D, *Stylonurus longicaudatus* Clk. & Rued., hind leg. E, *Pterygotus rhenaniae* Jaekel, reconstructed chelicera.

Lettering explained under figure 13.

segments are continuously sclerotized rings, is called the *metasoma.* In many species, however, the opisthosoma is abruptly narrowed between segments XIII and XIV (figs. 13 D, E, J; 14 A), so that the last five segments form a taillike part of the body, giving such eurypterids their characteristic scorpionlike appearance. The apical spine of the opisthosoma is usually straight and tapering in the eurypterids (figs. 13 A, B, J; 14 B) as in *Limulus,* but in the tailed species it may be curved and sharp-pointed (figs. 13 D, E; 14 A), suggesting the sting of a scorpion. In some forms, however, the apical structure is broad and fan-shaped (fig. 14 C).

In one respect the eurypterids differ conspicuously from the xiphosurids, which is the absence of exposed branchial organs on the opisthosoma. On the concealed dorsal surfaces of the opercular plates of the eighth segment, and on the inflected upper surfaces of the next four underlapping sternal plates, however, are paired structures that are interpreted as gills. In the Eurypterida, therefore, the five pairs of supposed branchiae occur on segments VIII to XII; in modern Xiphosurida the five pairs of gills are on segments IX to XIII.

The eurypterids undoubtedly must be in some way related to the xiphosurids. In the less elaborate development of the prosoma and in the segmentation of the opisthosoma they appear to be more generalized than the modern *Limulus,* but, as we have seen (fig. 8 E, F), the ancient fossil xiphosurids had a segmented opisthosoma. On the other hand, the scorpionlike form of some of the eurypterids has given rise to theories of relationship between the eurypterids and the scorpions. Probably the majority of zoologists with any opinion on the subject would regard the eurypterids as the ancestors of the scorpions, and therefore of the Arachnida in general, of which the scorpion is a member. It is certain, however, as Versluys and Demoll (1920) have emphatically asserted, that the scorpion is not a primitive arachnid; so these authors contend that the relationship is the other way around, namely, that the eurypterids and through them the xiphosurids have been evolved from scorpions, fossil scorpions being known from Silurian deposits. According to this theory the arachnids came from some independent source. Quite inconsistent with the scorpion ancestor theory, however, is the evident fact that the scorpionlike eurypterids are specialized forms, and that the earliest-known merostomes, from which the later eurypterids and modern xiphosurids must have descended, have no resemblance to scorpions, either ancient or modern. There is the Cambrian fossil *Strabops* (fig. 13 F), for example, which looks like a much-simplified eurypterid. Though paleontologists do not agree as to whether *Strabops* is a eurypterid or a xiphosurid, or whether it has 11 or 12 opisthosomatic segments, it is undoubtedly a primitive merostome, and has little to suggest that its ancestor was a scorpion. Moreover, historically the earliest eurypterids long antedate the first known scorpion. Neither theory of merostome-scorpion relationships, therefore, seems to be tenable. In the fourth chapter it will be shown that the scorpions differ in some important respects from the merostomes, and cannot be regarded even as primitive arachnids.

IV

THE PYCNOGONIDA

THE pycnogonids are strange-looking marine animals with small bodies and long, sprawling legs (fig. 15 A, C). Most of them measure a few centimeters across the outspread appendages, but some only 3 or 4 mm., while a deep-sea species has a spread of 50 cm., about 18 inches. The animals inhabit all parts of the ocean, but are most numerous in arctic and antarctic regions; they live chiefly along the shore, though a few species are deepwater inhabitants. Because of the number and length of their legs, the pycnogonids are known as "sea spiders." The name Pycnogonida refers to the crowded condition of the numerous "knees" when the legs are bent upward on their bases; but another group name is Pantopoda, which presumably expresses the fact that when the animals are crawling they appear to be "all legs." The number of appendages varies from four to eight or even nine pairs in different species.

As an example for a study of the general structure of a pycnogonid we may take one named *Nymphon hirtipes* Bell (fig. 15 A), a common subtidal species of the northern Atlantic coast of North America, having seven pairs of appendages. The trunk of the animal is divided into a partially segmented, limb-bearing prosoma (D, E), from which projects forward a large proboscis (*Prb*), and a very small, simple, unsegmented opisthosoma, or abdomen (*Ab*). The prosoma includes an anterior unsegmented section bearing the first four pairs of limbs and, on the posterior part of its dorsal surface (D), a tubercle with four very small eyes (*dO*); its posterior part is composed of three distinct segments, each of which carries a pair of legs. All the limbs are supported on large lateral lobes of the

48

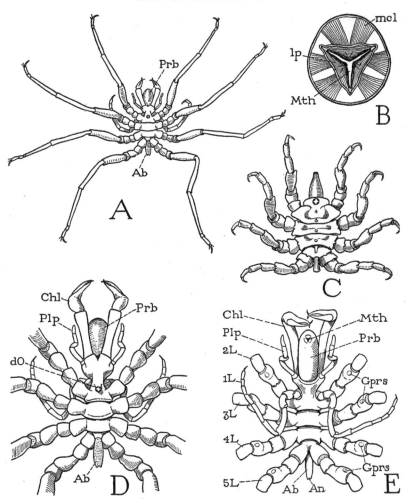

Fig. 15. Pycnogonida (Pantapoda).

A, *Nymphon hirtipes* Bell, female, dorsal. B, same, cross section of proboscis just behind the mouth. C, *Pycnogonum littorale* Ström., female, dorsal. D, *Nymphon hirtipes* Bell, female, body and bases of legs, dorsal. E, same, ventral.

Ab, abdomen; *An,* anus; *Chl,* chelicera; *dO,* dorsal ocelli; *Gprs,* gonopores; *L,* leg; *lp,* lip; *mcl,* muscle; *Mth,* mouth; *Plp,* palp; *Prb,* proboscis.

body. The minute abdomen (*Ab*) is a slender tube projecting posteriorly from between the bases of the last pair of legs and bears the anus at its extremity (E, *An*).

The appendages include a pair of relatively large, three-segmented *chelicerae,* or *chelifores* (fig. 15 D, E, *Chl*), embracing the probos-

49

cis, a pair of short, slender, five-segmented appendages, known as the *palpi* (*Plp*), closely associated with the bases of the chelicerae, and five pairs of many-jointed legs (E, *1L–5L*). The first legs (*1L*), however, differ from the others in their extreme slenderness and their ventral position. These appendages are termed the *ovigerous legs* because the male uses them for carrying the eggs (fig. 16, H, I), which are loaded upon him by the female at the time of mating. They are present in both sexes of the species here described, but in the females of some families they are absent. The following four pairs of appendages are the *walking legs* (fig. 15 E, *2L–5L*), and are all practically alike in size and structure. The segmentation of the legs will be discussed later. Inasmuch as there are seven pairs of appendages on the prosoma, this part of the body must include at least seven primitive segments. The anterior section bearing the chelicerae, the palps, the ovigerous legs, and the first pair of walking legs, therefore, is a composite tagma of four segments, and only the posterior three prosomatic segments remain ununited.

The huge proboscis is a characteristic feature of the pycnogonids, though it varies in relative size in different species; in *Pycnogonum*, for example (fig. 15 C), it is small compared with the proboscis of *Nymphon* (A, D, E, *Prb*), and in some forms it is almost as large as, or even larger than, the body itself. The mouth is situated on the anterior end of the proboscis (E, *Mth*) and is enclosed between three small lips, one dorsal, the other two ventrolateral, so that when the mouth is closed (B, E), the aperture is Y-shaped, but becomes triangular when the mouth is opened. The mouth leads directly into the cavity of a large sac occupying the entire length of the proboscis. In cross section the sac is triangular (B), its three sides corresponding with the three lips (*lp*) of the mouth (*Mth*). In the grooves between the sides are slender rods proceeding inward from the mouth angles. Each lip is provided with a pair of large muscles (*mcl*) arising behind the mouth on the walls of the proboscis, and throughout its length the walls of the sac are covered with radiating bundles of dilator muscle fibers with their peripheral attachments on the wall of the proboscis. The inner sac of the pycnogonid proboscis thus much resembles the sucking "pharynx" of an arachnid, but it has no constrictor muscles.

The lips of the mouth are strongly sclerotized along their edges and are fringed with very small hairs; in *Nymphon hirtipes* they have no armature other than two minute teeth just within the apex of each lip. As noted by Dohrn (1881) for the pycnogonids in general, the lips have no muscular closing apparatus; the mouth apparently shuts by its own elasticity. The inner sac also, though clearly dilatable by its radial muscles, has no antagonistic compressor muscles. The walls of the sac are smooth on their distal parts and devoid of spines, but on the inner half of each section are several broad, overlapping brushes of thickly set, extremely delicate hairs directed toward the mouth. The proboscis sac of *Nymphon hirtipes*, therefore, appears to be merely a sucking apparatus, and certainly can have no mechanical action on the ingested food. The brushes evidently serve as filters guarding the entrance to the oesophagus. The structure of the pycnogonid proboscis and its variations in different species have been fully described by Dohrn (1881). The distal end of the inner sac is surrounded by a ganglionated nerve ring, connected by a median dorsal nerve with the brain and by a pair of ventrolateral nerves with the suboesophageal ganglion. From the ring a nerve then proceeds posteriorly on the wall of each section of the sack, and the three are united by circular commissures, from which the proboscis muscles are innervated (see Wirén, 1918; Hanström, 1928). The proboscis sac leads into a short oesophagus, and the oesophagus discharges into a mesenteron stomach, from which are given off tubular diverticula that extend into all the walking legs and into some of the other appendages.

The pycnogonids feed on soft animals, particularly on coelenterate polyps. Their methods of feeding have been described by Cole (1906) and by Prell (1910). An individual of *Nymphon* feeding on the hydrozoan *Campanularia*, as observed by Prell, when it happens on a hydranth tentacle, grasps the stalk of the latter with one of its chelicerae and brings the hydrotheca against its mouth. A sucking action of the proboscis now begins, and the tentacle of the hydranth slowly disappears into the mouth of the pycnogonid. Finally the last cellular connection is broken, and the empty hydrotheca is cast away. The ingested material, Prell says, forms a conical mass within the proboscis, which in a short time, by the ceaseless action of the straining apparatus, is reduced to plasmatic lumps, which go into the oesophagus. The description by Cole of an *Anoplodactylus* feeding on *Eudendrium* is essentially the same, except that Cole says the hydranth is broken off and forced into the mouth by the

51

chelicerae, or sometimes broken up by the pincers and the pieces then apparently sucked into the mouth. A species of *Phoxichilidium* is recorded by Prell as feeding on a variety of hydrozoans, on the scyphozoan *Lucernaria*, and also on Bryozoa. In attacking *Lucernaria* it grasps the stalk of a tentacle group with one of its chelae, cuts it off, and brings the severed end to its mouth, whereupon the sucking action of the proboscis begins. *Pycnogonum littorale*, according to Prell, feeds on sea anemones (Actinozoa). *Pycnogonum* has no chelicerae (fig. 15 C), but it clings to the anemone by its feet and thrusts its proboscis into the body of the victim. The end of the proboscis is abruptly narrowed, but the mouth has no armature other than the three triangular, hard-edged, jawlike lips, which evidently are of sufficient strength to cut the skin of the anemone.

The food material ingested by the Pycnogonida, according to Schlottke (1933), contains no solid particles or remnants of tissue, indicating that it must be at once subjected to enzyme action, since it could not be reduced to this condition by the straining apparatus in the proboscis. Schlottke observes that the mesenteron discharges secretion into the lumen shortly before the beginning of food intake. The liquefied food is then absorbed and digested within cells of the mesenteron walls and diverticula; the digestive cells, finally containing only waste products, are cast off from the epithelium and discharged through the intestine. The whole process of feeding and digestion in the Pycnogonida is clearly very similar to that in Arachnida.

The appendages of the Pycnogonida are of special interest because of their numerous segments, their differentiated structure, and particularly their variable number in related species, a thing hardly to be expected in a group of animals otherwise so nearly alike. The large, three-segmented, chelate first appendages of *Nymphon* (fig. 16 C, *Chl*) and of most other genera appear to be identical with the chelicerae of *Limulus* and the arachnids, and they are said to be likewise innervated from the tritocerebral ganglia of the central nervous system. The chelicerae arise from anterior lobes of the body at the sides of the base of the proboscis. In a few families the chelicerae are absent, as in *Pycnogonum* (fig. 15 C), or, if present, they may be nonchelate. The second appendages, or palpi (fig. 15 D, E, *Plp*), may very well correspond with the pedipalps of the arachnids. In *Nymphon hirtipes* they are short, slender appendages of five segments each, arising from small body lobes at the sides of the cheliceral lobes (fig. 16 C, *Plp*). In other forms the palpi have a variable number of segments, and in many genera they are much reduced in size, or are absent as in *Pycnogonum*. The third appendages, or ovigerous legs (fig. 15 E, *1L*), arise ventrally on the anterior section of the body close in front of the bases of the first walking legs. They are well developed in both sexes of *Nymphon hirtipes*, though they differ somewhat in the male and the female; they appear to have 11 segments, counting the terminal claw, but their structure can be better described after a study of the walking legs. The presence or absence of the first three appendages, and their variable character in the different families and principal genera of the Pycnogonida are shown in tabular form by Hedgpeth (1947).

The four pairs of long walking legs do not differ essentially from one another in length, shape, or segmentation. Each leg (fig. 16 A) has nine apparent segments, but a study of the musculature will show that the two parts between the tibia (*Tb*) and the pretarsus (*Ptar*) are tarsal subsegments (*Tar*). The pycnogonid leg strikingly resembles an arachnid leg in having a patellar segment (*Pat*) that takes a horizontal position between the femur and the tibia, so that the leg might be said to have a double knee (*k, k'*). In the basal part of each leg, proximal to the femur (*Fm*), are three short segments that would appear to be the coxa (*Cx*), a first trochanter (*1Tr*), and a prefemur, or second trochanter (*2Tr*). The first segment has a horizontal hinge on the supporting body lobe, instead of the more usual vertical hinge of the coxa on the body; the second segment has a vertical hinge on the first, but the next two joints have horizontal hinges. Among the Arachnida three similar basal segments are present in some of the legs of Solpugida. The pycnogonid leg as a whole turns up and down in a vertical plane on its basal hinge with the body, and its only point of anterior and posterior movement is at the joint between the first and second segments. The musculature of each of the segments of the leg is appropriate to the movement at the joint with the segment proximal to it, the second segment having anterior and posterior muscles, while all the others have dorsal and ventral muscles, except the tarsus, which in the walking legs has only a ventral muscle. The movements of the legs of Pycnogonida have been described by Cole (1901) and by Prell (1910), and it is easily seen that the action of the legs at the joints is precisely adapted to the way the pycnogonids move their legs, which is principally up and down both in walking and in swimming.

The apical segment of the pycnogonid leg is a decurved claw (fig. 16 A, *Ptar*), which is clearly shown to be the pretarsus by the presence of a levator and a depressor muscle inserted on its base (B). In *Nymphon hirtipes* two small accessory claws (B, *Un*) arise from the base of the pretarsus, but in other species there may be only a single accessory claw or none. Intervening between the pretarsus and the end of the tibia are two segmentlike divisions of the leg, the first of which is usually termed the "propodus" and the second the "tarsus." Since no muscles connect these two parts, however, the latter are simply subdivisions of the tarsus (A, *Tar*). Attached on the base of the first tarsomere is a long depressor muscle of the tarsus arising in the tibia. The same tarsal structure and musculature are seen in the much shorter leg of *Pycnogonum* (E). The tibia and the patella have each two antagonistic muscles, a levator and a depressor. The simple musculature of the pycnogonid leg has little resemblance to the leg musculature of *Limulus*.

The first legs, or *ovigers* as they are called, in *Nymphon hirtipes* differ somewhat between the male and the female. In the female they are uniformly slender with no particular differentiation of the segments (fig. 15 E, *1L*). In the male (fig. 16 G) the patella (*Pat*) is disproportionately long and abruptly thickened at its distal end, where it bears a brush of long stiff hairs on the outer surface. In an egg-carrying leg (H) it is seen that the egg mass surrounds the slender part of the patella, and is held here by the distal enlargement of the segment and the bristlelike hairs. Between the tibia and the pretarsus of the ovigerous leg in each sex of *Nymphon* are four short divisions of the limb (G, *Tar*) in what appears to correspond with the two-part tarsal section of a walking leg (A, *Tar*). An examination of these four parts of the ovigerous leg, however, reveals the unexpected fact that they are individually musculated (J), each having a flexor muscle inserted on its base arising in the part proximal to it, while the first has two muscles arising in the tibia. The presence of intratarsal muscles in the ovigerous legs of the pycnogonids was noted by Börner (1903), but in no other arthropods are tarsal subdivisions known to be equipped individually with muscles. A segmentation of a different type appears to be present in the ovigerous leg of *Anoplodactylus lentus*. The oviger of this species consists of

54

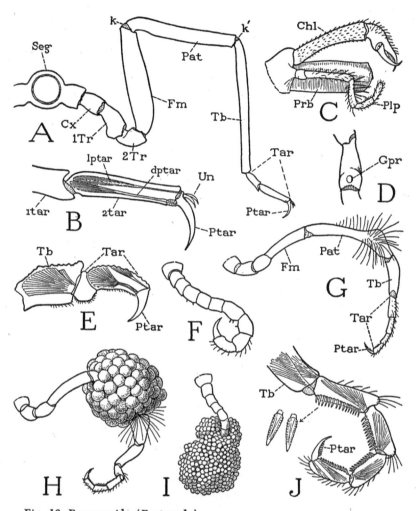

Fig. 16. Pycnogonida (Pantapoda).

A, *Nymphon hirtipes* Bell, female, section of body with leg. B, same, tarsus and pretarsus of a walking leg, showing pretarsal muscles. C, same, proboscis, chelicera, and palp of right side. D, same, female, first trochanter of a walking leg, with opening of oviduct on ventral surface. E, *Pycnogonum littorale* Ström., tarsus and pretarsus of a walking leg, showing muscles. F, same, ovigerous leg of male. G, *Nymphon hirtipes* Bell, ovigerous leg of male. H, same, ovigerous leg of male bearing eggs. I, *Pycnogonum littorale* Ström., ovigerous leg of male bearing eggs. J, *Nymphon hirtipes* Bell, distal segments of ovigerous leg of male, showing tarsal segments individually musculated.

Cx, coxa; *dptar,* depressor muscle of pretarsus; *Fm,* femur; *Gpr,* gonopore; *k, k',* double knee bend of leg; *lptar,* levator muscle of pretarsus; *Pat,* patella; *Ptar,* pretarsus; *Seg,* body segment; *Tar,* tarsus; *1tar,* first tarsal subsegment; *2tar,* second tarsal subsegment; *Tb,* tibia; *1Tr,* first trochanter; *2Tr,* second trochanter; *Un,* ungues.

only seven segments, the terminal claw segment being absent, and at each movable joint there are two antagonistic muscles, suggesting that the terminal segment alone is the tarsus. In *Pycnogonum littorale* ovigerous legs are present only in the male; they are relatively very small (F) and the segments are but little differentiated in size, but the number of them is the same as in *Nymphon*. An egg-bearing leg of *Pycnogonum* (I) has the claw-bearing tarsal segments simply hooked into the egg mass. In *Nymphon hirtipes* the ventral sides of the tarsal segments and the pretarsal claw are armed with combs of small denticulate spines (J), and these segments are usually strongly flexed. It is shown by Prell (1910) that the tarsi of the ovigerous legs are used to clean the other legs.

The exit apertures of the reproductive system of the pycnogonids are on the second segments of the walking legs, generally but not always on the undersurfaces (fig. 16 D, *Gpr*). In the female there is usually an opening on each leg (fig. 15 E, *Gprs*), but in the male the openings may be fewer, sometimes limited to a single pair on the last legs; in *Nymphon hirtipes* the male has apertures on the last two pairs of legs. The testes and the ovaries lie in the body above the alimentary canal, but they send branches into the walking legs; in a gravid female the femora may be seen to be full of developing eggs. The pycnogonids are unique among the arthropods in the possession of multiple genital openings, but it is not anomalous that the exits should be on the bases of the legs; this undoubtedly was their position in the primitive arthropods, since the genital ducts are to be traced back to coelomic ducts opening at the bases of the appendages, as in the modern Onychophora. In *Limulus* the genital apertures are on the opercular appendages of the eighth body segment, and in many Crustacea and most Diplopoda they are on the coxae of a specific pair of legs. The position of the gonopores in the Pycnogonida, therefore, is exceptional only in that it is on the second leg segment instead of the first.

It is impossible to arrive at any positive conclusion concerning the relationships of the Pycnogonida, except that the animals are arthropods. The absence of antennae and the presence of three-segmented, chelate first appendages might seem in themselves sufficient evidence for including the pycnogonids with the merostomes and arachnids in the large group of Chelicerata, but, as seen in the arachnids and crustaceans, chelae may be independently developed on almost any of the limbs, and in some other arthropods antennae are absent. However, the pycnogonids have the arachnid type of leg, and the eyes are said to have the "inverted" structure characteristic of the median eyes of arachnids (Wirén, 1918). A common feature of the Xiphosurida, Arachnida, and Pycnogonida is the branching of the stomach into radiating, tubular diverticula, in the walls of which, according to Schlottke (1933), the final phase of digestion takes place intracellularly. The inner sac of the proboscis that opens into the oesophagus resembles the sucking organ of the arachnids, and the proboscis itself might be comparable to the snout of the primitive Palpigradi among the Arachnida (fig. 23 D).

The pycnogonids, of course, differ from the other chelicerates, except Solpudiga, in that the leg-bearing part of the body is partly segmented and that the opisthosoma is reduced to a diminutive abdomen. The unsegmented anterior part of the body carrying the first four pairs of limbs might be likened to the "head" of a trilobite with its four pairs of legs, but the idea of a trilobite relationship for the pycnogonids is hardly to be entertained. All the pycnogonids have at least four pairs of walking legs, which gives them a spiderlike appearance, but since there are usually three pairs of appendages in front of the walking legs, the pycnogonids in general have seven pairs of limbs, and hence are more comparable to *Limulus* in the number of appendages, since the fourth pair of pycnogonid walking legs would correspond with the chilaria of the xiphosurids. In three families of the Pycnogonida, however, there are forms having five pairs of walking legs, and in one of these families a six-legged species is known. To keep the comparison with *Limulus*, then, it would seem that we should have to regard the extra legs as representing the first two pairs of opisthosomatic appendages; but, since the last leg-bearing segment and the abdomen retain the same structure regardless of the number of legs, it is probable that the extra pairs, when present, are interpolations somewhere between the other legs. It has been shown by Hedgpeth (1947) that the number of legs in the Pycnogonida can have no phylogenetic or taxonomic significance, since the five-legged forms occur in widely separated families and are closely related to four-legged forms in these same families, and the single species known to have six pairs of walking legs belongs to a family including four-legged and five-legged forms.

In conclusion, then, we can only repeat what was said at the beginning, that the pycnogonids are queer animals. But still, we can feel safe in claiming them for arthropods, and probably as a branch of the Chelicerata having more arachnid than merostome characters. Particularly arachnoid is their method of feeding and their sucking apparatus for the ingestion of food. For a full account of the pycnogonids the student should consult Helfer and Schlottke (1935).

V

THE ARACHNIDA

THE arachnids are more easily recognized than defined. They have so many features in common with *Limulus* that some zoologists have classed *Limulus* in the Arachnida. The essential differences between the Xiphosurida and the Arachnida are in the feeding organs and the organs of respiration. The arachnids feed on liquids extracted from their prey, which are ingested by a pharyngeal sucking pump; the xiphosurids feed on solid food, which is ground up in a proventricular grist mill. The arachnids are terrestrial and breathe by means of lungs or tracheae; the xiphosurids, being aquatic, have abdominal gills, and theoretical attempts to derive the arachnid lungs from gills are not convincing.

The most primitive of modern arachnids, the Palpigradi, are more generalized than *Limulus*. The Xiphosurida and the Arachnida, therefore, are two branches of the subphylum Chelicerata, but their common ancestors are not known. While there are paleontological reasons for believing that the xiphosurids and the trilobites had a common progenitor, the actual origin of the arachnids is obscure. However, as was noted in the last chapter, the pycnogonids have some surprisingly arachnoid characters. The scorpions have a superficial resemblance to the Eurypterida, but the scorpion, as compared with the Palpigradi, is not a primitive arachnid. However, it is not an object of the present text to discuss theoretical arthropod phylogeny. The student may learn the essentials of arachnid anatomy from a study of the scorpion, the spiders, and a tick, which are the principal subjects of this chapter.

THE SCORPION

The scorpion in appearance (fig. 17 A) is a highly distinctive arachnid that could not possibly be mistaken for any other member of the class; the combination of large, chelate pedipalps with a sting at the end of a segmented tail alone proclaims the animal to be a true scorpion. The so-called whip scorpion (fig. 23 E) and the pseudoscorpion have no sting. About 600 species of scorpions are known, most of them two to four inches in length, but there is one only half an inch long, while the huge African *Pandinus* attains a length of seven inches. Scorpions are widely distributed throughout the tropical parts of the earth and in most of the warmer regions of the temperate zones; 22 species occur in the United States. For a general account of the habits and modes of life of scorpions the reader is referred to the article "Scorpion" by Petrunkevitch in the *Encyclopaedia Britannica* (1947).

General Structure of a Scorpion

A scorpion at first glance (fig. 17 A) appears to have an elongate oval body, supported on four pairs of legs, and a thick, jointed tail bearing the sting at its extremity. The body, however, is divided into an unsegmented anterior part, which alone bears the appendages, and a larger segmented posterior part; the tail is a slender extension of the body, consisting of five segments, with the anus in the last segment. Morphologically considered, then, the trunk of the scorpion includes a *prosoma,* covered by an unsegmented plate, or *carapace,* and a segmented *opisthosoma,* or *abdomen,* which is differentiated into an anterior *mesosoma,* or *preabdomen,* and a posterior *metasoma,* or *postabdomen,* which is the tail. The prosoma bears the usual arachnid appendages, which are the *chelicerae* (A, B, *Chl*), the *pedipalps* (*Pdp*), and four pairs of *legs* (*1L–4L*); it therefore includes six primary postoral somites, and is thus comparable with the prosoma of *Limulus,* except that it lacks the chilaria and the corresponding seventh segment. The seventh body segment of the scorpion, in fact, is known to be suppressed during embryonic development, so that the first segment of the opisthosoma in the adult is the eighth. The mesosoma contains seven segments, which are segments *VIII–XIV,* and the tail has five segments (*XV–XIX*), not including the terminal sting. The adult scorpion, therefore, has 18 postoral

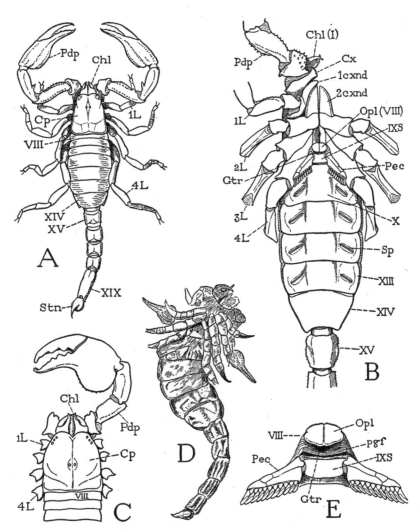

Fig. 17. Arachnida—Scorpionida.
A, *Chactas vanbenedeni* Gervais, Chactidae (2 lateral eyes on each side),
Colombia. B, *Pandinus* sp., Scorpionidae, Congo, ventral surface. C, same,
dorsal (3 lateral eyes on each side). D, *Palaeophonus hunteri* Pocock, Scottish
Silurian scorpion, 35.5 mm. long (from Pocock, 1901). E, *Pandinus* sp., genital
region of segment VIII, and pectines of segment IX, ventral.

For explanation of lettering see pages 126–127.

segments in all, which is the maximum number of segments possessed by any other arachnid, but if we count the suppressed seventh segment, it has 19 segments.

Studies by different writers on the correlation of the nerve centers of the scorpion with the body segmentation have given somewhat different results. McClendon (1904) described 20 pairs of neuromeres, or primary segmental ganglia, of which the first pair forms the brain, and the other 19 the ventral nerve cord. According to Buxton (1917), however, there are only 18 neuromeres in the nerve cord, which is one less than the number of body segments. Both Buxton and Petrunkevitch (1949), therefore, suggest that the first tail segment is a secondary subdivision of the last mesosomatic segment. Kästner (1940), on the other hand, re-examining the subject in species of several genera, finds that there are 19 pairs of primary ganglia formed in the postoral nervous system, of which the cheliceral ganglia unite with the brain, and the ganglia of segment VII disappear along with the segment itself; in other words, there is at first a pair of ganglia for each of the primary 19 postoral segments. In the adult, however, the five postcheliceral ganglia of the prosoma and the first four persisting opisthosomatic ganglia unite in a large nerve mass, or suboesophageal ganglion, lying in the prosoma, while the ganglia of segments XII to XVII remain separate, though the first two are displaced forward, and the ganglia of segments XVIII and XIX unite to form a double last ganglion lying in segment XVIII.

The carapace of the prosoma (fig. 17 A, C, *Cp*) is widest behind, somewhat narrowed anteriorly. On it are situated three groups of small, simple eyes, there being a pair of median eyes, and two groups of more anterior lateral eyes, each with from two to five eyes according to the species. No arachnid has compound eyes. The front margin of the carapace projects as a free fold over the bases of the chelicerae (fig. 21 D). From the inner end of the under lamella, or doublure, of the fold, the membranous anterior wall of the body goes downward to the base of a large median lobe (*Lm*), which is the *labrum*. The labrum lies between the bases of the pedipalps (B), but the chelicerae arise entirely dorsal to the labrum. At the base of the labrum is an irregular sclerotization of the body wall (D, *Epst*) representing the *epistome* of other arachnids, which is usually a horizontal plate forming a bridge between the upper surfaces of the pedipalp coxae and supporting the labrum. Though the adult arachnid has no distinct head, the labrum, the epistome, and the eye-bearing region of the carapace are derived from the cephalic lobe of the embryo. The lateral margins of the carapace are not extended beyond the leg bases (B), and a cross section of the prosoma of the scorpion (fig. 22 A), therefore, has quite a different shape from that of *Limulus* (fig. 7 A). The edges of the carapace are separated from the bases of the legs by narrow pleural folds of the integument (figs. 19 E, 22 A, *Pl*). The ventral surface of the prosoma is occupied almost entirely by the coxae of the legs (fig. 17 B), there being only a small median sternal plate between the posterior two pairs of coxae.

The fully segmented mesosoma, or preabdomen, is much broader than the prosoma. On the dorsum (fig. 17 A) each of its seven segments is covered by a distinct tergal plate. On the undersurface (B) there are likewise seven segmental divisions, but the sternal plates are not all so simple or uniform in size as the tergal plates. The venter of the first segment (*VIII*) is wedged between the anterior ends of the last leg coxae, where it is greatly reduced in size and has no connection with the tergum of its segment. On it, however, is situated the genital opening, or gonotreme, in each sex (B, E, *Gtr*), which is covered by a small plate or pair of plates forming an operculum (*Opl*), and is bordered behind by a transverse postgenital fold (E, *pgf*) of the integument. The second opisthosomatic sternum (B, *IXS*) is a small quadrate plate bearing a pair of comblike appendages known as the *pectines* (*Pec*), which will be described in connection with the other appendages. This sternum also is separated from its tergum, except for narrow pleural folds that run along the posterior margins of the hind coxae. The following five mesosomatic sterna are broad plates united laterally with the corresponding terga by infolded pleural membranes (fig. 22 B, *Pl*) that allow a considerable dorsoventral expansion of the abdomen. The tergal and sternal margins of the last preabdominal segment, however, come together posteriorly and are united at the posterior angles of the segment. On the lateral areas of sterna X to XIII are oblique slits (fig. 17 B, *Sp*), which are the apertures, or *spiracles,* of the internal respiratory organs known as *book lungs,* to be described later.

The five segments of the metasoma, or tail, are simple rings, the tergal and sternal arcs being entirely confluent. In cross section (fig. 18 B) a tail segment has an octagonal shape with unequal sides. The segments have no specific articulations on each other, but they are strongly connected, and in such a manner that their principal movement is in a vertical plane, most freely in a dorsal direction (D),

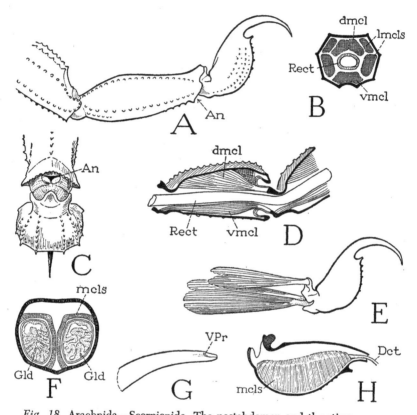

Fig. 18. Arachnida—Scorpionida. The postabdomen and the sting.

A, *Centruroides* sp., last two postabdominal segments and sting. B, same, cross section of a postabdominal segment, showing position of muscles. C, *Pandinus* sp., end of last postabdominal segment, with sting, ventral. D, *Centruroides* sp., longitudinal section of first postabdominal segment and base of second. E, same, the sting and its muscles. F, same, cross section through base of sting. G, same, terminal part of sting, showing aperture of left venom duct. H, same, right half of base of sting, mesal, showing muscles covering right venom gland, and exit duct.

For explanation of lettering see pages 126–127.

though lesser sidewise movements also are permitted. On the base of each segment are attached six muscles arising in the preceding segment, one dorsal (B, *dmcl*), four dorsolateral and ventrolateral (*lmcls*), and one ventral (*vmcl*). The basal segment has a corresponding but stronger musculature from the last mesosomatic segment. The dorsal flexion of the segments on one another allows the tail to be freely turned forward over the back of the scorpion and the inverted point of the sting to be thrust upward into the prey held in the jaws of the pedipalps. The tail is traversed by the long intestine (D, *Rect*), which opens ventrally in the end of the last segment (C, *An*) beneath the base of the sting. The segmentlike sting of the scorpion, as the tail spine of *Limulus,* is generally regarded as the telson, but in the mandibulate arthropods, especially in the Crustacea, the anus is situated *on* the apical telson, which fact might suggest that the sting-bearing segment of the scorpion is the true telson, and that the sting is a specialized appendage of it. On the other hand, the presence during embryonic development of a nerve center in the sting-bearing segment would appear to be conclusive evidence that this segment is a true somite and not the telson. Possibly, then, the anus has been transposed secondarily from the telson into the intersegmental membrane before it.

The sting of the scorpion (fig. 18 A) consists of a large, bulbous basal part and of a long, sharp, decurved distal spine that contains near its tip the outlets (G, *VPr*) of the venom glands in the bulb. The base of the organ is movably articulated on the end of the tail and is provided with four muscles (E) arising in the last tail segment, two dorsal and one ventrolateral on each side. In the reversed position of the tail, the smaller dorsal muscles depress the point of the sting, and the larger ventrolateral muscles give the upward thrust of the weapon into the body of the prey. Probably, acting as antagonists, the ventral muscles produce also lateral movements of the sting. The venom of the sting is secreted in two saclike glands contained in the swollen base of the organ (fig. 18 F, *Gld*). The glands have individual ducts opening separately near the point of sting through two lateral pores (G, *VPr*), from which grooves extend to the tip. Each gland is closely invested along its mesal and dorsal surfaces by a thick muscular sheath of several layers of semicircular fibers (F, H, *mcls*), attached dorsally above the gland on the outer wall of the containing capsule and ventrally on the lower wall. Contraction of the muscles evidently compresses the gland sacs against the rigid capsule walls and expels the venom through the ducts. The effect of the scorpion's poison on insects captured for food is to kill them, but the effect on man is highly variable according

65

to the species of scorpion. Some species produce symptoms no worse than those of a bee's sting, while with others the results may be serious, and even fatal.

A cuticular endoskeleton is but little developed in the scorpion. Such as there is consists of a pair of epistomal apodemes, a small median apodeme of the prosomatic sternum (fig. 22 A, S), and apodemal extensions from the bases of the chelicerae and the coxae of the legs. Within the prosoma, however, is a noncuticular plate corresponding with the endosternum of *Limulus,* which gives attachment to numerous muscles but has no other connection with the body wall. The endosternum of the scorpion is much more complex than the simple endosternal plate of *Limulus;* its structure has been described and illustrated in detail by Lankester and Beck (*in* Lankester, Benham, and Beck, 1885).

The Appendages

The segmental appendages of the scorpion include the chelicerae, the pedipalps and the legs of the prosoma, and the pectines of the abdomen. The prosomatic appendages will best be studied beginning with the legs, because the chelicerae and the pedipalps have fewer segments and their reduced segmentation can be more easily interpreted after a study is made of the legs and their musculature.

The Legs— The leg of the scorpion (fig. 19 A) is divided by flexible joints into eight podomeres (leg sections), but since the two parts in the tarsal region (*Tar*) are not connected by muscles, they are to be regarded as subsegments of the tarsus, or tarsomeres. The seven true leg segments, therefore, beginning at the base, are the *coxa* (*Cx*), *trochanter* (*Tr*), *femur* (*Fm*), *patella* (*Pat*), *tibia* (*Tb*), *tarsus* (*Tar*), and *pretarsus* (*Ptar*). The pretarsus is visible principally as a pair of claws, the body of the segment being mostly concealed in the end of the distal tarsomere (F, *Ptar*).

The coxae, as already noted, occupy most of the undersurface of the prosoma (fig. 17 B). Those of the first legs (*1L*) are short, but the others become successively longer and more oblique, the last pair being particularly long and slender. The coxae of the first and second legs have each a large mesal lobe, or endite (*cxnd*), projecting forward beneath the mouth. The coxae have no points of articulation on the body, and, except those of the first pair, are but little movable; yet, as shown by Beck (1885), they all have a strong

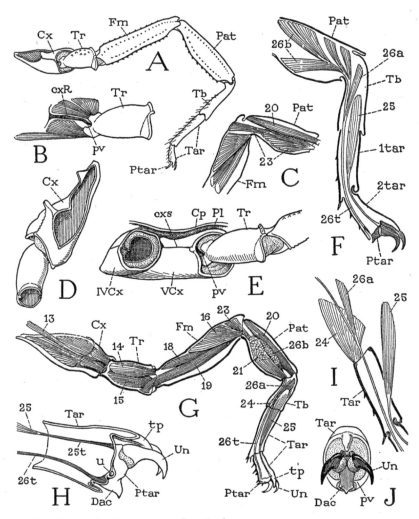

Fig. 19. Arachnida—Scorpionida. The legs.

A, *Centruroides* sp., third left leg, anterior. B, same, trochanter and its muscles. C, same, patella and distal half of femur (muscles 21 and 26 removed, see G). D, coxa and trochanter of third left leg, dorsal. E, same, base of third leg and coxa of second leg, lateral. F, *Pandinus* sp., leg segments beyond femur, showing distribution of pretarsal muscles. G, same, third left leg, showing muscles visible in anterior view. H, same, pretarsus and its muscle tendons. I, *Centruroides* sp., base of tarsus, with single muscle (*24*) attached ventrally on posterior side of base. J, same, end view of tarsus and pretarsus.

For explanation of lettering see pages 126–127.

musculature, including muscles from the carapace and from the endosternum, inserted on their bases or on basal apodemes. The long ventral surfaces of the second, third, and fourth coxae reach to the mid-line of the body, but the dorsal and lateral surfaces are relatively short (fig. 19 A, D, *Cx*). On its outer side the coxal wall is marked by a longitudinal groove (E, *cxs*) forming a strong internal ridge (B, *cxR*) that runs out posteriorly in a pivotlike process (B, E, *pv*), on which alone the trochanter is articulated. Because of the immobility of the coxae, the principal basal movement of the scorpion's legs is at the coxotrochanteral joints.

The trochanter is movable on the coxa by a great mass of muscle fibers inserted on the entire periphery of its base (fig. 19 G). The fibers, however, can be separated into at least 12 distinct muscles taking their origins in the coxa. The trochanter, therefore, is movable in any direction, but the lateral or forward movement is restricted by the position of the articular pivot and is dependent on the small muscles of the short anterior wall of the coxa (B) inserted above and below the articulation. The forward reach of the scorpion's legs is due to the more or less horizontal position of the extended limbs, by which flexion at the joints becomes anterior instead of ventral. A slender muscle from the body (G, *13*) that runs through the coxa is probably the "plastro-deutomeral" muscle of Beck (1885), but it tapers distally into a tendon that traverses the trochanter and is attached on the dorsal rim of the base of the femur.

The elongate femur is hinged to the trochanter by a strong, horizontal, dicondylic articulation that permits movement only in a dorso-ventral direction. It is accordingly provided with a dorsal levator muscle (fig. 19 G, *14*), and a two-branched ventral depressor muscle (*15*) arising in the trochanter, in addition to the slender levator (*13*) arising in the body, apparently on the endosternum.

The patella is an important segment of the scorpion's leg; its interpolation between the femur and the tibia gives the leg its "double knee." The femoropatellar articulation is a dicondylic hinge with a horizontal axis, but the active movement of the patella is downward on the end of the femur, since it has only depressor muscles. A wide, fan-shaped anterior depressor (fig. 19 G, *16*) and a similar posterior depressor (*17*, not seen in the figure) arise in the distal part of the femur; a long ventral depressor (*18*) comes from the base of the femur, and another (*19*) takes its origin ventrally in the end of the trochanter.

The patellotibial joint again has a transverse hinge line between anterior and posterior articulations, and the short tibia turns abruptly downward from the end of the patella, but it has both flexor and extensor muscles. A single long extensor (fig. 19 G, *20*) arises dorsally in the base of the patella; two lateral flexors (*21*, and *22* not seen in the figure) arise on opposite sides of the patella, while a ventral flexor (C, *23*) has spreading fibers on the ventral wall of the patella and a median bundle arising in the distal end of the femur.

The long basal subsegment of the tarsus in the scorpion here described, *Centruroides*, has an anterior articulation on the tibia and is provided with but one muscle, a depressor (fig. 19 G, I, *24*), which arises posteriorly in the tibia and is inserted ventroposteriorly on the base of the tarsus. No muscles are present between the two tarsomeres, though the latter are united by a flexible joint. From the end of the short distal tarsomere a fingerlike process (G, *tp*) projects over the pretarsus.

The claw-bearing pretarsus is often not recognized to be a true segment of the limb, but its structure as seen when fully exposed (fig. 19 H) leaves no doubt of its segmental nature. The body of the segment is a short but complete ring with anterior and posterior articular sockets that receive pivotal processes of the tarsus (J, *pv*), and it is provided with antagonistic dorsal and ventral muscles attached on its base by strong tendons (H, *25t*, *26t*). Ventrally the pretarsus is produced into a short median claw, or *dactyl* (*Dac*), which suggests that it corresponds with the terminal dactylopodite of a crustacean leg. The curved paired claws, which may be designated the *ungues* (*Un*), are set on membranous bases and are therefore flexible on the pretarsus, but they have no means of independent movement. The levator, or extensor, muscle of the pretarsus arises in the tibia (F, G, *25*), the depressor, or flexor, has a branch in the tibia (*26a*), but the tendon continues into the patella and gives attachment to several fiber bundles (F, *26b*) arising in the patella. The pretarsus thus has a rocking movement in a vertical plane on its tarsal articulations, which turns the claws up and down. The large size of the depressor muscle gives the claws the necessary strength on the downstroke. It seems remarkable that 26 muscles should be required to operate a single leg of the scorpion, in addition to those inserted on the coxa.

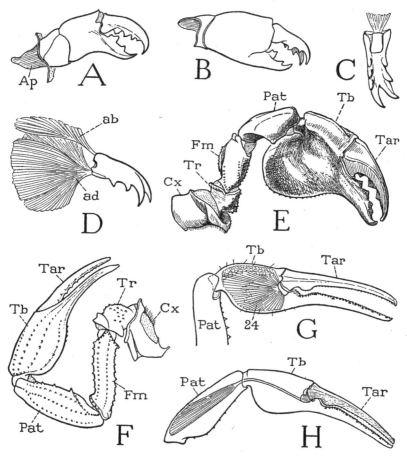

Fig. 20. Arachnida—Scorpionida. The chelicerae and the pedipalps.
A, *Pandinus* sp., chelicera. B, *Centruroides* sp., chelicera. C, same, movable finger of chelicera, ventral. D, same, movable finger of chelicera and its muscles. E, *Pandinus* sp., right pedipalp, ventral. F, *Centruroides* sp., left pedipalp, dorsal. G, same, chela of pedipalp, showing tibial muscles of movable finger (*Tar*). H, same, showing patellar muscle of movable finger.
For explanation of lettering see pages 126–127.

The Pedipalps— The scorpion's pedipalp has only six segments, including the movable finger of the chela (fig. 20 E, F), and thus lacks a segment present in the legs. It is therefore of interest to determine what segment is missing. By comparison with a leg the first three segments of the pedipalp must be the coxa (Cx), the trochanter (Tr), and the femur (Fm). If we next examine the musculature of the movable finger within the "hand" of the chela (G), we find that the latter is filled with a great mass of fibers all attached on a ventral process of the base of the movable finger. If the finger represents the pretarsus of a leg, it should have an opening as well as a closing muscle. The segment of the leg that has a single muscle, and that one a flexor, is the tarsus (fig. 19 I, 24). We must conclude, therefore, that the "hand" of the pedipalp chela is the tibia (fig. 20 E, F, G, Tb), the movable finger the tarsus (Tar), and the segment that supports the chela the patella (Pat). The movable finger, of course, may be supposed to include the pretarsal claw in its tip, there being evidence of this in some other arachnids with chelate pedipalps. The strength of the movable finger is reinforced by another muscle arising in the patella (H) and inserted by a strong tendon on the basal knob of the finger. It is surprising perhaps that the chela should not have an opening muscle, but evidently the elasticity of the hinge of the finger on the hand keeps the forceps open until closed by the finger muscles. In contrast to the scorpion chela, the similar chela of the crayfish (fig. 45 B) has both an opening and a closing muscle, and the movable finger is the pretarsus, or dactylopodite.

The Chelicerae— The chelicerae of the scorpion (fig. 20 A, B) are relatively short but powerful pincers. They are three-segmented, and since the chelicerae of none of the Chelicerata have more than three segments, there is no apparent way of identifying the segments. The strongly toothed movable finger of the cheliceral appendage has both an opening and a closing muscle arising in the hand (D), and hence is comparable to the finger of a crustacean chela, which is the dactylopodite. The structural uniformity of the chelicerae in all the chelicerate arthropods would indicate that these appendages have been handed down in their present form from some remote ancestor of the group.

The Pectines— The comblike organs borne on the base of the undersurface of the abdomen (fig. 17 B, Pec), known as the pectines, are movably attached by their bases to the sternal plate of the ninth segment (B, E, IXS), and are regarded as the appendages of this segment. They vary in form in different genera of scorpions, but a

71

typical example of their structure is that shown at E in *Pandinus*. Each pecten bears an anterior row of long teeth, and a posterior row of small teeth. The pectines are peculiar to the scorpions; their specific function is not known, but they are unquestionably important sensory organs, since the teeth bear numerous innervated sensilla. Immediately in front of them is the genital opening, or gonotreme (E, *Gtr*).

The Feeding Apparatus, Digestion, and Excretion

The mouth of the scorpion is concealed within a large, open *preoral cavity* between the broad, soft inner surfaces of the pedipalp coxae (fig. 21 A, *PrC*). The cavity is overhung dorsally by the chelicerae (*Chl*), but ventrally it is closed by a wide underlip formed of the closely approximated endites of the coxae of the first and second legs (*1cxnd, 2cxnd*). When the chelicerae and the pedipalps are removed (B), the mouth is seen to be a small aperture (*Mth*) beneath the base of the large, laterally compressed labrum (*Lm*) that projects forward from between the bases of the pedipalps. Extended anteriorly from below the mouth is the basinlike underlip, the upper part of which is formed of the concave dorsal surfaces of the first coxal endites (B, C, *cxnd*). The opposing edges of these two endites are not in contact but leave between them a median groove, which is closed below by the long, rigid supporting endites of the second leg coxae (A, *2cxnd*). The gutterlike groove leads directly into the mouth. The relation of these preoral structures to one another and to the mouth is seen in the longitudinal section shown at D. The labrum (*Lm*) has a rounded dorsal surface terminating in a fringe of long hairs, below which the anterior wall slopes back to the short ventral surface that overhangs the mouth (*Mth*). Crossing the inner part of the cavity of the labrum are two bundles of transverse compressor muscle fibers (*tmcl*). At the base of the labrum, dorsally, is the thick, irregular epistomal sclerotization of the head integument (*Epst*); on each side it gives off into the body cavity a long epistomal apodeme (*eAp*, only the one on the right seen in the figure). Beneath the labrum and the mouth is the floor of the preoral cavity composed of the endites of the first and second leg coxae.

The mouth leads into a small pear-shaped pouch (fig. 21 D, *Phy*) that enlarges upward from its narrowed entrance at the mouth. This pouch is the sucking organ known as the *pharynx* in arachnology. It is somewhat compressed laterally and rounded at its inner end; the slender oesophagus (*Oe*) departs from its lower wall at the end of a ventral channel from the mouth. The dorsal wall is deeply in-folded lengthwise, and the trough of the invagination is strength-

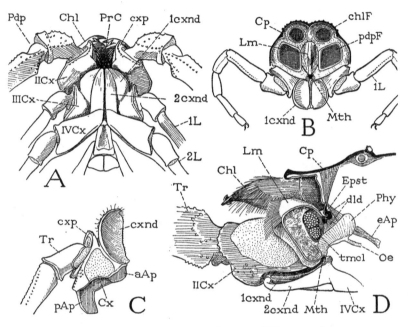

Fig. 21. Arachnida—Scorpionida. Mouth parts of *Centruroides* sp.

A, ventral side of anterior part of body, showing mouth parts surrounding a large preoral food cavity (*PrC*). B, anterior end of body, chelicerae and pedipalps removed, exposing the mouth (*Mth*) and coxal endites of first legs. C, base of first left leg with coxal endite, dorsal. D, longitudinal section through anterior end of body and labrum, right chelicera and base of right pedipalp in place, showing mouth leading into pharynx.

For explanation of lettering see pages 126–127.

ened by an elastic rod. Compressor muscles cover the walls of the pharynx; dilator muscles (*dld*) attached on the concave dorsal wall arise on the epistomal sclerotization (*Epst*) at the base of the labrum, and lateral dilators have their origins on the epistomal apodemes (*eAp*).

The preoral cavity of the scorpion, being open dorsally directly under the short chelicerae, serves as a receptacle for liquids or fragments of the prey held and crushed in the cheliceral pincers. There is no straining apparatus guarding the mouth, such as is present in many other arachnids, but the small size of the scorpion's mouth precludes the entrance of large pieces of food, and the scorpion, in common with all other arachnids, has only a sucking apparatus for the ingestion of food.

In each of the orders of Arachnida the mouth parts are different, as shown elsewhere by the writer (1948), but in all except the Palpigradi they are simply modifications of the same structures that compose the feeding organs of the scorpion. On the other hand, there is a radical difference between the arachnids and *Limulus* in the structure of the oral region of the body, the ingestion apparatus, and the manner of feeding. The arachnids feed on liquids extracted from the prey either mechanically or by extraoral digestion, and their organ of ingestion is a sucking pump; *Limulus* devours pieces of animal food, which are ground up in a proventricular gizzard.

Scorpions in captivity will eat any kind of small arthropod. The prey is seized by the chelae of the pedipalps, which in large species are able to crush hard-shelled beetles, but if the victim is not killed by crushing, it is subdued by the abdominal sting. From the pedipalps the food is passed to the chelicerae, one of which holds it while the other rips open the body and pulls out the viscera. The extracted material collected in the preoral cavity is thoroughly cut up by the chelicerae, then reduced to a pulp by digestive juices discharged upon it, probably from the stomach, and finally in liquid form it is sucked into the mouth by the pharynx. A detailed account of the feeding of a scorpion is given by Kästner (1940, pp. 154–158).

The stomodaeal oesophagus leads into the mesenteron, which consists of a tubular stomach section lying in the prosoma and mesosoma, and of a long intestinal section that begins in segment XII and extends into the last tail segment, where it joins the very short proctodaeum that opens through the anus. From the sides of the axial tube of the stomach are given off six pairs of diverticula that expand into large sacs with soft, infolded walls, all of which are closely packed along the sides of the body and bound together by a covering tunic of connective tissue. The first pair of diverticula is in the prosoma, the others arise in the first five segments of the mesosoma, but those of the last pair are branched and extend into the base of the metasoma. These diverticular sacs of the stomach form the major part of the alimentary system of the scorpion and occupy most of the space in the mesosoma. The digestive processes of the scorpion have been described by Schlottke (1934). The epithelial walls of the stomach sacs include secretory cells and digestive cells. The first produce enzymes that are given off into the lumina of the sacs and accomplish a preliminary digestion of the food pulp received from the pharynx; according to Pavlovsky and Zarin (1926), the digestive enzymes of the scorpion include amylase, lipase, and proteinases. The digestive cells absorb the products of enzyme action, and within them takes place the final stage of digestion. The cells at last become filled with excretory granules, which are thrown out and discharged through the intestine.

Connected with the mesenteron are two pairs of excretory tubes, known as Malpighian vessels, which remove waste matter from the blood; one pair branches in the mesosoma, the other goes into the prosoma. In addition, the scorpion has a single pair of coelomic excretory glands lying in the posterior part of the prosoma, the ducts of which open in the grooves between the coxae of the third legs and the prosomatic sternal plate.

The Respiratory Organs

The breathing apparatus of the scorpion consists of four pairs of respiratory organs located inside the abdomen above the sterna of segments X to XIII (fig. 22 B, *bl*) and opening by slitlike apertures, the *spiracles*, on the lateral parts of the sterna of these same segments (C, *Sp*). The organs are known as *book lungs* because the essential parts of them consist of numerous thin, hollow, leaflike lamellae attached on a common base like the leaves of a book (D). The spiracle of each organ (E, *Sp*) opens into an obliquely elongate atrial chamber (D, E, *Atr*), which is produced beyond each end of the spiracle in a tapering extension. The anterior and posterior walls of the atrium are membranous and flexible, but the arched dorsal wall is crossed by closely set, longitudinal bars (F, *s*), which are the septa between narrow slits (*o*) opening into the lumina of the lamellae (*lam*). The lamellae are somewhat triangular in shape (E), set vertically on the atrium, and extend anteriorly from it. The atrium and the leaflets are ingrowths of the body wall, and are lined with a delicate cuticle which is drawn out and renewed at each moult. In the species illustrated the leaflets appear to be entirely free from one another and can be readily spread apart (F), but in some scorpions they are said to be united by protoplasmic strands of their epidermal walls.

Each lung is enclosed in a pulmonary cavity, or sinus, of the haemocoele covered by a sheet of connective tissue. The lumina of the leaflets contain the respiratory air derived from the atrium, and

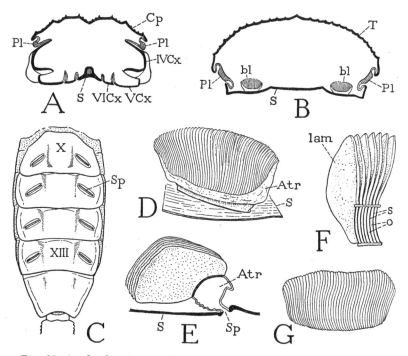

Fig. 22. Arachnida—Scorpionida.
A, *Centruroides* sp., outline of cross section of prosoma through coxae of second legs (*IVCx*), showing pleural folds (*Pl*) between carapace and coxae. B, same, outline of cross section through a mesosomatic segment, showing position of book lungs. C, *Pandinus* sp., ventral surface of mesosoma. D, same, right book lung of segment X, dorsoposterior view. E, same, vertical section of book lung, showing spiracular entrance to atrium. F, same, part of anterodorsal wall of atrium, showing slitlike openings (*o*) into lung lamellae. G, same, book lung, dorsal.
For explanation of lettering see pages 126–127.

the blood circulates in the spaces between the lamellae, the gas exchange taking place through the very thin walls of the latter. Air enters the leaflets probably by diffusion from the atrium, but the atrium is said, in some species at least, to have a ventilating action by means of muscles. According to Fraenkel (1929), in a species of *Buthus* there are two muscles attached on the posterior wall that produce an opening of the spiracle and an expansion of the atrium, the closing being automatic on relaxation of the muscles. The opening of the spiracles, Fraenkel observes, takes place only when the scorpion is active. The wall of each pulmonary sinus is connected with the pericardium in the dorsal part of the abdomen by a strand called the "pericardio-pulmonary muscle" by Lankester (1885), but which Fraenkel says is not muscle tissue; it transmits to the sinus, however, the vibrations of the heart beat, which cause a rapid pulsation of the lung in the sinus, and probably thus aids the circulation of blood between the lung lamellae.

The respiratory organs of Arachnida in general include book lungs and tubular tracheae. The scorpion has the greatest number of lungs; some other arachnids have not more than two pairs and, where only one pair is present, the lungs are supplemented by tracheae; but some arachnids have tracheae only. Those zoologists who formerly contended that the arachnids are derived from merostomes attempted to explain the arachnid lungs as invaginated gills such as those of *Limulus*. The theory, however, would have to assume that the individual lamellae of the gill have been turned outside-in to form the leaflets of the lung, an assumption that in itself is enough to discredit the theory, and, moreover, the gill leaflets of *Limulus* are transverse, while the lamellae of the arachnid lung are longitudinal and vertical.

A comparison of one of the tailed eurypterids (fig. 13 D, E) with a scorpion shows a rather striking superficial resemblance between the two, which is accentuated when the comparison is with a Silurian fossil scorpion (fig. 17 D), since the short legs of the latter much resemble those of the eurypterid. A close relationship between scorpions and eurypterids, therefore, has been almost taken for granted, but with a difference of opinion as to which is the ancestral form. The scorpion, however, is by no means a primitive arachnid, as Versluys and Demoll (1920) have emphatically stated; its feeding organs are specialized in the arachnid manner, and very probably those of the eurypterids were quite different, certainly the method of feeding and the structure of the mouth parts of *Limulus* are not arachnoid. Considering these points and others of equal importance, such as the radical differences between external gills and internal lungs, the idea of a close relationship either way between the merostomes and the arachnids is difficult to maintain. The two groups may be regarded as members of the subphylum Chelicerata, but their ancestry cannot be traced to any known common progenitor.

Comparison of a Scorpion with the Palpigradi

The most primitive arachnid structure known is to be seen in the members of the Palpigradi, a group of minute spiderlike forms in-

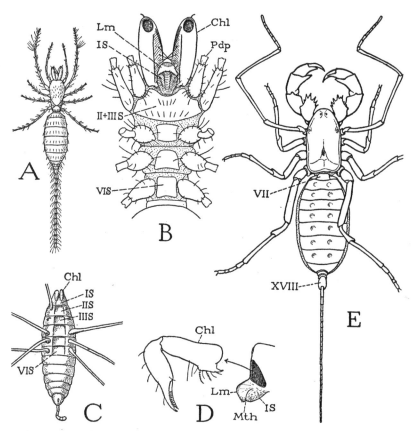

Fig. 23. Arachnida—Palpigradi (*Koenenia*) and Pedipalpida (*Mastigoproctus*).

A, *Koenenia mirabilis* Grassi (from Hansen and Sörenson, 1897). B, same, prosoma and bases of appendages, ventral (from Börner, 1901). C, *Sternarthron zitteli* Haase (var. *minus* Oppenheim), a Jurassic fossil palpigrade (from Haase, 1890). D, *Koenenia mirabilis* Grassi, anterior end of body with mouth cone and detached left chelicera (from Hansen and Sörensen, 1897). E, *Mastigoproctus giganteus* (H. Lucas).

For explanation of lettering see pages 126–127.

cluding the genus *Koenenia* (fig. 23 A) found in the Mediterranean region of Europe, *Prokoenia* from Texas and California, and a fossil Jurassic species named *Sternarthron* (C).

A palpigrade at first sight (fig. 23 A) has a superficial resemblance to a scorpion in the presence of a long jointed tail; but a closer inspection shows that the rings of the tail are not body segments, and that the tail itself is an appendage of the last segment of the body proper, being similar to the caudal flagellum of the whip scorpion (E), which is borne on a short three-segment stalk comparable to the postabdomen of the scorpion. It is in the anterior part of its body that the palpigrade shows its primitive features. The appendages corresponding with the huge chelate pedipalps of the scorpion are slender legs (A) arising entirely behind the mouth (B, *Pdp*) from a large sternal plate (*II + IIIS*) that carries the next long leglike appendages and is followed by three separate sternal plates of segments IV, V, and VI. The "pedipalps" of the palpigrades thus form no part of the feeding apparatus; in all other arachnids their coxae are intimately associated with the mouth. The mouth of the palpigrade is situated on a small, snoutlike projection of the head end of the body (D, *Mth*), and the chelicerae arise just above the base of the snout.

The dorsal surface of the snout of *Koenenia* is evidently the labrum (*Lm*), the ventral surface would appear to be, as Börner (1901) has interpreted it, the sternum of the cheliceral segment (*IS*). The organ contains the sucking pharynx. In the fossil *Sternarthron* (C), as illustrated by Haase (1890), there are shown six distinct sternal plates on the venter of the prosoma. In no other arachnid order is a cheliceral sternum present, or recognizable as such in the adult, though it is present in the embryo. There is no question with arachnologists that the Palpigradi present the most primitive structure of the prosoma known among the arachnids. It follows, then, that the scorpion is entirely too specialized to qualify as a modern representative of the arachnid ancestors.

SPIDERS

The arachnids commonly known as spiders belong to the order Araneida, or Araneae. In some respects the spiders are the most remarkable product of the arthropod phylum. In their instincts they equal or surpass the insects, and as spinners and weavers of silk they have no rivals. The cocoon of a silkworm and the unsightly domiciles of the webworms and tent caterpillars are but crude things compared with the geometric webs of the orb-weaving spiders. "The orb web," says Gertsch (1949) "would seem to stand alone as a glorious creation, an incredible novelty designed by superior artisans." Yet, there is no evidence that the intricate activities of the spiders in the construction of their silken snares and webs for catching prey and in their extraordinary modes of mating are guided by any faculty other than that of "blind instinct"; two spiders of the same species are never known to do the same thing in different ways.

Anatomically the spiders are equally remarkable for the structural adaptations that subserve their instincts. In no other arachnid is the body so narrowly constricted between the leg-bearing prosoma and the abdomen, which carries the spinning organs. No other arachnid has abdominal silk-producing glands; the spinnerets have been evolved from a pair of segmental appendages, with accessory structures between them. In the male spider the apical segment, or the last two segments, of the pedipalps have been elaborated, often to an extreme degree, into a complex structure for transferring sperm to the female, who, in turn, is provided with receptacles that in intricacy match the intromittent organs of the male. On the other hand, in most other respects, the spiders are simply arachnids: their feeding aparatus is in no way specialized; the alimentary canal and the respiratory, circulatory, and reproductive organs are essentially those of the arachnids in general.

General External Structure

In a typical spider the prosoma is relatively small and depressed as compared with the rotund abdomen, which is attached to the prosoma by such a narrow pedicel that it is freely movable in all directions. The pedicel is traversed by the alimentary canal, the aorta, tracheae, and nerve trunks. The nerve centers of the spiders all lie in the prosoma, where the ganglia of the nerve cord are condensed in a large suboesophageal nerve mass closely united with the brain around the sides of the oesophagus. The abdomen contains the heart, most of the stomach and its diverticula, the intestine, the Malpighian tubules, the respiratory organs, the silk glands, and the reproductive organs.

The prosoma (fig. 24 A), as in other arachnids, carries the chelicerae

(*Chl*), the pedipalps (*Pdp*), and the four pairs of legs (*L*). Its dorsal surface is covered by a leathery carapace, on which indistinct lines radiating from the center suggest a division into a "head," bearing the eyes, chelicerae, and pedipalps, and four segments corresponding to the legs. The V-shaped anterior line is called the "cervical groove," and the part behind it the "thorax," but there is nothing in the development of any arachnid to indicate that the spider was ever divided in this manner into a head and a thorax. The embryonic head of an arachnid lies in front of the cheliceral segment, and the eyes pertain to it, but all the prosomatic appendages belong to postcephalic body segments. However, inasmuch as these segments in the adult are intimately united with the embryonic head, the prosoma of the spider may appropriately be termed a cephalothorax. Most spiders have eight eyes distributed across the anterior part of the carapace (D), but in some all the eyes are grouped on a median tubercle (F, G). The number of eyes, however, may be reduced to six, four, or two, and some cave-inhabiting spiders have no eyes. The number, position, and relative size of the eyes serve as diagnostic characters for classification.

The body wall in front of the carapace abruptly descends as a wide membranous area from which the chelicerae arise (fig. 24 E), and ends below between the bases of the pedipalps, where it supports the pendent labrum (*Lm*). In most other arachnids the labrum is attached on a sclerotic bridge, the epistome, uniting the pedipalp coxae. A small epistomal plate is present in some of the mygalomorph spiders (fig. 31 B, *Epst*), but in most species the epistomal region above the labrum is unsclerotized or not distinctly separated from the labrum (fig. 27 F, G). Araneologists commonly call the labrum the "rostrum," and give the name "labrum" to the anterior part of the carapace before the eyes, but this usage of the term *labrum* is clearly a misapplication, since the labrum of all arthropods is a free lobe overhanging the mouth, though in ordinary spiders it is concealed behind the chelicerae, which hang downward from beneath the front of the carapace (fig. 24 D). The lumen of the labrum is crossed by two compressor muscles (fig. 31 A, M, *tmcls*); within it also is a gland or pair of glands. According to Petrunkevitch (1933), there are probably two labral glands, or so-called "rostral" glands, present in all spiders, but in some they are so closely united as to appear to be a single organ; the two ducts discharge into a wide,

81

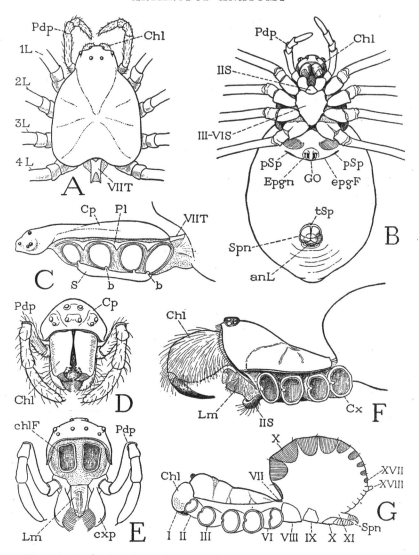

Fig. 24. Arachnida—Araneida. General Structure.

A, *Argiope trifasciata* Forsk., female, prosoma and pedicel, dorsal. B, same, entire body, ventral. C, same, prosoma with appendages detached, lateral. D, same, anterior end of body, showing eyes, chelicerae, and pedipalps. E, same as D with chelicerae removed. F, *Eurypelma hentzi* Chamb., female, prosoma and base of opisthosoma, pedipalps removed, legs cut off beyond coxae, lateral. G, *Liphistius desultor* Schiödte, female, showing complete arachnid body segmentation (from Bristowe and Millot, 1932).

For explanation of lettering see pages 126–127.

slitlike atrium that opens on the anterior surface of the labrum. The structure of the labral glands in *Atypus piceus* Sultzer is described in detail by Bertkau (1885).

The undersurface of the prosoma of most spiders, as seen in *Argiope* (fig. 24 B), is formed of a small sternal plate between the pedipalp coxae and of a large, usually heart-shaped plate between the leg coxae. The small anterior plate (*IIS*) is the sternum of the pedipalp segment; the second plate bears on each margin (C) four small knobs on which the leg coxae are articulated, and therefore represents the combined sterna of the four leg-bearing segments (B, *III–VIS*). Since the pedipalp sternum lies immediately below the mouth (fig. 31 A, *Mth*) and serves the spider as an underlip, it is commonly called the "labium," though it has no homology with the labium of an insect, which is formed of the appendages corresponding with the second pair of legs of an arachnid. In some spiders the pedipalp sternum is united with the leg sternum (fig. 27 I). The legs of the spider arise from the membranous lateral walls of the prosoma between the carapace and the sternum (fig. 24 C, F). Intervening between the carapace and the coxae there may be a narrow sclerotic band (C, *Pl*), which, though it does not give articular supports to the coxae, may be regarded as a pleural sclerotization.

The araneid opisthosoma, or abdomen, varies in shape from globular to elongate, or takes on irregular and sometimes bizarre forms. In the Liphistiomorpha the abdomen is shown definitely, by the presence of distinct tergal plates (fig. 24 G), to be composed of 12 segments, including the anal lobe (*XVIII*), but in the other spiders the abdominal segmentation is obscured or obliterated in the adult, though certain external features are always associated with specific segments. The first opisthosomatic segment is the abdominal pedicel (A, C, *VIIT*). The dorsum of the abdomen has no special characters; on the venter (B) are located the genital opening, the apertures of the respiratory organs, the spinnerets, and, in the female, the orifices of the sperm receptacles.

Crossing the anterior part of the ventral surface of the abdomen is a groove known as the *epigastric furrow* (fig. 24 B, *epgF*). In the middle of the furrow is the simple genital opening (*GO*) in both sexes, and in the lateral parts, in the majority of spiders, the slitlike apertures (*pSp*) of a pair of book lungs. Before the genital opening of the female of most spiders is a strongly sclerotized plate termed the *epigynum* (*Epgn*), which contains the openings of the sperm receptacles. In the four-lunged Liphistiomorphae, Mygalomorphae, and Hypochilomorphae, a second pair of pulmonary spiracles lies a short distance behind the first pair (fig. 34 B, *2Sp*), but in the two-lunged spiders the second lungs are replaced by tracheae. The tracheal spiracles may lie in the position of the second lung spiracles, or more centrally on the abdominal venter, but usually they come together in a common posterior opening just before the spinnerets (fig. 24 B, *tSp*). In a few spiders the first lungs also are replaced by tracheae. The first pair of respiratory organs pertains to the second abdominal segment (*VIII*), the second pair, whether lungs or tracheae, belongs to segment IX, which on the venter extends back to the spinnerets.

The spinnerets (fig. 24 B, *Spn*) are a group of small appendages, usually six of them, in most spiders lying close before the anal lobe (*anL*). The spinnerets pertain to segments X and XI, but the anal lobe represents segment XVIII; the ventral arcs of the six intervening segments, therefore, are compressed in the narrow space between the spinnerets and the anal lobe, except in *Liphistius* (G), in which there is a long segmented area of the abdominal venter behind the spinnerets. The abdomen of most spiders projects more or less beyond the spinnerets, but the surface seen from below behind the anal lobe (B) is a part of the dorsum.

The respiratory organs, the epigynum, and the spinnerets will be more fully described in following sections.

The Legs

The legs of spiders are so attached on the sides of the prosoma between the carapace and the sternum that they turn anteriorly and posteriorly. There may be no specific articulation of the leg base on the body, but usually the coxae are articulated ventrally on marginal knobs of the sternum (fig. 24 C, *b*). The extrinsic leg muscles include dorsal muscles arising on the carapace, and ventral muscles from the endosternum.

The spider leg (fig. 25 A) consists of seven true segments, but the small terminal segment, or pretarsus (*Ptar*), bearing the claws is mostly concealed by hairs, or by retraction into the end of the tarsus. The other segments, beginning at the base of the leg, are the coxa (*Cx*), a single trochanter (*Tr*), the femur (*Fm*), a short patella

(*Pat*), the tibia (*Tb*), and the tarsus (*Tar*). The tarsus, however, is distinctly subdivided into a long basal part (*1tar*) and a shorter distal part (*2tar*). Arachnologists commonly term the basal tarsomere the "metatarsus" and the distal tarsomere the "tarsus," though by analogy with vertebrate anatomy the two names should be re-

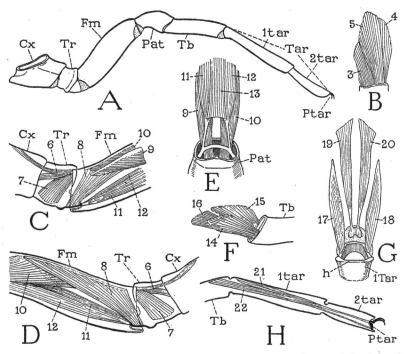

Fig. 25. Arachnida—Araneida. Segmentation and musculature of the legs of *Eurypelma hentzi* Chamb.
A, second left leg, anterior. B, ventral muscles of trochanter arising in coxa. C, muscles of femur (*6, 7, 8*) arising in trochanter and coxa, and proximal ends of patellar muscles (*9, 10, 11, 12*), anterior. D, same part of leg, posterior. E, base of patella and its muscles, ventral. F, base of tibia and its muscles from patella, anterior. G, base of tarsus and its muscles, dorsal. H, distal part of leg, showing subsegments of tarsus and distribution of pretarsal muscles.
For explanation of lettering see pages 126–127.

versed, and in entomology "metatarsus" would refer to the tarsus of a metathoracic leg. To avoid confusion, therefore, it will be better to call the basal tarsomere of the spider leg the *basitarsus* (*1tar*) and the distal one the *telotarsus* (*2tar*). That the two tarsal parts are not true segments is shown by the consistent absence of interconnecting muscles.

Differences in the articulations at the joints between the leg segments give to the leg a variety of movements. The trochanter, and
therefore the telopodite as a whole, turn up and down on the end
of the coxa; between these two segments there is a strong anterior
articulation, but no specific posterior articular point. The trochantero-
femoral joint is dicondylic with both anterior and posterior articulations, so that the femur moves on the trochanter in a vertical plane,
though its principal flexion is in a dorsal direction. The patella is
joined to the femur by a strong, transverse, dicondylic dorsal hinge;
the femoropatellar joint is the "knee" of the spider leg, with a principal
ventral flexion. The patellotibial joint differs from the other joints
in that its axis is obliquely vertical with a dorsal point of articulation between the adjoining segments. Movement at this joint, therefore, is transverse to the axis of the limb, and the nature of the joint
enables the patella in its up and down movement on the femur to
carry the distal part of the leg with it. The basitarsus again moves
in a vertical plane since it has a strong transverse dorsal hinge (fig.
25 G, h) on the end of the tibia. The distal tarsomere is freely flexible
on the basitarsus, but there are no controlling points of articulation
and no muscles at this intratarsal joint. At the end of the tarsus are
the apical claws of the leg, commonly termed the "tarsal claws." An
examination of the foot of the spider (fig. 26), however, shows that,
as in other arthropods, the claws pertain to a small end segment of
the limb, which is the pretarsus, or dactylopodite, having its own
muscles. The body of the pretarsus is set vertically in the articular
membrane at the end of the tarsus, and in most spiders is produced
in a median claw, or *dactyl* (C, D, E, *Dac*); the paired claws, or
ungues (C, *Un*), are flexibly attached to the upper end of the
pretarsus (*Ptar*).

The intrinsic musculature of the leg appears to be essentially the
same in all spiders. In a recent study of the leg muscles of the
tarantula *Eurypelma,* Dillon (1952) finds 31 muscles in all for each
leg, 11 of which are in the coxa. This account is much more complete
and accurate than any other hitherto published on the arachnid leg
muscles. It is often difficult, however, to decide how many individual
muscles may be represented in a compact mass of fibers having a
common insertion. The writer, for example, has enumerated only
five major groups of fibers in the coxa of *Eurypelma,* two being
dorsal and three ventral (fig. 25 B). In the following descriptions

of the muscles of the telopodite 17 muscles have been recognized instead of 20 as given by Dillon, but the 17 muscles shown on figure 25 will sufficiently illustrate the mechanism of the arachnid leg.

The coxal muscles operate directly the trochanter, but they serve as levators and depressors of the telopodite as a whole. The femur is individually movable in a vertical plane on the trochanter. Its levator muscles, attached dorsally on the base, include two groups of fibers, one a horizontal dorsal muscle (fig. 25 C, D, 6) with a short branch arising in the base of the trochanter, and a longer branch from the coxa, the other (7) is a thick bundle of obliquely dorsoventral fibers from the ventral wall of the trochanter. The depressor of the femur is a large muscle (8), the fibers of which arise on the extended lower lip of the trochanter and spread distally in the posterior part of the femur (D) to be attached dorsally in the proximal two-thirds of the segment. This muscle is present in dipneumone spiders examined, though more weakly developed than in *Eurypelma*. The patella has only depressor muscles (E). Two of them are large anterior and posterior muscles (9, 10) arising dorsally in the proximal part of the femur (C) and inserted directly on the lower lip of the base of the patella (E). Traversing the femur ventrally is a compact bundle of fibers attached anteriorly on the lower lip of the trochanter, which is separable into two thick lateral muscles (C, D, E, 11, 12) and a thin, flat ventral muscle (E, 13). Posteriorly the lateral muscles are attached separately by a pair of tendons to an arcuate bar in the ventral articular membrane at the base of the patella (E); the ventral muscle (13) is attached on the membrane itself. The lateral muscles of this group evidently are depressors of the patella; the median ventral muscle possibly pulls on the infolding membrane. The "knee" joint of the limb between the femur and the patella is an important point of ventral flexure.

The patella is fully occupied by three short muscles of the tibia (fig. 25 F), one anterior (14), one posterior (15), and one ventral (16), but the last has an anteroventral insertion; 14 and 16, therefore, are productors, 15 a reductor, the movements of the tibia on the patella being transverse to the axis of the leg. The basitarsus has four muscles (G), all effective as depressors because they are attached below the dorsal tibiotarsal hinge (h). An anterior and a posterior muscle (17, 18) are inserted on the base of the tarsus; a pair of ventral muscles (19, 20) is attached by tendons on a small

87

plate in the ventral articular membrane of the joint. Between the two parts of the tarsus (H, *1tar, 2tar*) there are no muscles, but the tarsus as a whole is traversed by the muscles of the pretarsus (*Ptar*), a dorsal muscle (*21*) arising proximally in the basitarsus, and a ventral muscle (*22*) arising dorsally in the distal end of the tibia. The mechanism of the pretarsus will be discussed in a following paragraph.

A study of the musculature of the spider's leg, as has been noted by other writers, shows that there are no levator (extensor) muscles of the patella, the tibia, or the tarsus; the tibia has no upward movement by reason of the nature of its connection with the patella. Ellis (1944) has given reasons for believing that extension of the leg at the femoropatellar and tibiotarsal joints is produced by blood pressure, there being no evidence of elasticity at the joints, since the legs of a dead spider are always flexed. "Experimental evidence," he says, "demonstrates that extension of the leg is intimately associated with changes in the volume and pressure of the blood in the leg." In a freshly killed spider compression of the basal part of a leg at once extends the distal part.

A comparison of the leg musculature of the spider, the scorpion (fig. 19 G), and *Limulus* (fig. 11 A) shows numerous differences among the three. A common feature, however, not found in the mandibulate arthropods, is the presence of a ventral muscle or muscles in the femur extending from the lower lip of the trochanter to the base of the patella, represented by muscle 10 in *Limulus,* muscle 19 in the scorpion, and muscles 11, 12, and 13 in the spider (fig. 25 E).

The pretarsus of the legs of most spiders (fig. 26 C, D) has the same structure and mechanism as that of the scorpion (fig. 19 H). It is a much-shortened apical segment of the limb attached by membrane within the end of the tarsus (fig. 26 C, *Ptar*), and rocks on a transverse axis by the action of its antagonistic muscles attached by tendons dorsally and ventrally on its base (fig. 25 H). Usually the pretarsus is produced into a solid median claw, or dactyl (fig. 26 C, D, E, *Dac*), and is thus seen to represent the dactylopodite of a generalized arthropod limb. The lateral claws, or ungues (C, *Un*), however, are flexibly attached by basal membranes on the upper part of the pretarsus and are evidently secondary outgrowths having no independent movement of their own. In different spiders all the

88

claws of the foot differ much in shape, and the ungues are usually armed below with teeth or comblike rows of spines (C, F).

A type of pretarsal mechanism somewhat different from that of ordinary spiders occurs in the Mygalomorphae. The pretarsus of Euryphelma, for example, is a small vertical plate without a median

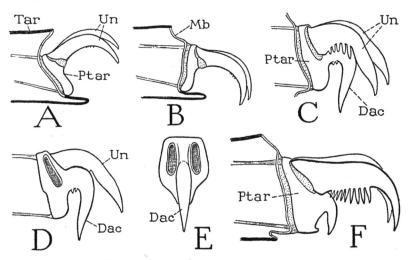

Fig. 26. Arachnida—Araneida. The pretarsus.

A, *Eurypelma hentzi* Chamb., pretarsus and claws retracted. B, same, pretarsus and claws protracted. C, *Argiope trifasciata* Forsk., pretarsus with median dactyl and lateral ungues. D, same, pretarsus with anterior claw removed, lateral. E, same, pretarsus with both lateral claws removed, end view. F, *Ancylometes* sp., pretarsus and claws.

For explanation of lettering see pages 126–127.

claw (fig. 26 A, *Ptar*) bearing dorsally a pair of ungues (*Un*). Its lower end is set in a deep median notch in the lower lip of the tarsus, and is here firmly but flexibly attached, so that it rocks back and forth on the tarsal support; its proximal movement elevates and retracts the claws (A), its distal movement protracts and deflects them (B). The usual pretarsal muscles are attached by tendons dorsally and ventrally on the base of the pretarsus, and the dorsal muscle is clearly a retractor of the claws. The tendon of the ventral muscle, however, being attached just above the tarsal fulcrum (A, B), this muscle in *Eurypelma* would appear to be merely an inefficient accessory to the dorsal retractor muscle. The fibers of the two muscles are closely adherent in the basitarsus (fig. 25 H), but proximally they are separated at their respective origins.

89

The Pedipalps

The pedipalps of the spiders have the same segmentation as the legs (fig. 27 A), but the tarsus is undivided, and the pretarsus has no lateral claws. In the male spider the pretarsus of the palpus is variously developed as a sperm-carrying and intromittent organ; in the female it is usually a simple, dactylopoditelike claw (B, *Ptar*), mostly concealed in the retracted condition by hairs on the end of the tarsus (C). On the base of the claw are attached the tendons (E) of the usual levator and depressor muscles of the pretarsus, the first, in the female (D, *lvptar*), arising in the distal end of the tibia, the second (*dpptar*) in the base of the tibia. The pretarsal claw of the female as seen in *Eurypelma* (B) is attached by membrane on the end of the tarsus and probably is protractile by blood pressure. Though it has no articulation on the tarsus, it rocks up and down because its open basal connection with the supporting membrane is shorter than the base of the claw (E), the lower angle of which is produced downward as a lever giving attachment to the tendon of the depressor muscle. The intrinsic musculature of the pedipalp is essentially the same as that of the legs; the trochantero-femoral depressor of the femur (fig. 25 D, 8) is well developed in the pedipalp, lying posterior to the other muscles in the femur.

The coxae of the pedipalps form a part of the feeding apparatus insofar as they constitute the side walls of the entrance passage to the mouth. In most of the Mygalomorphae the pedipalp coxa is produced distally mesad of the trochanter in a small coxal process (fig. 27 A, *cxp*), but in *Atypus* the coxal process is a large lobelike extension of the coxa (I, *cxp*), as it is in some of the other spiders (J, *cxp*). In a majority of the spiders, however, the coxal lobes become more differentiated and individualized structures, having the appearance of a pair of jaws appended to the coxae (F, G, *cxp*). The coxal lobes are commonly termed the "maxillae," but they have no independent movement on the coxae, and according to Kästner (see Gerhardt and Kästner, 1937, '38), the mechanical treatment of the prey is done entirely with the chelicerae, not with the coxal lobes of the pedipalps. The lobes generally bear dense brushes of hairs that curve together over the mouth entrance and serve to strain the liquid food sucked into the mouth. The term "maxilla," therefore, is doubly inappropriate as applied to the coxal lobes of the pedipalps,

90

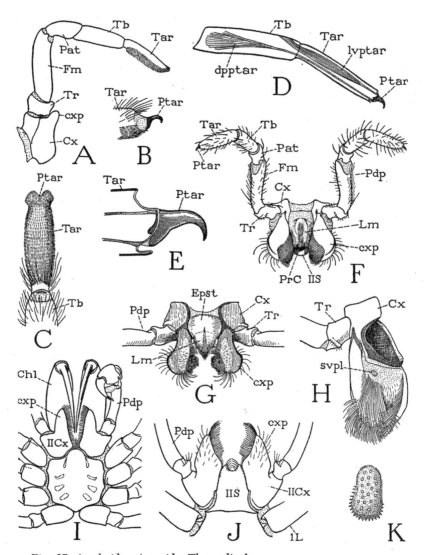

Fig. 27. Arachnida—Araneida. The pedipalp.

A, *Eurypelma hentzi* Chamb., pedipalp of female. B, same, pretarsus of pedipalp exposed by removal of tarsal hairs. C, same, undersurface of pedipalp tarsus. D, same, distal segments of pedipalp, showing pretarsal muscles. E, same, pretarsus. F, *Argiope trifasciata* Forsk., female pedipalps, labrum, and pedipalp sternum, anterior. G, *Ancylometes* sp., female, epistome, labrum, and bases of pedipalps, anterior. H, *Argiope trifasciata* Forsk., mesal surface of coxal lobe of right pedipalp. I, *Atypus bicolor* Lucas, male, prosoma and bases of appendages, ventral. J, *Dysdera crocata* C. Koch, female, bases of pedipalps and pedipalp sternum. K, *Argiope trifasciata* Forsk., sieve plate of pedipalp coxal lobe.

For explanation of lettering see pages 126–127.

because the arachnid pedipalps are appendages homologous with the mandibles of mandibulate arthropods. The coxal lobes of the araneid pedipalps, moreover, evidently do not correspond with basal lobes of the coxa, such as those of the scorpion (fig. 21 C, *cxnd*) and other arachnids, which have been termed coxal "endites," since a distal coxal process (*cxp*) may be present also.

Glands contained in the pedipalp coxae and their lobes are said to be present in all the Araneida; they open into the preoral cavity between the coxal lobes and are known as *salivary glands,* or "maxillary glands." According to Petrunkevitch (1933), these glands are unicellular in *Hypochilus,* but in all other genera they are multicellular saclike organs, the number in each coxa varying with the species of spider. In *Liphistius* and the Mygalomorphae the glands are shown by Bertkau (1885) to be distributed along the entire length of the coxa and to open irregularly on the upper surface near the inner edge. In other spiders the glands open in a small oval area near the base of the mesal surface of the coxal lobe (fig. 27 H, *svpl*), known as the *sieve plate* because of its perforation by the duct orifices. In *Argiope trifasciata* the sieve plate (K) is a somewhat convex oval membrane with a dark border partly fringed with minute spines, and perforated by about 20 pores.

A male spider is usually known at a glance to be a male by the enlarged ends of his pedipalps, the terminal segments of which are elaborated into organs for the transfer of sperm to the sperm receptacles of the female. The palpal intromittent organ varies in different spiders from a relatively simple structure to one of extreme complexity, and its characters are of much importance in taxonomy for the identification of species. The segments of the limb involved are mainly the tarsus and the pretarsus, and to a lesser degree the tibia.

A relatively generalized structure of the intromittent organ is shown by Comstock (1910) and Comstock and Gertsch (1949) to be present in the genus *Filistata* (fig. 28 A). The organ here consists of the pretarsus alone, which is differentiated into an enlarged, subdivided basal part called the *bulb* and a slender, somewhat twisted terminal neck termed the *embolus.* The bulb arises from an alveolar depression in the end of the tarsus (*Tar*), which segment shows no special modification. At the apex of the embolus is the opening of an internal canal, coiled in the bulb and ending with a vesicular enlargement, which is the receptacle in which the male spider carries the sperm.

A study of the musculature of the male palpal organ leaves no doubt that the organ is the apical segment, or pretarsus, of the pedipalp, represented by the simple, dactylopoditelike claw of the

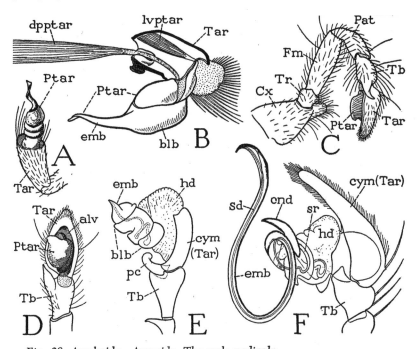

Fig. 28. Arachnida—Araneida. The male pedipalp.

A, *Filistata* sp., end segments of the pedipalp (from Comstock, 1910). B, *Eurypelma hentzi* Chamb., the pretarsal intromittent organ and section of tarsus of left pedipalp, showing pretarsal muscles. C, right pedipalp of a gnaphosid spider, posterior. D, distal segments of same, ventral. E, *Erigone autumnalis* Emerton, distal segments of pedipalp (from Nelson, 1909). F, *Agelena naevia* Walck., distal segments of pedipalp (from Petrunkevitch, 1925).

alv, alveolus; *blb,* bulb; *cnd,* conductor; *Cx,* coxa; *cym,* cymbium; *dpptar,* depressor muscle of pretarsus; *emb,* embolus; *Fm,* femur; *hd,* haematodocha; *lvptar,* levator muscle of pretarsus; *Pat,* patella; *pc,* paracymbium; *Ptar,* pretarsus; *sd,* seminal duct; *sr,* seminal reservoir; *Tar,* tarsus; *Tb,* tibia; *Tr,* trochanter.

female palpus (fig. 27 B, E, *Ptar*). As shown here in *Eurypelma* (fig. 28 B) the structure of the organ is even simpler than in *Filistata* (A); the bulb is supported on the end of the short tarsus (*Tar*), though it is flexed proximally, and on its base are attached the tendons of the usual two muscles of the pretarsus. The levator muscle (*lvptar*) arises dorsally in the base of the tarsus, the large depressor (*dpptar*) takes its origin in the tibia. The same structure is shown in *Eurypelma californica* by Barrows (1925, fig. 13), who identifies the palpal organ of the male as a hypertrophied claw, representing the dactylopodite of Crustacea. The large depressor muscle of the pretarsus in *Eurypelma* evidently causes a proximal ventral flexion of the intromittent organ on the tarsus. The tarsus of the male palpus, as that of the female, is a single, undivided segment; on its base is attached the usual flexor muscle of the arachnid tarsus.

In most male spiders the pedipalp tarsus itself becomes modified in connection with the pretarsus. The pretarsus shifts to the undersurface of the tarsus and takes a position near its base (fig. 28 C, D); the tarsus develops a depression, or *alveolus* (D, *alv*), for the reception of the intromittent organ and is now termed the *cymbium* (E, *cym*). Furthermore, the articular membrane between the tarsus and the pretarsus becomes enlarged in the form of a vesicle, called the *haematodocha* (E, *hd*), which may be distended by blood pressure, and is supposed to be effective in forcing the embolus into the female receptacle during mating.

The palpal organ of *Erigone autumnalis* (fig. 28 E), as shown by Nelson (1909), is more complex than that of the gnaphosid given at D, but it is still relatively simple. The short embolus (*emb*) projects from a three-part bulb (*blb*), which is supported on a large, inflated haematodocha (*hd*), and the tarsus, or cymbium (*cym*), has an accessory branch (*pc*) termed the *paracymbium*. An example of extreme elaboration of structure in the palpal organ is given at F, as figured by Petrunkevitch (1925) for *Agelena naevia*. The embolus (*emb*) is here a long, slender, doubly curved process containing the seminal duct (*sd*), and is accompanied at its base by an accessory process (*cnd*) termed the *conductor*, which in some spiders is extended the full length of the embolus. The pretarsal part of the organ, based on the haematodocha (*hd*), arises from a deep alveolus in the base of the elongate tarsal cymbium (*cym*). The palpal organ attains an even greater complexity in some other spiders, but for further examples of its variable structure the student may refer to the comparative studies by Comstock (1910) and by Osterloh (1922) or, for a concise account of the essential nature of the organ, to the paper by Nelson (1909).

Preliminary to mating, the male spider of most species spins a small, flat web on which he discharges a drop of sperm from his genital opening on the abdomen. Then, applying the tips of the palpal organs to the *under* surface of the web beneath the drop of sperm, the latter is taken into the sperm canals, presumably by capillary attraction. This act is called *sperm induction.* During mating the emboli of the male organs are inserted into the apertures of the seminal receptacles of the female, either both at the same time or alternately, and are forced into the ducts by blood pressure in the haematodochae. The ejection of the sperm is generally attributed also to blood pressure, but Osterloh (1922) suggests that the intrusion of secretion from epithelial gland cells surrounding the seminal canals may drive the sperm out, perhaps in combination with blood pressure. Numerous observations on sperm induction by the male, courtship, and mating among spiders are recorded by Montgomery (1903, 1909b); Baerg (1928) gives an account of sperm induction and mating by the tarantula; Ewing (1918) covers the life history of the house spider; Gertsch (1949) fully reviews the whole subject of courtship and mating.

The Chelicerae

The chelicerae in all spiders are two-segmented (fig. 29), and only rarely does the basal segment have a process opposing the fang-like apical segment. The appendages arise from the anterior membranous wall of the body (fig. 24 E) between the carapace above and the pedipalp coxae and labrum (*Lm*) below, but this supraoral position they assume secondarily during embryonic growth, as in other arachnids, from a primitive ventral position behind the mouth. In the Mygalomorphae and Liphistiomorphae, the basal segments of the chelicerae project forward from the body (fig. 24 F, G), and the fangs turn downward and posteriorly; by contrast, in the typical spiders the basal segments hang downward (fig. 29 D) and the fangs close against their mesal surfaces, where generally they are received in grooves, which may be armed on one or both margins with small spines or teeth. The fang in all cases is strongly movable by antagonistic muscles (A, B) arising in the basal segment. The chelicerae do not vary much from the typical structure, but in some of the ant spiders they attain an enormous relative length by elongation of both segments, and the basal segment is armed below with a row of slender spines (see Millot, 1949, fig. 369).

95

In all the araneid families but the Uloboridae the chelicerae contain poison glands. The gland is an elongate sac (fig. 29 C, E) with
a duct traversing the fang to open on the convex side of the latter near
the tip (C, *VPr*). The gland of the mygalomorph spiders is contained
in the basal segment of the chelicera; in other spiders it may extend
into the body cavity as far as the prosomatic nerve mass, or beyond
it. The gland is covered by a layer of muscle fibers, said by Millot

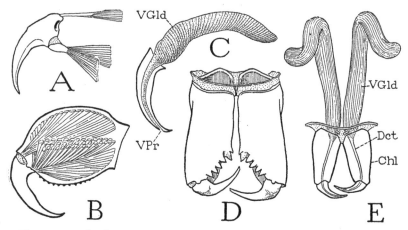

Fig. 29. Arachnida—Araneida. The chelicerae.

A, *Argiope trifasciata* Forsk., cheliceral fang and its muscles. B, *Eurypelma
hentzi* Chamb., chelicera, showing muscles of the fang. C, same, venom gland
of chelicera and duct. D, *Ancylometes* sp., chelicerae, anterior. E, *Latrodectus
mactans* (F.), chelicerae and venom glands.

For explanation of lettering see pages 126–127.

(1931) to be generally arranged spirally along the length of the sac
(fig. 29 C) but to present variations and irregularities; in the highly
venomous "black widow" spider, *Lactrodectus mactans*, the muscles
as shown by Reese (1944) run longitudinally on the gland (E).
In one spider, *Scytodes thoracica* (Latr.), as described by Millot
(1931), the cheliceral gland is bilobed; one lobe secretes venom,
the other a silk liquid which the spider ejects on its prey to entangle
it before killing it with venom from the poison lobe. The chelicerae
are the most essential external organs the spiders possess, since without them they could neither capture nor kill their prey.

The Eyes

The eyes of the spiders have each a single lens and are therefore of the kind known as simple eyes, or *ocelli*. The two median anterior eyes differ from the others in their mode of development, which turns the retinal layer upside down, and for this reason these eyes are said to be *inverted*.

The structure of a noninverted eye is shown at B of figure 30. Beneath the thick corneal *lens* (*Ln*) is a deep layer of translucent *corneagenous cells* (*CgCls*), which are the epidermal cells that generated the lens but which form a *vitreous body* in the mature eye. Beneath the corneagenous cells is the *retina* (*Ret*), composed of numerous light-sensitive cells, the nuclei of which are in the outer ends of the cells, and the sensory, receptive zones (*rz*) on the parts proximal to the nuclei (C, D, *rz*). Eyes of this kind are termed "erect" or "converted" eyes, or "prebascillary" eyes in reference to the nuclei being distal to the rodlike sensory parts of the retinal cells. A spider eye of this type is similar to the eye of a scorpion (A) except that the corneagenous cells of the scorpion eye do not intervene between the lens and the retina and the sensory zones of contiguous retinal cells form intercellular rods, or *rhabdoms* (*Rhb*). In the scorpion eye, as in the simple eye of *Limulus* (fig. 9 D), the retinal nuclei lie in the inner parts of the cells, as they do in the ocelli of insects.

A modification of the erect type of eye occurs in the web-spinning spiders (fig. 30 E), in which the nucleated ends of the retina cells (F) diverge from beneath the lens (E, *Ret*) and thus expose the receptive surfaces (*rz*) more directly to the light. The outer ends of the cells are imbedded in a dark pigment (*Pig*), and the sensory zones are surrounded by a sheath of light-reflecting cells forming a *tapetum* (*Tap*) that throws the light back into the retina.

The inverted eyes have the retinal nuclei in the inner parts of the cells (fig. 30 L) behind the light-sensitive rods (*r*), and are hence termed "postbascillary" eyes. In the development of these eyes, as described by Locy (1886), the prospective corneagenous layer and the retinal layer are derived from contiguous areas of the surface epidermis (H) and become superposed by an involution and consequent inversion of the retina beneath the corneagenous layer (I, J, K). The inner wall of the retinal pocket (I, K, *prl*) becomes a

97

postretinal layer of the mature eye (L, *prl*); the retina closes against the corneagenous layer, and the opposing basement membranes, or included connective tissue, form a *preretinal membrane*. At the original point of involution (H) the corneagenous cells, which secrete

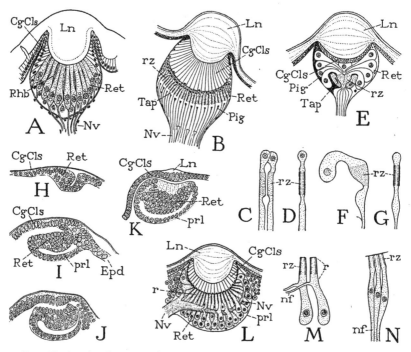

Fig. 30. Arachnida. Eyes of a scorpion and of spiders.

A, lateral eye of *Euscorpius italicus* (Hbst.) (from Lankester and Bourne, 1883). B, erect eye of *Pardosa monticola* C. L. Koch (*Lycosa agricola* Thorell) (simplified from Widmann, 1908). C, D, retinal cells from B. E, erect eye of *Tegenaria derhamii* (Scopoli) (*domestica* Clerck) (from Widmann, 1908). F, G, retinal cells from E. H, I, J, K, development of inverted eye of *Agelena naevia* Walck. (from Locy, 1886). L, inverted eye of *Tegenaria derhamii* (Scopoli) (from Widmann, 1908). M, retinal cells from L. N, examples of inverted retinal cells with secondarily proximal nerves (from Widmann, 1908).

CgCls, corneagenous cells; *Epd*, epidermis; *Ln*, lens; *nf*, nerve fiber; *Nv*, nerve trunk; *Pig*, pigment; *prl*, postretinal layer; *r*, optic rod; *Ret*, retina; *Rhb*, rhabdom; *rz*, receptive zone of retinal cell; *Tap*, tapetum.

the cuticular lens (K, L, Ln), become continuous with the epidermis (I, *Epd*). As a consequence of the inversion of the retina, the retinal nuclei come to be at the inner ends of the cells, while the nerve fibers issue distally; and usually the optic nerve trunk departs from one side of the eye (L, *Nv*). In the mature eye the outer ends of the retinal cells are extended beyond the nerve roots (M, *nf*) to form light-receptive, rodlike distal processes (*r*). In some spiders, however, the nerves are transposed to the inner ends of the cells (N, *nf*). In the mature eye, therefore, little evidence of inversion may remain, except for the presence of a preretinal membrane, remnants of the postretinal cell layer, and the inner position of the retinal nuclei. The eyes of the Pycnogonida are said by Wirén (1918) to resemble the inverted eyes of the Araneida.

The inverted eyes, at least in some spiders, are provided with muscles, a most unusual thing for arthropod eyes. The ocular musculature of the spiders has been described by Widmann (1908) and by Scheuring (1914). According to Widmann, in *Lycosa* dorsal and ventral muscles from the head wall that are attached on the eye appear to regulate the focal depth of the eye by compressing the corneagenous layer. In other species only a dorsal or a ventral muscle may be present, which, by tilting the eye up or down, changes the direction of the lens with regard to external objects. Scheuring (1914) describes in *Salticus scenicus* a much more complex ocular musculature, consisting of six muscles attached on each eye, to which he attributes eight different accommodation movements.

The Organs of Ingestion, Digestion, and Excretion

The spiders subsist entirely on liquid exudates extracted from the animals, mostly insects, on which they feed. The prey is seized, held, killed, punctured, lacerated, or crushed by the chelicerae, but extra-oral digestion plays an important part in rendering the food available for ingestion, a highly potent digestive liquid from the stomach being discharged on or into the prey which completely liquefies the soft tissues. So copious and effective is this digestive fluid that some of the large spiders are able to consume even small vertebrates, which they kill with the venom of the chelicerae. It is probable that the salivary glands opening on the inner faces of the pedipalp coxal lobes contribute their secretion to that from the stomach.

The mouth parts of the spider, except the chelicerae, form merely a receptacle before the mouth, from which the liquid food is ingested by the sucking action of the pharynx. The preoral food chamber (fig. 31 A, *PrC*) is enclosed by the labrum above (*Lm*), the pedipalp coxal lobes on the sides, and the pedipalp sternum (*IIS*)

99

below (see fig. 27 F). The coxal lobes, as already noted, are not jaws, or "maxillae" as they are often called, since they have no independent movement on the coxae; their distal ends and inner surfaces are furnished with large brushes of hairs that close over the mouth and form an effective strainer that prevents the ingress of hard pieces of the prey or food particles too large to be swallowed. The mouth (fig. 31 A, *Mth*) lies beneath the base of the labrum and leads directly into the sucking organ known as the *pharynx* (*Phy*). The continuity of the pharyngeal walls with the undersurface of the labrum and the upper surface of the suboral sternum might suggest that the pharynx itself is an external derivative such as the cibarial pump of sucking insects, but the arachnid pharynx is generally said to be a part of the stomodaeum. It is an elongate, flattened pouch that turns upward and posteriorly from the mouth (A, M, *Phy*), and the narrow oesophagus (*Oe*) dips downward from its upper end.

The anterior, or dorsal, wall of the pharynx is formed of a dorsal pharyngeal plate (fig. 31 A, *dpl*); the posterior, or ventral, wall contains a ventral plate (*vpl*) or pair of plates, the two being connected along the sides by flexible membranes allowing expansion and contraction. The dorsal plate is usually the more strongly developed: in the mygalomorph *Eurypelma* it is roundly convex ventrally (C, D), presenting a wide median lobe and a pair of lateral folds; in other spiders it is usually flat or slightly concave (F, G, I, J). Along the middle of the dorsal plate runs a narrow dorsal channel (*dc*), extending from the mouth to the opening of the oesophagus (C, F, J), and the lateral parts of the plate are finely and closely ridged transversely. The specific function of the dorsal channel of the pharynx is not definitely known, but Bartels (1930) suggests that the grooves between the lateral ridges of the dorsal plate direct the liquid food into the median channel, through which it is conducted to the oesophagus, while undissolved particles are retained by the ridges and grooves. The main lumen of the pharynx, then, according to Bartels, is the conduit of the digestive liquid discharged on the prey. This conclusion Bartels deduces from the observation that in spiders allowed to drink water containing a suspension of India ink or carmine particles, the particles are found in the transverse grooves and massed along the sides of the dorsal canal. The ventral plate of the pharynx may be more or less divided into lateral halves (fig. 31

100

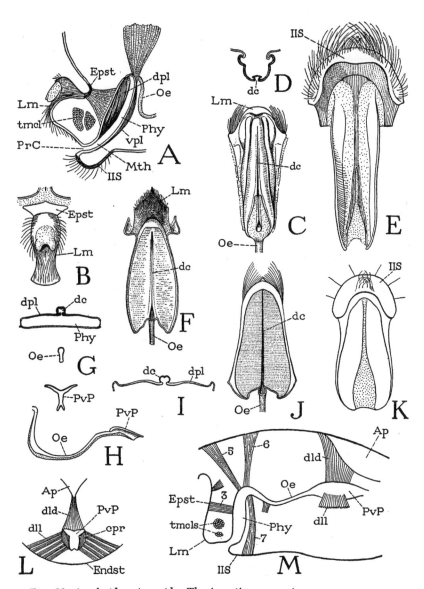

Fig. 31. Arachnida—Araneida. The ingestion apparatus.

A, *Eurypelma hentzi* Chamb., longitudinal vertical section of pharynx, epistome, and labrum. B, same, epistome and labrum, anterior. C, same, dorsal plate of pharynx, ventral. D, same, cross section of dorsal plate of pharynx. E, same, ventral plate of pharynx and end of suboral sternum, dorsal. F, *Ancylometes* sp., dorsal plate of pharynx and labrum, ventral. G, same, cross sections of pharynx, oesophagus, and proventricular pump. H, same, oesophagus and proventricular pump. I, *Argiope trifasciata* Forsk., cross section of dorsal plate of pharynx. J, same, dorsal plate of pharynx, ventral. K, same, ventral plate of pharynx and end of suboral sternum, dorsal. L, *Agelena naevia* Walck., cross section of proventricular pump and muscles (from Brown, 1939). M, same, stomodaeum and its muscles (from Brown, 1939).

For explanation of lettering see pages 126–127.

E, K) and is either concave or flat according to the shape of the over-
lying dorsal plate.

The pharyngeal musculature of the spiders is simpler than that
of most other arachnids; that of *Agelena naevia,* as described by
Brown (1939), includes the following muscles (fig. 31 M): a pair of
median dilator muscles (3) arising at the base of the labrum and
converging to the front of the pharynx (in most arachnids these
muscles arise on an epistomal plate above the base of the labrum, as
seen at A in *Eurypelma*); a pair of dorsal dilators from the pedipalp
coxae to the edges of the anterior wall of the pharynx; a dorsal
dilator (5) from the carapace to the upper end of the pharynx; a
dorsal retractor (6) from the carapace attached on the pharynx behind the last muscle; and a ventral retractor (7) from the pedipalp
sternum ("labium") to the upper end of the posterior wall of the
pharynx. In other arachnids the pharynx is closely surrounded by
constrictor muscles; in the spiders apparently compression of the
pharynx must result from the pull of the dorsal retractor muscles.

The oesophagus is a narrow tube (fig. 31 H, M, *Oe*) that goes
posteriorly from the pharynx through the central nerve mass to the
posterior part of the prosoma, where it expands before joining the
stomach to form a proventricular pump (*PvP*). The walls of the
pump are expanded by strong dorsal muscles (L, M, *dld*) from an
apodeme (*Ap*) of the carapace and by lateral muscles (*dll*) from
the endosternum (L, *Endst*); compression is effected by short compressor muscles (L, *cpr*) on the side walls of the organ. The mesenteron section of the alimentary canal includes an axial stomach
tube with diverticula, and most of the intestine. The first diverticula
are a pair of tubes from the anterior end of the stomach in the
prosoma, which give off lateral branches extending into the leg bases.
The abdominal diverticula are mostly branched into many small
lobes that form a large mass of soft tissue filling the upper part of
the abdomen. Secretion and the final stages of digestion take place
in the cells of the gastric diverticula. Beyond the stomach the mesenteron is continued as an intestinal tube ending with a large saclike
expansion known as the cloaca, into which open the Malpighian
tubules. The cloaca discharges through a very short ectodermal
proctodaeum opening on the anal lobe of the abdomen.

The excretory organs of the spiders are nephidial coxal glands
and the Malpighian tubules. In Liphistiomorphae and Mygalo-

102

morphae there are two pairs of coxal glands, those of the first pair opening on the first-leg coxae, those of the second pair on the third-leg coxae; the other spiders have only the pair opening on the first-leg coxae. An extensive comparative account of the arachnid coxal glands is given by Buxton (1913). The Malpighian tubules are a pair of slender tubes arising from the cloaca and branching among the stomach diverticula in the abdomen.

The Sperm Receptacles of the Female and the Epigynum

The paired ducts of the ovaries of the spiders unite anteriorly in a common part of the female genital tract, termed the *uterus* because it is of mesodermal origin. The uterus discharges through a short ectodermal exit passage commonly called the *vagina*, but termed also the "uterus externus" because in most spiders it has no copulatory function. Associated with the female genital opening are special sperm receptacles, or *spermathecae*. In many, mostly primitive, genera of spiders the spermathecae open directly from the vagina (fig. 32 F, *Spt*). There may be two, three, five, or more of these vaginal spermathecae, which are relatively simple structures. In the majority of spiders the sperm receptacles, two in number, except when secondarily divided, open independently in the neighborhood of the genital outlet, and usually the apertures are contained in a sclerotic plate on the anterior margin of the epigastric furrow known as the *epigynum* (fig. 24 B, *Epgn*). In this case, the emboli of the male intromittent palpal organs carrying the sperm are thrust into the entrance canals of the receptacles (fig. 32 G, *ic*), and each spermatheca (*Spt*) has a *fertilization canal* (*fc*) connecting it with the uterus, in which the eggs are inseminated.

The epigynum of different spiders is highly variable in size and form, in ways that have little relation to general classification but are characteristic of species. A very simple structure occurs in the common house spider, *Theridion tepidariorum*, in which the epigynum is little more than a basinlike impression of the abdominal wall with a sclerotic rim, containing the orifices of the intromittent canals of the spermathecae. A more typical example of the epigynum is shown here in a member of the Pisauridae (fig. 32 A), in which the epigynum is a large, strongly sclerotized plate with a pair of lateral cavities, from the inner ends of which open the spermathecal canals. A variant of this type of epigynal structure is seen in *Argiope tri-*

103

fasciata (B) and in many other spiders. Again, however, the epigynum takes the form of a long, projecting process, as in *Neoscona benjamina* (C, D), with the intromittent openings on the undersurface (C, *io*). A reference to books and taxonomic papers on spiders will show the endless variety of forms that the epigynum assumes.

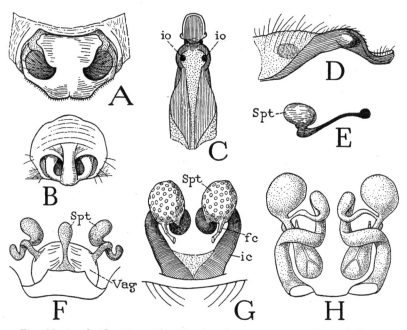

Fig. 32. Arachnida—Araneida. The female sperm receptacles and the epigynum.

A, *Ancylometes* sp., epigynum. B, *Argiope trifasciata* Forsk., epigynum. C, *Neoscona benjamina* (Walck.), epigynum, ventral. D, same, epigynum, lateral. E, same, sperm receptacle and intromittent duct of left side. F, *Tetragnatha solandri* Scopoli, vagina and connected spermathecae, internal view (from Engelhardt, 1910). G, *Theridion tepidariorum* (C. L. Koch), spermathecae with independent openings. H, *Clubiona neglecta* Cambridge (*montana* L. Koch), spermathecae and canals (from Engelhardt, 1910).

fc, fertilization canal; *ic*, intromittent canal; *io*, intromittent orifice; *Spt*, spermatheca; *Vag*, vagina.

The sperm receptacles connected with the epigynum in their simplest development are a pair of globular or oval, usually hard-walled vesicles at the ends of the intromittent canals (fig. 32 E). Generally, however, the organs are more complex owing to a lengthening and coiling of the canals, or also to a differentiation of the vesicles into two or more parts (G, H). An example of a highly com-

104

plicated spermathecal structure is that described and figured by Petrunkevitch (1925) in *Agelena naevia* Walck. Here the inner ends of the intromittent canals expand into large copulatory bursae connected by ducts with the sperm vesicles, which themselves are produced into tubular diverticula, and the fertilization canals wrap themselves around the intromittent canals on their way to the uterus. A comparative study of the sperm-receiving organs will be found in the papers by Engelhardt (1910) and Osterloh (1922).

The structure of the female receptive organs and that of the male intromittent organs are intimately correlated, but it is difficult to see a reason for the great diversity in form and complexity of the copulatory apparatus. Since the structure is characteristic of species, it is sometimes explained as a "lock-and-key" device to prevent crossbreeding, though probably there are few observed cases of a male spider attempting to mate with a female not of his own species.

The Respiratory and the Circulatory Organs

The breathing organs of the araneids include *book lungs* like those of the scorpion, and tubular ingrowths of the integument known as *tracheae*. The tracheae, however, are not all of the same origin, some being primarily respiratory in function, others primarily apodemal. There are never more than two pairs of original respiratory organs, both of which may be lungs or both tracheae. They open on the venter of the abdomen and pertain to the second and third abdominal segments (segments VIII and IX). In the Liphistiomorphae, Mygalomorphae, and the Hypochilidae among the Araneomorphae both pairs of respiratory organs are lungs; in the families Caponiidae, Thelmidae, and Symphytognathidae both pairs are tracheae. In most of the rest of the araneids the organs of the first pair are lungs and those of the second pair tracheae, though in some Pholcidae tracheae are absent. The tracheal spiracles of Dysderidae, Segestriidae, and Oonopidae lie close behind the apertures of the lungs (fig. 33 D, *tSp*); in other families the paired tracheae open medially, sometimes near the middle of the venter of the abdomen but usually close before the spinnerets (fig. 24 B, *tSp*). When the spiracles have a posterior position, they lie at the sides of the sternal apodemes of segment IX, and in most such cases the hollow apodemes themselves become trachealike organs.

The position of the book lungs is generally evident because of the

105

slightly raised and darker areas of the integument beneath them, known as the *opercula* (fig. 33 D, *Opl*). The lung openings are transverse clefts behind the opercula, those of the anterior pair being in or close to the lateral parts of the epigastric furrow (figs. 24 B, 33 D, *pSp*). Each aperture leads into a vertical atrial chamber (fig. 33 A, *Atr*), the ends of which are extended on each side beyond the spiracular cleft (B). In the anterior wall of the atrium are numerous very narrow parallel slits (A, *o*), which are the mouths of leaflike air pouches (*lam*). The lung structure of the spider differs in no way from that of the scorpion (fig. 22 D, E, F). The lung lamellae in different spiders may be vertical, oblique, or nearly horizontal, and are said to vary in number from 25 to 100 for each lung. In *Eurypelma* (fig. 33 C) the lamellae are triangular and are attached below on a supporting wall, so that they are free only along their dorsal edges, and it is here that the outer surfaces of adjacent lamellae are held apart by short spacing rods between them. Each lung lies in a ventral blood sinus of the abdomen, which is connected by a venous channel with the anterior part of the pericardium (I, *Vp*), the aerated blood being thus conveyed directly to the heart. (In most spiders the pericardial cavity around the heart is much narrower than that of *Liphistius* shown in the figure.)

The mechanism of respiration in spiders by the lungs has been investigated by several writers, whose work is reviewed by Kästner (1929) along with his own studies on the subject. The spiders make no pulsatory breathing movements of the body wall, and the lungs themselves have no capacity for expansion and contraction. The only lung muscle is an opener of the spiracle arising on the body wall behind the spiracle and inserted on the posterior spiracular lip. This muscle effects also an expansion of the atrium, but the closing action depends on elasticity. Kästner observes, however, that the spiracular slit is opened only when the spider has been active or is in some way artificially stimulated to activity. At other times it appears that enough air has access to the atrium to satisfy the needs of the spider. An expansion and contraction of the lung lamella was postulated by Weiss (1923) as resulting from the action of associated body muscles, other writers have attributed it to the pulsation of the blood, and it has even been supposed that the rods connecting the lamellae are elastic; but Kästner concludes that gas exchange between the atrium and the narrow cavities of the lamellae can be attributed only to

106

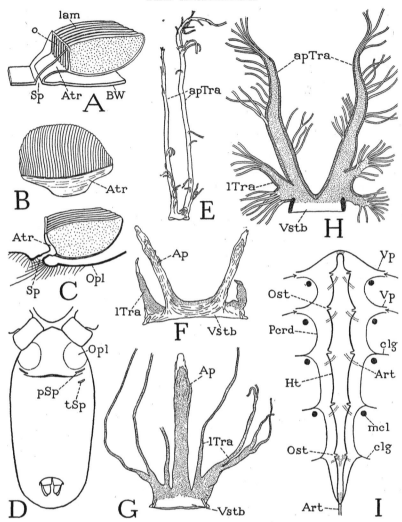

Fig. 33. Arachnida—Araneida. Respiratory organs and the heart.

A, diagram of a book lung of *Eurypelma*. B, *Eurypelma hentzi* Chamb., book lung, dorsal. C, same, vertical section of book lung and sublying operculum. D, *Segestria senoculata* L., abdomen, ventral (from Purcell, 1909). E, *Miagrammops* sp., apodemal posterior tracheae (from Lamy, 1902). F, *Filistata capitata* Hentz., nonrespiratory apodemes and small tracheal pouches of segment IX (from Lamy, 1902). G, *Loxosceles rufescens* L., single ventral apodeme of segment IX, and lateral tracheae opening from a common vestibule (from Lamy, 1902). H, *Attus floricola* C. Koch, apodemal and tracheal respiratory organs of segment IX (from Purcell, 1909). I, *Liphistius malayanus* Abr., pericardium and heart, anterior aorta not shown (from Bristowe and Millot, 1932).

For explanation of lettering see pages 126–127.

diffusion. Kästner's explanation of the respiratory mechanism of the lung books of spiders closely agrees with that of Fraenkel (1929) on the scorpion (pp. 76–77).

The tracheae of the spiders are far more variable in their structure than are the lungs. They are branched or unbranched tubes, and the inner walls of the larger trunks are strengthened by thickenings in the form of reticulations and anastomosing processes. Some writers have described the smaller tubes as having spiral taenidia in the intima, but Richards and Korda (1950) report that electron microscope studies show no circular or helical thickenings in species they examined (*Theridion, Tetragnatha, Neoscona*).

In the Caponiidae, one of the families having two anterior lateral pairs of tracheae but no lungs, the first spiracle on each side leads into an atrial pouch from which are given off numerous capillary tubules into the anterior blood sinus of the abdomen. The structure of these tracheae, therefore, suggests a lung in which the air pockets are tubular instead of lamelliform, and, as in the case of the lungs, it is the blood that is aerated. In those spiders in which the single pair of tracheal spiracles lies close behind the pulmonary spiracles (fig. 33 D, *tSp*), each spiracle opens into a wide tubular atrium from which a bundle of capillaries is given off as from the first tracheae in Caponiidae. In other spiders these spiracles have a more median and posterior position. In the Filistatidae they lie on the venter of the abdomen midway between the epigastric furrow and the spinnerets, and open into the ends of a transverse groove, or vestibulum (fig. 33 F, *Vstb*), laterad of the pair of ventral apodemes of segment IX (*Ap*); the tracheae themselves, however, are but little developed. When, as in most spiders, the spiracles are transposed to a position immediately in front of the spinnerets (fig. 24 B), the apodemes are carried back with them by the elongation of sternum IX, and the three associated structures open from a single vestibule (fig. 33 G). The apodemes may be united in a single median process as at G, but more commonly they are developed as a pair of long, tapering, hollow tubes, giving off numerous small branches (H, *apTra*), and thus take on the structure and supposedly the function of tracheae. United with the bases of these apodemal tracheae are the primary lateral tracheae (*lTra*), which vary much in their extent and branching in different spiders. Finally, in some spiders the lateral tracheae of segment IX are suppressed, and the apodemal tubes remain as the only tracheal organs (E). The conversion of hollow apodemes into respiratory organs is not confined to the spiders; in the diplopods all the tracheae arise from similar sternal apodemes serving also for muscle attachments.

Considering the fact that the araneid respiratory organs, whether lungs or tracheae, occur always on the same body segments and that either the second pair or both pairs of lungs may be replaced by tracheae, the question arises as to whether the two kinds of organs are really distinct structures, or merely different forms of development of the same thing, the ingrowths from the atrium in one case being lamellar, in the other tubular. The question has been much discussed, and both sides have been advocated by different writers, but neither side seems to offer a conclusive answer. From studies on the development of the lungs by Purcell (1909, 1910) and by Montgomery (1909a), it is evident that the lungs are intimately associated with the appendage rudiments of segments VIII and IX. The atrium of the lung is formed as a depression of the integument immediately behind the base of the appendage rudiment, and the latter itself then sinks into the cavity. According to Montgomery, the lung lamellae grow into the body cavity from the sunken part of the appendage, while the flattened external part becomes the operculum. Purcell contends that the tracheae are formed in the same way as the lungs, but Montgomery says that in the case of the tracheae the appendage rudiment disappears and the tracheae arises as an independent ingrowth behind the site of the appendage. When we note that the prosomatic spiracles of the Solpugida, Ricinulei, and Acarina are associated with the bases of the appendages, it might be deduced that the respiratory ingrowths of the Arachnida in general have a postcoxal position. The structure of the first tracheae in Caponiidae and that of the postpulmonary tracheae in Dysderidae, which are comparable to that of a lung with capillary tubes instead of hollow lamella, and the position of the single pair of tracheal spiracles in Phalangida, which is at the site of the first lungs in Araneida, suggest that the primary respiratory organs of the Arachnida, regardless of their form, are serially homologous structures. The antiquity of the scorpions gives weight to the idea that lungs have priority over tracheae, but it is a curious fact that lung spiracles have not been observed with certainty in any Paleozoic scorpion (see Petrunkevitch, 1949, pp. 134–135). There can be no question,

109

as shown by Ripper (1931), that the arachnid tracheae have no homology with the tracheae of myriapods and insects.

The improbability of book lungs being derived from gill plates has been discussed in the section of Scorpionida. Purcell (1909) contends that the leaflets of the spider's lung appear first as folds on the base of the associated appendage, but Montgomery (1909a) says these folds disappear and that the air pockets of the definitive lungs are formed secondarily after the insinking of the appendage rudiment. Purcell offers no explanation of how external gill plates may have been turned outside in to form internal air pouches.

The heart of the spider is a median muscular tube lying within a pericardial sinus in the dorsal part of the abdomen. The blood enters the heart through the lateral ostia; it is discharged into the abdomen through paired arteries and into the prosoma through a median aorta that divides into numerous branches. The terminal branches of the arteries and the aorta end in blood sinuses, through which the blood flows to the region of the lungs, and thence through the "pulmonary veins" to the pericardial cavity. The heart is most highly developed in *Liphistius* (fig. 33 I), in which there are five pairs of ostia (*Ost*) and five pairs of arteries (*Art*) besides a median posterior artery. In other spiders the heart has become shortened from behind, and the ostia reduced to four or two pairs.

The Spinnerets

Abdominal spinnerets are organs peculiar to the Araneida; they pertain to the fourth and fifth segments of the abdomen (segments X and XI). In most spiders, because of the great posterior extension of segment IX and the compression of the segments between segment XI and the anal lobe, the spinnerets come to lie just before the anus (fig. 24 B, *Spn*), which is at the posterior end of the body except when the abdomen takes on some unusual shape (fig. 34 K). In the Liphistiomorphae, however, a more primitive condition of the abdominal segmentation is retained, and the spinnerets arise near the middle of the venter well forward of the anal lobe (fig. 24 G).

Three pairs of spinnerets are commonly present in a compact group (fig. 24 B, *Spn*); their relative positions can be seen when the six are separated, as shown at D of figure 34. Two pairs are lateral, those of segment X being known as the *anterior spinnerets* (*aSpn*),

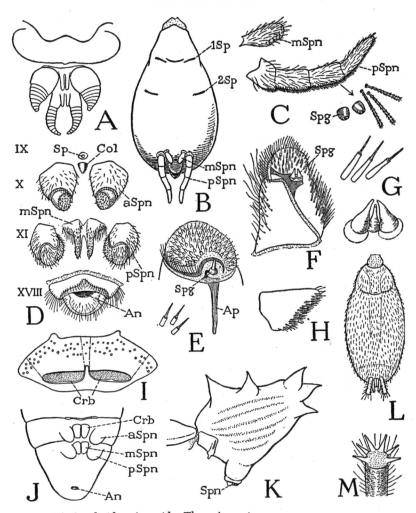

Fig. 34. Arachnida—Araneida. The spinnerets.

A, *Liphistius birmanicus* Thorell, spinnerets (from Bristowe and Millot, 1932). B, abdomen of a mygalomorph spider, ventral. C, *Eurypelma hentzi* Chamb., spinnerets of left side, with examples of ventral hairs and spigots. D, *Argiope trifasciata* Forsk., colulus, spinnerets, and anal lobe, separated. E, same, end of left anterior spinneret, mesal, and examples of spinning hairs. F, same, left posterior spinneret, mesal. G, same, spines and spigots of posterior spinneret. H, same, median spinneret. I, *Filistata* sp., cribellum (from Montgomery, 1909a). J, same, posterior end of embryo with rudiments of spinnerets (from Montgomery, 1909a). K, *Micrathena gracilis* (Walck.), abdomen, lateral. L, abdomen of male *Zelotes rusticus* L. Koch, ventral. M, lateral spinneret of same with spigots extruded by pressure.

For explanation of lettering see pages 126–127.

those of segment XI as the *posterior spinnerets* (*pSpn*). The third pair, or *median spinnerets* (*mSpn*), lie between the bases of the posterior spinnerets. It is probable, however, that originally there were four pairs of spinnerets, since in *Liphistius* (A) there is a pair of slender processes between both the anterior and the posterior lateral spinnerets. In most of the spiders having six spinnerets a small conical process termed the *colulus* (D, *Col*) lies between the bases of the anterior spinnerets and is regarded as representing the first median pair of *Liphistius*, since in its development the colulus of some spiders is formed of two rudiments. The colulus is apparently a vestigial organ having no special function. A number of spider families lack the colulus but have in place of it a plate or pair of plates perforated by numerous silk-duct openings, known as the *cribellum* (I, *Crb*). The cribellum in its development, as shown by Montgomery (1909a), is formed from two median spinneretlike processes of the ninth segment (J, *Crb*), so that there is little doubt that the cribellum also is a derivative of a pair of anterior median spinnerets of more primitive spiders. The glands of the cribellum produce a special kind of gluey silk, which is combed out by a row or double row of spines on the upper surface of the basitarsus of the hind legs, termed the *calamistrum*. The calamistrum is thus a character of the cribellate spiders.

The close association of the median spinnerets with the bases of the lateral spinnerets has given rise to the idea that the spinnerets represent four biramous appendages, each composed of an exopodite and an endopodite comparable with the two rami of a crustacean limb. According to Montgomery (1909a), however, the development of the organs shows that only the larger, outer spinnerets are the true appendages of segments X and XI, the median spinnerets being merely secondary outgrowths of the body wall between the others. In most of the Mygalomorphae, which have only the two pairs of spinnerets of segment XI (fig. 34 B, C), the long, four-segmented outer spinnerets are much suggestive of a pair of legs, and the small median spinnerets are entirely unconnected with them. The lateral spinnerets are movable by muscles inserted on their basal segments, and each of the other segments is individually musculated.

A typical structure of the spinnerets in spiders having the usual three pairs is shown here in the female of *Argiope trifasciata* (fig. 34 D). The four outer spinnerets (*aSpn, pSpn*) are thick conical lobes each with a small apical segment; the two median spinnerets (*mSpn*) are simple triangular lobes (H), compressed from side to side. The flattened end of the apical segment of each anterior spinneret is turned somewhat mesally on the basal segment and is covered (E) with minute, upstanding cylinders bearing tapering spines, three of which are shown enlarged below in the figure. These structures are the spinning tubes of the silk glands and are known as the *fusules;* the gland ducts open at the tips of the spines. In a notch on the mesal margin of the segment is a much larger gland outlet distinguished as a *spigot* (*Spg*); connected with its base is a long apodeme (*Ap*) for muscle attachments. In some specimens there are two spigots here. The posterior spinneret (D, *pSpn*) is similar in shape to the anterior spinneret, but the apical segment extends up the mesal surface of the basal segment (F) and is covered with fusules somewhat longer than those of the anterior spinneret (G); on a mesal sclerotization it bears a pair of spigots (F, *Spg*), or in some specimens three (G). The median spinnerets of *Argiope* (D, *mSpn*, H) are fringed with small fusules on their ventral margins and have brushes of long slender ones on their expanded bases (D); near the middle of the ventral margin of each is a conspicuous spigot.

The spinnerets of different spiders differ in shape and relative size. In the male gnaphosid shown at L of figure 34 the anterior spinnerets are long and cylindrical; when one of them is compressed, six fingerlike spigots are projected from the distal end (M), from each of which a strand of silk may be drawn out. The anterior and posterior spinnerets of *Liphistius* (A) are multisegmented. The four-segmented outer spinnerets of *Eurypelma* (C) have numerous small conical spigots (two shown enlarged below) on their ventral surfaces in a dense clothing of slender, club-shaped spines. The simple median spinnerets (*mSpn*) have similar ventral spigots.

The silk-producing glands of the spiders are contained in the abdomen. As classified by Gertsch (1949), there are at least seven kinds of silk glands differing in size, shape, and numbers, but they do not all occur in any one spider or family of spiders. The glands that open through the small spinning tubes are the most numerous since there is a gland for each tube. These glands, either aciniform or pyriform in shape, occur in dense masses and are present in all spiders. Similar to them are the cribellar glands of the cribellate spiders. The other glands open through the spigots; they are long

113

tubular glands, relatively few in number, cylindrical, ampullate, lobed, or branched in form.

The production of silk is a faculty peculiar to the arthropods, and one that evidently has been independently acquired in different groups. The silk glands of the pseudoscorpions are in the chelicerae and discharge through the movable fingers of these appendages. Among the diplopods silk glands are present in the Nematophora opening through spines on the end of the body; in Symphyla silk glands open through the abdominal cerci. The silk glands of insects are mostly labial glands, but the Embioptera have silk glands in the fore tarsi, and in certain beetles and neuropterans the Malpighian tubules produce silk discharged through the anus. For quantity production the lepidopterous silkworms probably rank first, but for architectural achievements in the art of spinning the spiders excel all their arthropod relatives.

A TICK

The ticks and the mites constitute the arachnid order Acari, or Acarina. The number of species, genera, and families is so large that acarologists find it necessary to divide the order into six or more suborders, and under each suborder to recognize several subsidiary ranks, one under the other, before coming even to the families (see Vitzthum, 1931, 1940–1943). An important character used in classifying the larger groups is the position of the spiracles, or the presence or absence of spiracles. The ticks belong to the acarine group called the Ixodides, and most of them are included in two families, the Ixodidae and the Argasidae.

In external appearance the Acarina have little resemblance to other arachnids, but fundamentally in both their outer and their inner organization they show that they are merely arachnids that have developed a highly specialized type of structure, mostly correlated with parasitic habits. Two distinctive acarine characters are the absence of any constriction or other separation between the prosoma and the opisthosoma (fig. 35 A) and the presence of a discrete head structure (*Capt*), known as the *capitulum*, or *gnathosoma*. As in other arachnids, the adult has six pairs of appendages, including a pair of chelicerae, a pair of palps representing the pedipalps, and four pairs of legs. The food is generally a liquid, either blood or body juices extracted directly from the host, body tissue liquefied by predigestion, or the sap of plants. Some species of mites, however, feed on dry material, and ingest spores or particles of the food, but it is questionable whether they swallow such material in a dry form or suspended in an ejected liquid. The chelicerae are the cutting or piercing instruments, the organ of ingestion is a sucking pharynx as in other arachnids.

For a study of the structure of a tick we may take the common "dog tick," *Dermacentor variabilis* (Say), of the eastern and central states, a member of the family Ixodidae. It is closely related to the western *Dermacentor andersoni* Stiles, and both species carry the Rickettsiae of Rocky Mountain spotted fever. Other common genera of the same family are *Ixodes, Amblyomma, Boophilus, Ripicephalus,* and *Haemaphysalis.* The ticks are blood feeders at all stages of their postembryonic development. From the egg hatches a six-legged young tick called the *larva,* or *seed tick;* it feeds on small rodents, and when replete with its first meal it drops off the host and undergoes a moult into a second-stage form known as a *nymph,* which has eight legs. The nymph, after feeding usually on some larger mammal, drops off and moults into the *adult.* In the tick embryo rudiments of eight legs are present, but those of the fourth pair, as described by Falke (1931) in *Ixodes ricinus,* are reduced to small masses of cells that remain latent beneath the cuticle of the larva and redevelop into the fourth pair of functional legs during the quiescent stage that precedes the moult to the nymph after the larva has engorged. The reappearance of the fourth pair of legs in the nymph after their suppression in the larva, therefore, is not literally a case of reacquisition of a lost organ.

The active nymph, which resembles the adult female but is much smaller, finds a suitable animal, usually a field mouse, to which it attaches itself, and fills with blood in from four to eight days. It then detaches from the host, and, after a few days of activity, the skin splits and an adult tick, either a male or female, crawls out. The adults find a new host, almost any common mammal, including man, and engorge themselves with blood. An unfed female of *Dermacentor variabilis,* measuring about 5 mm. in length, after feeding becomes an oval, turgid bag 13 mm. long and 10 mm. wide, with a thickness of 5 mm. The increase in body size is accompanied by the flattening out of innumerable small wrinkles in the denser, outer layer of the cuticle, and probably by a stretching of the soft inner

layer. The male tick does not become distended with feeding, but
the body thickens slightly. Mating takes place on the host, and the
fully fed female drops off to lay her eggs on the ground. There is
no fixed relation between the developmental stages of the tick and
the yearly seasons, as there is in most insects; overwintering stages
of ticks may include adults, nymphs, and larvae, or even eggs. The
results of an elaborate study of the seasonal stages of the dog tick
on the island of Martha's Vineyard, Massachusetts, are given by
Smith, Cole, and Gouck (1946).

General Structure of the Body— The body of an unfed *Derma-*
centor variabilis (fig. 35 A) is ovate in outline, much flattened
dorsoventrally, and deeply incised anteriorly for the reception of
the capitulum (*Capt*). The broadly rounded posterior end is margined with 11 quadrate divisions of the integument termed *festoons*
(*fst*). A female (A) is readily distinguished from a male by the
presence of a shieldlike area on the anterior part of the dorsum, called
the *shield*, or *scutum* (*Shld*), which is distinctly defined by a thickened margin and is variously mottled with white pigment. The male
has no such shield, or, as it is said, the shield covers his entire back.
Located laterally on the dorsum just behind the second legs, in the
lateral angles of the shield of the female, is a pair of ocellar eyes
(*O*). Near the middle of the back in each sex, shortly behind the
shield of the female (fig. 35 A), are two clear, circular areas, each
with a small central body bristling with minute peglike processes,
presumably sensory in function. These organs are larger in the female
than in the male. The integument otherwise is spotted with the alve-
oli of very short, blunt setae directed forward, and is everywhere
speckled with cuticular pores said to be the outlets of dermal glands.
In the unfed female tick the surface of the integument, except that
of the shield, is marked by fine, closely set, parallel striations running in highly irregular, zig-zag bands from one side to the other.
These striations are the minute wrinkles seen in sections on the
surface of the cuticle; they entirely disappear with the stretching of
the skin in the engorged female. The shield of the female, which
gives attachment to the leg muscles, remains unaffected by the increase in size of the body. The cuticular striations are not present in
the male. The surface features of the integument are best seen by
cutting the dorsal and ventral walls apart, so that they can be examined individually by transmitted light.

116

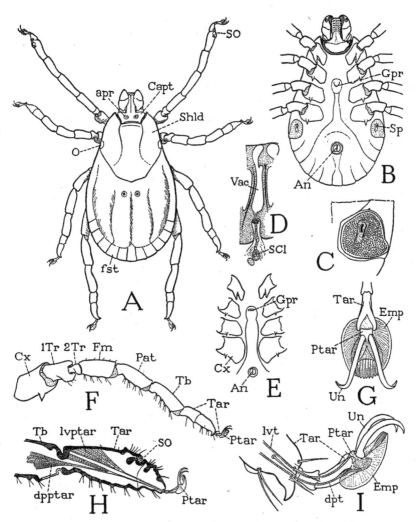

Fig. 35. Arachnida—Acarina. General external structure of a tick.

A, *Dermacentor variabilis* (Say), adult unfed female, dorsal. B, same ventral. C, same, left spiracle and surrounding plate. D, *Ixodes reduvius* L., section of a large vacuole of spiracular plate, with apparent sense organ in basal canal (from Nordenskiöld, 1909). E, *Dermacentor variabilis* (Say), male, details of undersurface. F, same, third left leg of female, ventral. G, same, tarsus and pretarsus with empodium spread out, dorsal. H, same, optical section of distal part of first leg, showing tarsal sense organs and pretarsal muscles. I, same, tarsus and pretarsus of third leg as seen by transmitted light.

For explanation of lettering see pages 126–127.

On the underside of the body (fig. 35 B) the flat coxal segments of the legs form a double row of lateral plates, separated by a wider median space in the female (B) than in the male (E). Between the coxae of the second pair is the gonopore (*Gpr*), a V-shaped aperture in the female (B), a transverse slit in the male (E). From the sides of the gonopore two lines, the *genital grooves*, run backward and widely diverge behind the fourth coxae to the posterior margin of the body. In the male (E) the parts of the grooves between the coxae are usually farther apart than in the female. Although the genital opening of the tick appears to be on the prosomatic part of the body, it is probable that the abdominal sterna have been extended forward between the legs and that the gonopore is really on segment VIII as in other arachnids. The anus (B, E, *An*) lies at about the center of the undersurface of the abdominal region of the tick, but, as with the spiders, its position represents the posterior pole of the body.

The Spiracles— The tick has only a single pair of spiracles (fig. 35 B, *Sp*), which in *Dermacentor* lie laterally on the venter of the abdominal region behind the last legs. Each spiracle of *Dermacentor* is contained in a large spiracular plate, or peritreme (C). The surface of the plate is smooth, but it has a finely punctate appearance, and in transmitted light the punctations appear as minute bright spots of two sizes. Sections show that the interior of the plate is vacuolated by larger and smaller cavities corresponding to the bright spots seen on the surface. The small vacuoles lie in a middle layer of the plate between cuticular strands that connect a thinner outer layer with a thick inner layer of the plate. The large vacuoles extend to the outer surface, where they are said to have minute openings, and they open on the inner surface of the plate through narrow basal canals. The large vacuoles of *Dermacentor* are described by Stiles (1910) as goblet-shaped, the stems being the basal canals; those of *Haemaphysalis punctata* are said by Nuttall, Cooper, and Robinson (1908) to be pear-shaped; in *Ixodes reduvius* they are shown by Nordenskiöld (1909) as more slender cavities (fig. 35 D, *Vac*); and Falke (1931) describes those of *Ixodes ricinus* as tubular pores opening externally on the surface of the plate and internally into the wider basal canals. Both Nordenskiöld and Falke find within the basal canals structures that appear to be sense organs (D), each, according to Nordenskiöld, being connected with a sense cell (*SCl*), or sometimes with two cells, in the sublying epidermis. On the other hand, Mellanby (1935) says the spiracular plate of *Ornithodoros moubata* is pierced by minute pores opening into the spiracular atrium, and in *Dermacentor andersoni* Douglas (1943) asserts that the atrium is surrounded by a periatrial space with which the goblet vacuoles are connected, "thus establishing a passage for air via the area porosa." Yet Douglas notes that the stems of the goblets (opening through the inner wall of the cuticle) "contain protoplasmic strands from the hypodermis." The structure of the spiracle plate clearly needs further study. The plate, however, must have some function in connection with respiration; if it is sensory, Falke suggests that it possibly gives the tick a warning to close the spiracles against some harmful condition of the air. Stiles (1910) gives a large number of illustrations showing the form of the spiracular plates and their surface appearance in species of *Dermacentor*. The spiracular orifice is a small crescentic slit in an elongate oval depression of the plate; it opens into an atrial chamber, from which branching tracheae are given off to all parts of the body.

The Legs— The legs of the tick have each eight segments (fig. 35 F). The coxae, as already noted, have the form of plates on the undersurface of the body (B); those of the first legs are slightly movable, but the others are adnate on the body wall, and the principal movement of all the legs is at the coxotrochanteral joints. A very small second trochanter (F, *2Tr*) is firmly fixed to the base of the femur in each leg. The next three segments, the femur, patella, and tibia, are of approximately equal length. The tarsus (*Tar*) is differentiated into a thick, strongly sclerotic proximal section of two subsegments and a slender, delicate distal part that forms a stalk supporting the pretarsal foot (*Ptar*). The footstalk itself is divided into three subsegments (I), the short basal one being mostly retracted into the end of the preceding section of the tarsus, which, in all the legs but the first, is produced into a strong, posterior, hooklike process.

The pretarsus is a small sclerotic body (fig. 35 G, I, *Ptar*) articulated on the end of the footstalk. Dorsally it bears a pair of long, slender decurved claws (I, *Un*), and ventrally a large pad (*Emp*), which when spread out (G) has an oval outline and a flat, smooth undersurface. The pad is evidently an adhesive organ. To be consistent with insect nomenclature, the footpad of the tick must be termed a pulvilliform *empodium*, unless it is composed of two lateral

119

parts united, "pulvilli," so called, being paired lobes beneath the bases of the claws. In some Acarina the body of the pretarsus is produced into a median claw between the bases of the two articulated lateral claws, or ungues, while in others there may be only a median claw. On the base of the pretarsus are attached the tendons (I, *lvt*, *dpt*) of the usual two pretarsal muscles, the levator (H, *lvptar*) arising on the dorsal wall of the proximal part of the tarsus, the depressor (*dpptar*) arising mostly in the tibia but with a small bundle of fibers from the base of the tarsus.

On the tarsi of the forelegs of the ticks are located important sense organs (fig. 35 A, *SO*), known as Haller's organs from their discoverer. At the distal end of the basal subsegment of the tarsus, on the dorsal surface, is a capsulelike cavity with only a small opening to the exterior, containing a number of sensory setae (H, *SO*). Just beyond it on the base of the second subsegment of the tarsus is a shallow, open depression also containing sensory setae. Each group of setae is innervated from its own sublying sense cells. The structure of the organs has been described by Schulze (1941) and by Lees (1948). The capsular organ is said by Schulze to be fully developed in all ticks, the open organ is variable and may be absent in some species. It has long been known that the capsular organ is an odor receptor; experimental evidence presented by Lees suggests that the open organ is responsive to humidity. Probably by means of these tarsal organs the ticks recognize their prospective hosts. When a tick has ascended a blade of grass or the stem of a bush, it clings to the support by the third legs, keeping the others free for grasping, while the long first legs are extended to catch the odor of a passing animal. The sensory reactions of ticks have been described by Totze (1933) and by Lees (1948).

The Capitulum and the Organs of Feeding— The headlike capitulum is the most distinctive feature of the Acarina; in the ticks it is a strongly sclerotized structure (fig. 36 A, B), with a thick base bearing laterally a pair of palps and produced medially into a cylindrical rostrum directed forward. The basal part, or *basis capituli* (A, *Bcp*) as it is called, has particularly strong walls and is rectangular in cross section; a narrower, posterior necklike extension fits into the receptive cavity of the body between the bases of the first legs, where it is attached by a tough, flexible integument. On the dorsal surface of the basis capituli of the female are two porous areas, the

120

areae porosae (fig. 35 A, *apr*), which, according to Falke (1931), contain sense organs like those of the spiracular plates. The rostrum is composed of three parts: dorsally are two closely adjacent tubular extensions of the basis (fig. 36 A, *cSh*), which ensheath the long slender chelicerae (C, *Chl*); ventrally (B) a prolongation of the lower wall of the capitulum forms a long, somewhat spoon-shaped underlip, known as the *hypostome* (*Hst*), armed in D. *variabilis* with six rows of strong teeth on its undersurface. Above the base of the hypostome is the mouth (C, *Mth*), which leads into the pharynx (*Phy*). Projecting over the mouth, beneath the cheliceral sheaths, is a conical lobe ending in a slender, tapering stylet, which is the labrum (*Lm*). On the dorsal surface of the hypostome a narrow median groove, or gutter, runs back beneath the labrum to the mouth and is the food conduit (*fc*) leading into the pharynx, which is the sucking organ. The distal end of the hypostome (B) is delicately membranous and is covered by numerous minute papillae bearing short hairs, presumably having a sensory function. The rest of the hypostome is rigid, and its teeth are directed toward the base. The chelicerae are the cutting organs that make an incision into the skin of the host; the hypostomal teeth retain the rostrum when the latter penetrates the wound.

The palps are movably attached to the basis capituli at the sides of the rostrum (fig. 36 A, B, *Plp*). In *Dermacentor variabilis* they extend somewhat beyond the end of the rostrum, and each is four-segmented, though the basal segment is firmly united with the second. The apical segment is a very small lobe bearing sensory hairs, located in a membranous ventral area on the end of the third segment (B). Dorsally each palp is expanded mesally in a broad flange that, in the usual position of the palps, overlaps the upper surface of the rostrum. When, during feeding, the rostrum penetrates the skin of the host, the palps spread out to the sides.

The chelicerae are long, slender, cylindrical rods with their bases deeply sunken into the upper part of the capitulum and, when fully retracted, extending into the anterior part of the body (fig. 36 C, *Chl*). The shaft of each organ is indistinctly divided into two segments and bears at the distal end a small, movable apical segment. The base of the apical segment (H) is produced into an elongate, toothed mesal process and bears a shorter, strongly toothed, movable lateral process. On the base are attached the tendons of two

121

antagonistic muscles arising proximally in the shaft; in action the apical segment as a whole turns laterally and somewhat downward. The lateral dentate lobe of the apical segment of the tick's chelicera is clearly not the "movable finger" of the chelicerae of other arachnids, since the tendons of both muscles are attached on the common base of the segment. The entire segment, therefore, appears to be the "movable finger" with an accessory lateral lobe. It is guarded mesally by a liplike extension of the end of the shaft, and, covering the toothed processes dorsally, is a delicate membranous fold (H, I). The chelicerae can be protracted from their sheaths and retracted; retraction is produced by muscles attached on the cheliceral bases, protraction is thought to be produced by a bulblike contraction of the body.

The cheliceral sheaths are tubular extensions of the capitular integument that entirely surround the retracted chelicerae (fig. 36 C). Their dorsal walls are weakly sclerotized, the ventral walls are membranous. At the distal end the outer wall of each sheath (*ocSh*) is inflected to form an inner membranous tube, or inner sheath (*icSh*), that more closely invests the cheliceral shaft and is attached on the proximal segment of the latter. The free part of the inner sheath becomes everted with the protraction of the chelicera. On the basal part of the rostrum the outer sheaths are united (A, *cSh*), and their cavities are separated by only a median septum.

In the more generalized Acarina there is a free epistomal region proximal to the labrum, beneath the chelicerae, on which arise dorsal dilators of the pharynx. In the argasid ticks as shown by Robinson and Davidson (1913–1914) and by True (1932), and in *Ixodes ricinus* as described by Arthur (1946), a similar plate, termed the subcheliceral plate, lies beneath the chelicerae and gives attachment to the dorsal dilator muscles of the pharynx. The subcheliceral plate of these ticks, therefore, must be the *epistome*. In *Argas persicus* and *Ixodes ricinus* the plate is thin medially but thickened along the margins (fig. 36 F, *Epst*) where the dorsal muscles of the pharynx (*dld*) are attached on it. With the plate apparently is united the lower walls of the outer sheaths (*cSh*) of the chelicerae. In *Dermacentor* a distinct epistomal sclerotization extends a short distance proximally from the upper side of the base of the labrum, but otherwise the epistome (C, *Epst*) is not distinguishable from the under

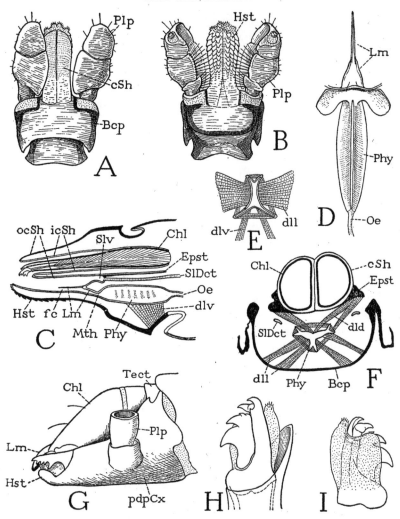

Fig. 36. Arachnida—Acarina. The capitulum and mouth parts.

A, *Dermacentor variabilis* (Say), capitulum, dorsal.   B, same, capitulum, ventral.   C, same, diagrammatic longitudinal section of capitulum, reconstructed from dissections.   D, same, pharynx and labrum, dorsal.   E, same, cross section of pharynx and its muscles.   F, *Argas persicus* (Oken), transverse section of capitulum through pharynx (from Robinson and Davidson, 1913–1914, relettered).   G, *Opiliacarus segmentatus* With, Notostigmata, capitulum, lateral (from With, 1904).   H, *Dermacentor variabilis* (Say), apical segment of left chelicera, ventral.   I, same, apical segment of left chelicera, dorsal.

For explanation of lettering see pages 126–127.

walls of the cheliceral sheaths. In *Dermacentor* there are no dorsal dilator muscles of the pharynx.

The extraoral space between the bases of the cheliceral sheaths and the hypostome has been called the "buccal cavity" of the tick, but since it is entirely outside the mouth it is properly a *preoral cavity*. The ducts of the salivary glands, one on each side (fig. 36 C, *SlDct*), open into a pocket of the preoral cavity (*Slv*) above the base of the labrum.

The pharynx of *Dermacentor* is an elongate wedge-shaped sac (fig. 36 D, *Phy*) lying in the ventral part of the capitulum (C). In the contracted condition, as seen in cross section (E), it is triradiate, being very narrow above and expanded ventrally, with the four walls incurved toward the lumen. On each of the side walls are attached about seven winglike bundles of dilator muscle fibers (*dll*), and on the ventral wall a double row of ventral dilators (*dlv*), all of which arise on the walls of the capitulum. Between the four angle ridges of the pharynx are stretched small constrictor muscles alternating with the bundles of dilator fibers. In most other Acarina that have been studied, dorsal dilator muscles of the pharynx arise on the epistomal plate, as shown in the argasid tick *Argas persicus* (F, *dld*) by Robinson and Davidson (1913–1914). In *Argas* and *Ixodes*, however, the dorsal wall of the pharynx is wide, and the ventral wall narrow. Douglas (1943) records the presence of only lateral and ventral dilators of the pharynx in *Dermacentor andersoni*, and in *D. variabilis* the writer has observed no muscles attached on the narrow upper wall of the pharynx. The anterior end of the dorsal pharyngeal wall in *Dermacentor* is produced into a pair of lateral plates (D). Douglas has shown that transverse muscles on the dorsal surfaces of these plates apparently serve to close the mouth by depressing a valvelike lobe of the pharynx lumen between them. From the inner end of the pharynx the slender oesophagus (C, *Oe*) proceeds back to the stomach.

A more detailed description of the structure of the capitulum and the feeding mechanism of *Ixodes hexagonus* Leach is given in a recent paper by Arthur (1951a).

At the base of the dorsal wall of the capitulum of the female tick, in the anterior end of the body, is an eversible wax-secreting structure known as the *organ of Géné*. It is used for coating the eggs with a protective covering of wax. The structure of the organ in *Ixodes*

124

ricinus and the method by which the female tick brings the eggs
into contact with it are fully described by Lees and Beament (1948)
and by Arthur (1951a). At the time of oviposition the capitulum is
turned downward and backward against the undersurface of the
body in front of the gonopore. The vagina, containing an issuing egg,
is then everted in the form of a tube, and the egg is pressed against
the organ of Géné, which, when fully extruded, has a globular form
with a pair of bifid horns on its outer end. Finally, the organ is retracted, leaving the egg on the hypostome, and the capitulum reverts to its usual position. Most eggs that are not waxed shrivel
quickly, and "few hatch even in a humid atmosphere."

Inasmuch as no other arachnid has a head structure comparable
with the capitulum of the Acarina, it becomes somewhat of a problem to understand how the acarine capitulum has been evolved from
ordinary arachnid parts. The generalized notostigmatid mite *Opiliacarus* (fig. 36 G), as figured by With (1904), however, readily
gives the clue to the capitular composition. The chelicerae (*Chl*) of
Opiliacarus are fully exposed, except at their bases, which are slightly
retracted into the capitulum. The large pedipalp coxae (*pdpCx*) are
united below and extended upward on the sides to a small dorsal
plate (*Tect*) over the cheliceral bases. It is clear, then, that a union
of the dorsal plate with the upper ends of the coxae would produce
an annular basis capituli. Distally the united coxae are produced
into a small hypostome (*Hst*), and their upper margins beneath the
chelicerae are united by an epistomal bridge giving attachment to
dorsal dilator muscles of the pharynx. In the formation of the
notostigmatid capitulum, therefore, the only part added to the usual
arachnid structure is the dorsal plate connecting the upper ends
of the pedipalp coxae. This plate can be nothing more than a secondary sclerotization of the integument over the bases of the chelicerae, which in the ticks is produced into the cheliceral sheaths.
Acarologists commonly call the dorsal wall of the capitulum, together
with the upper walls of the cheliceral sheaths, either the "rostrum"
or the "epistome." The term "epistome" in this connection is clearly
misapplied. Since the surface in question is the roof of the basis
capituli, the writer (1948) has named it the *tectum capituli* (fig.
36 G, *Tect*). The functional rostrum of the tick is the composite feeding organ composed of the cheliceral sheaths, the chelicerae, and
the hypostome.

Explanation of Lettering on Figures 17–27, 29, 31, 33–36

aAp, anterior apodeme of coxa.
ab, abductor muscle.
ad, adductor muscle.
An, anus.
anL, anal lobe.
Ap, apodeme.
apr, area porosa.
apTra, apodemal trachea.
Art, artery.
aSpn, anterior spinneret.
Atr, atrium of spiracle.

b, sternal articulation of coxa.
Bcp, basis capituli.
bl, book lung.
BW, body wall.

Capt, capitulum.
Chl, chelicera.
chlF, cheliceral foramen.
clg, cardiac ligament.
Col, colulus.
Cp, carapace.
cpr, compressor muscle.
Crb, cribellum.
cSh, cheliceral sheath.
Cx, coxa.
cxnd, coxal endite (*1cxnd,* of first leg; *2cxnd,* of second leg).
cxp, coxal process.
cxR, coxal ridge.
cxs, coxal sulcus.

Dac, dactyl, median claw of pretarsus.
dc, dorsal canal of pharynx.
Dct, duct.
dld, dorsal dilator muscle.
dll, lateral dilator muscle.
dlv, ventral dilator muscle.
dmcl, dorsal muscle.
dpl, dorsal plate of pharynx.
dpptar, depressor muscle of pretarsus.
dpt, tendon of depressor muscle.

eAp, epistomal apodeme.
Emp, empodium.
Endst, endosternum.
epgF, epigastric furrow.
Epgn, epigynum.
Epst, epistome.

fc, food canal.
Fm, femur.
fst, fastigium.

Gld, venom gland.
GO, genital opening.
Gpr, gonopore.
Gtr, gonotreme.

h, dorsal hinge of tarsus on tibia.
Hst, hypostome.
Ht, heart.

icSh, inner cheliceral sheath.

L, leg.
lam, lamella of lung.
Lm, labrum.
lmcls, longitudinal muscles.
lTra, lateral trachea.
lvptar, levator muscle of pretarsus.
lvt, tendon of levator muscle.

Mb, membrane.
mcl, muscle.
mSpn, median spinneret.
Mth, mouth.

o, atrial openings into lung lamellae.
O, ocellus.
ocSh, outer cheliceral sheath.
Oe, oesophagus.
Opl, operculum.
Ost, ostium of heart.

pAp, posterior apodeme of coxa.
Pat, patella.
Pcrd, pericardium.

Pdp, pedipalp.
pdpCx, coxa of pedipalp.
pdpF, foramen of pedipalp.
Pec, pecten.
pgf, pregenital fold.
Phy, pharynx.
Pl, pleural fold.
Plp, palp.
PrC, preoral food cavity.
pSp, pulmonary spiracle.
pSpn, posterior spinneret.
Ptar, pretarsus.
pv, articular pivot.
PvP, proventricular pump.

Rect, rectum.

s, septa between openings of lung lamellae.
S, sternum.
SCl, sensory cell.
Shld, shield of female tick.
SlDct, salivary duct.
Slv, salivary pocket.
SO, sense organ of tarsus.
Sp, spiracle (*1Sp*, first spiracle; *2Sp*, second spiracle).
Spg, spigot.

Spn, spinneret.
Stn, sting.
svpl, sieve plate, opening of salivary glands.

t, tendon.
T, tergum.
tar, tarsomere (*1tar*, *2tar*, first and second tarsomeres).
Tar, tarsus.
Tb, tibia.
Tect, tectum capituli.
tmcl, transverse muscle of labrum.
tp, tarsal process.
Tr, trochanter.
tSp, tracheal spiracle.

u, tarsal articulation of pretarsus.
Un, unguis, lateral claw of pretarsus.

Vac, vacuole.
VGld, venom gland.
vmcl, ventral muscle.
Vp, vena pulmonaris.
vpl, ventral plate of pharynx.
VPr, venom pore.
Vstb, vestibule.

⁂ VI ⁂

THE CRUSTACEA

THE Crustacea introduce us to the arthropod group known as the Mandibulata, which includes also the chilopods, the diplopods, the pauropods, the symphylans, and the hexapods, or insects. The Mandibulata are so named from the fact that their principal organs for biting and chewing are a pair of jaws, the *mandibles,* fashioned from a pair of segmental appendages that correspond with the pedipalps of the Chelicerata. In the Crustacea the mandibles are preceded by two pairs of antennae. The second antennae evidently are represented by the chelicerae in the Chelicerata, but they are absent in mandibulate forms other than the Crustacea, except for transient embryonic vestiges. The first antennae, or antennules, however, are characteristic head appendages of all the mandibulate arthropods and are present in the trilobites. Since the antennules do not have the structure of the following appendages, it is probable that they represent very primitive head appendages of the ancestral arthropods, which have been lost only in the chelicerates and in a few forms here and there among the mandibulates. The second antennae, on the other hand, show by their structure that they belong to the series of body limbs.

The postantennular appendages of the Crustacea have typically a biramous structure, owing to the frequent presence of a lateral branch on the second segment from the base of the limb. The branch is termed the *exopodite,* and the main shaft of the limb beyond it the *endopodite.* The biramous limb is probably a primitive crustacean character, but by a suppression of the exopodite the appendage in many cases reverts to a simple uniramous form. Since the Crustacea

128

are generally regarded as the most generalized of the mandibulate arthropods, the biramous limb has been thought to be the primitive form of the arthropod appendages. Occasional branched limb structures in other mandibulates, therefore, have been interpreted as retentions of, or reversions to, the ancestral structure. Such interpretations, however, are in no case necessary, and are not supported by specific evidence; the exopodite branch of the limb is a crustacean specialty. Besides the exopodite, there may be an epipodite borne on the outer face of the coxopodite, and median lobes, or endites, on the coxopodite and basipodite. A determination of the homologies of the limb segments and their accessories is one of the interesting and often perplexing problems in carcinology.

The Crustacea include many diverse forms; taxonomically they have been divided into two major groups, the Entomostraca and the Malacostraca. The term Entomostraca, however, is just a convenient name for a large number of small crustaceans such as the branchiopods, the ostracods, the copepods, and the barnacles that may have no close relationships to one another. The Malacostraca, on the other hand, including such forms as the shrimps, crayfish, lobsters, and crabs, are a more homogeneous group, but included in it are the amphipods and the isopods, which have distinctive characters of their own. A chapter on the Crustacea in general would expand to the size of a book. Hence, we shall have to omit the entomostracans entirely and limit a discussion of the Malacostraca to three examples, *Anaspides,* a crayfish, and an isopod, representing the principal types of malacostracan organization.

ANASPIDES

Anaspides is a small, shrimplike crustacean known only from Tasmania, where it lives in pools of running water on Mount Wellington, mostly above 1,400 feet. A related form, *Paranaspides,* inhabits the Great Lake of Tasmania at a height of 3,700 feet, and another, *Koonunga,* has its home in Australia. A minute crustacean named *Bathynella,* found in springs and caves of central Europe, is usually classed with the Anaspidacea, but it has a number of quite distinctive features. The Tasmanian and Australian species are nearest related to certain fossil crustaceans from Permian and Carboniferous strata of Europe and North America, and they are the only known living representatives of this group, called the Syncarida, which includes several families comprised in the single order Anaspidacea. For earlier discussions on the syncarids, the student is referred to Smith (1909) for the structure of the Anaspidacea, living and fossil, to Calman (1896) for a demonstration of the affinities of *Anaspides* with the fossil syncarids, and to Manton (1930) for an account of the habits and feeding mechanisms of *Anaspides* and *Paranaspides. Koonunga* is described by Sayce (1908), *Bathynella* by Chappuis (1915) and by Calman (1917).

Anaspides tasmaniae Thomson (fig. 37 A) has a slender, uniformly segmented body, two pairs of long antennae, and a series of segmental limbs that are mostly locomotor in function. Ordinary specimens are about an inch and a half in length, but exceptional individuals are said to exceed two inches. While *Anaspides* cannot be regarded as a primitive crustacean, it is unquestionably the most generalized of modern Malacostraca, and its simplicity of structure suggests that in some respects it preserves features more primitive in form than those of the Entomostraca. Hence, though specimens may not be available for class use, a description of *Anaspides* will give the student an outline of the essential crustacean characters in their simplest available form.

The body of *Anaspides* (fig. 37 A) is segmented throughout its length and bears 17 pairs of appendages, not including the two pairs of antennae. There appear to be, however, only 15 segmental divisions of the body as indicated by the number of tergal plates, but it is found that the second apparent tergum corresponds to three pairs of appendages, so that it must be a composite of three primitive segments (*III + IV + V*), making 17 in all (*II–XVIII*). The large first tergum is enumerated as belonging to segment II because in the crustacean embryo there is a preceding segment of the second antennae, corresponding with the cheliceral segment I of the Arachnida. Anterior to the antennal segment of the embryo is a primitive head lobe that bears the eyes and the first antennae. This head lobe was perhaps itself segmented at some early stage of arthropod evolution, but arthropodists do not agree as to this, or as to how many segments may be represented in the cephalic lobe of modern embryos. Merely for purposes of comparative enumeration, therefore, the crustacean segments will be numbered the same as those of the Arachnida, beginning with the segment of the second antennae as segment I. The body of *Anaspides* terminates posteriorly with a

130

flat lobe, known as the *telson* (*Tel*), which contains the anus on its undersurface (fig. 39 B, *An*). The telson, however, has no appendages and is not regarded as a true segment, or, more properly, a somite.

The first tergal plate of the body (fig. 37 D, *II*) projects anteriorly in a short *rostrum* (*R*) over the eyes and the bases of the antennae.

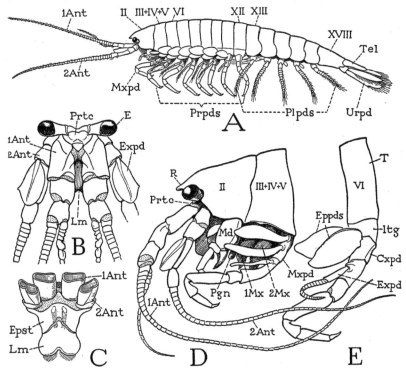

Fig. 37. Crustacea—Anaspidacea. *Anaspides tasmaniae* Thomson.
A, entire animal, male, with appendages of left side. B, head and its appendages, anterior. C, epistome and labrum, with bases of antennae, ventral. D, head and anterior part of body, with appendages, showing body segments of first and second maxillae and maxillipeds combined under one tergal plate (*III* + *IV* + *V*). E, body segment of first legs (first free segment).
For explanation of lettering see pages 190–192.

Concealed beneath the rostrum and the free edge of the tergum is a small head structure, which, when detached (B), is seen to carry the stalked eyes (*E*), the first and the second antennae (*1Ant*, *2Ant*), and the labrum (*Lm*); it, therefore, must include the embryonic head lobe and at least a part of the second antennal segment. Since no simpler cephalic structure than this headpiece of *Anas-*

131

pides is known in any other adult arthropod, it may be designated
a *protocephalon,* implying that it represents the *first* adult head in
the evolution of the arthropods. The protocephalon recurs as a discrete head in many other crustaceans, but it is to be noted that it
is principally a sensory tagma, since it does not carry the organs of
feeding. In various crustaceans, however, and in all the other mandibulate arthropods, the protocephalon is united with some of the
anterior body segments to form a secondary head of more complex
structure that combines in one unit both the sensory and the feeding functions. Though the protocephalon is a distinct anatomical
unit in the adult structure of most of the malacostracan Crustacea,
and always carries the second antennae, its segmental composition
is not clear, since, in the decapods at least, the dorsal muscles of
the second antennae take their origins on the anterior part of the
mandibular region of the carapace.

The head of *Anaspides,* as above noted, carries the eyes, the first
antennae, and the second antennae; the stalked compound eyes (fig.
37 B, *E*) project laterally from the dorsal wall, the antennae arise
beneath them. The first antennae, or antennules (*1Ant*), consist
each of a thick basal stalk composed of three large segments and of a
pair of slender, multiarticulate flagella, of which the outer flagellum
is much longer than the inner. Though the first antennae are two-
branched, the branches do not conform with those of the second
antennae and the legs known as the exopodite and the endopodite,
since they arise from the third and not the second segment of the
appendage, and neither of them is truly segmented. The eyes and
the first antennae pertain to the primary cephalic lobe of the embryo.

The second antennae (fig. 37 B, D, *2Ant*) have each a basal stalk
of four segments, the last of which bears a single, long, multiarticulate
flagellum. From the second segment, however, arises laterally a large,
flat lobe (B, *Expd*), which thus in position corresponds with the
exopodite of a leg. The second antennae, therefore, appear to be
truly biramous appendages serially homologous with the body limbs;
though they are carried by the head of the adult, their rudiments in
the embryo are formed behind the mouth, the appendages later becoming secondarily preoral, just as do the chelicerae in Arachnida.

The under wall of the head is formed by a wide plate beneath
the bases of the antennae (fig. 37 C, *Epst*). This plate is the *epistome*
of the Crustacea, and, as in the arachnids, it carries the labrum

(*Lm*). Its basal angles are produced into narrow arms that go up-ward behind the antennal bases to the dorsum of the head. The labrum represents the anterior pole of the arthropod; its ventral position behind the antennae is secondary.

The trunk segments of *Anaspides* are not entirely uniform: there is an abrupt change in the shape of the tergal plates between segments XII and XIII (fig. 37 A). The last six segments of the body (*XIII–XVIII*) constitute the *abdomen*, or *pleon*, of malacostracan Crustacea; the preceding part of the body is commonly termed the *cephalothorax*. In *Anaspides*, however, the latter includes the proto-cephalon, the first four postcephalic segments of the body (*II–V*), which carry the manibles (D, *Md*), two pairs of maxillae (*1Mx*, *2Mx*), and a pair of leglike maxillipeds (*Mxpd*), while a third region, composed of segments VI–XII (A), bears the seven pairs of walking legs, or *pereiopods* (*Prpds*), and may be termed specifically the *thorax* to distinguish it from the *gnathal region* of the feeding appendages, though generally the maxilliped segment is reckoned as a part of the thorax. The appendages of the abdomen are termed *pleopods* (*Plpds*), but the broad, leaflike appendages of the last segment are distinguished as *uropods* (*Urpd*).

The tergal plate of the dorsum of each body segment in the thoracic region comes down on the sides to the bases of the legs, but shortly above each leg base it is crossed by a faint longitudinal groove that sets off from the main plate of the back a small *latero-tergite* (fig. 38 A, *ltg*), on which the coxa of the leg is weakly articulated (*a*). The laterotergites of *Anaspides*, therefore, are evidently homologous, by position at least, with the limb-supporting sclerites of other arthropods, called the *pleurites*, such as the small pivotal sclerites of the legs in *Limulus* (fig. 7 F, *Pl*) and the segmental components of the highly developed pleural plate of the thorax of the crayfish (fig. 43 B). The pleuron of the crustaceans thus appears to be a derivative of the dorsal skeleton.

The ventral surfaces of the first six leg-bearing segments of *Anaspides* are mostly membranous, but in each segment (fig. 38 B) there is a narrow, transverse sternal sclerite (*S*) behind the leg bases, which is continuous on each side with the laterotergite (*ltg*) of its segment by a postcoxal arm (*pcx*). Between the sternum and each coxa is a small pit (*inv*), which marks the root of a short internal apodeme. The sternum of the last leg-bearing segment in the male is a large

133

plate lying before the coxae, and behind it are the paired, slitlike male genital apertures (fig. 39 D, *Gprs*). On the venter of the same segment in the female, between the bases of the legs, is the single aperture of the sperm receptacle. The female genital openings are on the mesal surfaces of the coxae of the fifth legs (segment X). The sterna of the maxillary segments and of the abdomen are somewhat differently developed, and will be described in their appropriate places.

The body appendages of the higher Crustacea are so diversified in form by adaptation to different functions that no general description can be made to fit them; yet it is to be presumed that they have all evolved from a common generalized type of limb. The dominance of the leg type of structure in all arthropods and the fact that the trilobite limbs are legs of uniform pattern and function suggest that in the Crustacea the typical ambulatory pereiopods have deviated least from the original limb structure. In studying the body appendages of *Anaspides*, therefore, we may proceed best by first understanding the structure and segmentation of the walking legs.

Any one of the first five pereiopods of *Anaspides* will serve to illustrate the typical structure of a crustacean appendage. The limb consists of a main shaft of seven segments (fig. 38 A). In the terms of carcinology, the first segment is the *coxopodite* (*Cxpd*), the second the *basipodite* (*Bspd*), the third the *ischiopodite* (*Iscpd*); then comes an elongate segment known as the *meropodite* (*Mrpd*), followed by a short *carpopodite* (*Crppd*), a slender *propodite* (*Propd*), and finally the clawlike *dactylopodite* (*Dactpd*). The knee bend of the leg is at the joint between the meropodite and the carpopodite. These two segments of the crustacean leg correspond with the femur and tibia of the arachnid leg (fig. 25 A), and consequently a patella is absent in Crustacea, as it is in the legs of all the other mandibulate arthropods. The coxopodite of *Anaspides* (fig. 38 A) is weakly articulated (*a*) on the small laterotergal plate of the dorsum (*ltg*), and bears on its outer surface a pair of large, flat, leaflike lobes (*Eppds*), which functionally are probably gills; but, in general, appendicular structures arising from the outer side of the coxopodite are termed *epipodites*. From the basipodite arises laterally a long slender branch of the limb (*Expd*), consisting of a two-segmented basal stalk and an annulated flagellum. This lateral branch of the basipodite is termed the *exopodite*, and the shaft of the limb beyond the basipo-

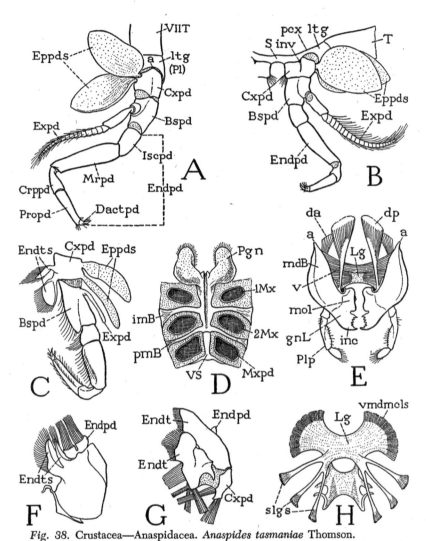

Fig. 38. Crustacea—Anaspidacea. *Anaspides tasmaniae* Thomson.

A, third left pereiopod and adjoining part of tergum, lateral. B, third right pereiopod and supporting parts of body segment, posterior. C, left maxilliped, anterior. D, ventral surfaces of maxillary and maxilliped segments, with paragnaths, showing intersegmental sternal brachia. E, mandibles and their musculature, posterior. F, left second maxilla, posterior. G, left first maxilla, posterior. H, intergnathal ligament and suspensory branches (flattened under cover glass).

For explanation of lettering see pages 190–192.

dite is then called the *endopodite* (*Endpd*). The pereiopods of the sixth pair in the male (fig. 37 A) differ from those preceding in that the gill lobes are relatively small and the exopodite is a simple short, unjointed appendage. The last pereiopods in each sex have neither epipodites nor an exopodite.

Though each leg of *Anaspides* is clearly divided into only seven interarticulated segments, some carcinologists contend that the leg really contains eight segments, or even nine. One of these supposed segments is seen in the demarked area of the basipodite in *Anaspides* that bears the exopodite (fig. 38 A); but there is nothing to indicate that this part was ever an independently movable section of the leg. The idea that the primitive arthropod limb included a "subcoxal," or "precoxal," segment has been much exploited, and in *Anaspides* the small laterotergal plate of the dorsum, on which the coxa is articulated, has been interpreted by Hansen (1925) as a remnant of this hypothetical basal segment. It is on the coxa, however, as in the arthropods generally, that the body muscles of the limb are inserted, indicating that the coxa is the true base of the appendage. In *Anaspides* the dorsal muscles of the legs pass over the laterotergites to insert on the coxae, and there is nothing to indicate that the laterotergites are anything else than lateral subdivisions of the tergum, representing the pleural sclerites more elaborately developed in some other forms.

The maxillipeds of *Anaspides* (fig. 38 C) resemble the legs and have the same number of segments. The coxopodite of each maxilliped also has two epipodites (*Eppds*), but they are much smaller than those of the legs, and the exopodite (*Expd*) is reduced to a small, simple, slender appendage of the basipodite. The maxilliped coxa, however, bears two well-developed endites (*Endts*) fringed with long hairs. Anatomically, the maxillipeds of *Anaspides* are the first pair of legs, but they are turned forward at the sides of the mouth parts in front of them (fig. 37 D) and are functionally a part of the feeding apparatus; according to Manton (1930), they are never used for walking or digging. In some Crustacea the next pair of legs, or the next two pairs, may also be transformed into maxillipeds.

In front of the maxillipeds are the two pairs of maxillary appendages (fig. 37 D, *1Mx, 2Mx*), which, together with the maxillipeds (*Mxpd*), are suspended from the third, fourth, and fifth segments that have in common the single composite second tergal plate

($III + IV + V$). Both maxillae of *Anaspides* are small, much simplified appendages having no resemblance at all to a leg. Anatomically they are of interest chiefly as examples of the extent to which
a segmental appendage may be reduced in size and simplified in
structure. The second maxilla (fig. 38 F) consists principally of a
large basal part, from the mesal margin of which projects a pair of
hair-fringed endites (*Endts*), and which bears distally a movable,
bilobed segment (*Endpd*) with two long brushes of hairs, which possibly represents the endopodite. The first maxilla, or maxillule (G),
has a small basal segment (*Cxpd*) that appears to be the coxopodite,
but the main body of the appendage presents on its posterior surface
two large plates lying side by side, each of which is extended mesally
into a broad endite (*Endt*). A small lobe (*Endpd*) on the outer
side of the lateral plate may possibly be a remnant of the endopodite.
It is evident that the parts of appendages such as the maxillae of
Anaspides can be identified only when we have a series of forms
that show the steps in the modification from a typical biramous limb.
In some of the Crustacea the maxillae retain more distinctly a structure comparable with that of a leg, as will be seen in the crayfish,
while in others they may be even more reduced than are those of
Anaspides. The crustacean maxillae serve principally for passing food
forward to the mouth.

Finally we come to the mandibles, which are the biting jaws of
the animal. They are the first of the series of segmental appendages
behind the head (fig. 37 D, *Md*) and correspond with the pedipalps
of the Arachnida. The conversion of these appendages into jaws is
distinctive of the crustaceans, the myriapods, and the insects, and
separates these groups from the chelicerate arthropods, which have
no true biting or masticatory organs.

The mandibles of *Anaspides* (fig. 37 D, *Md*) hang vertically from
the tergum of the first body segment (*II*), on which each jaw has a
single point of articulation. The body of each mandible (fig. 38 E,
mdB) is broadly attached to the membranous lateral wall of its segment, and thus is freely movable on its tergal articulation (*a*). Visible
evidence that the mandible is a modified leg is seen in the presence
of a three-segmented palpus (*Plp*) arising from its outer surface,
which evidently represents in reduced form the distal part of an
ordinary limb. On the large basal part of the mandible (*mdB*) are
attached the body muscles that move the appendage as a whole;

137

these muscles correspond with the coxal muscles of a leg and thus show that the mandibular base is the coxopodite of the mandibular appendage. At its lower end, mesad of the palpus, the mandible is produced into a strong, free *gnathal lobe* (*gnL*), which is a specially developed coxal endite and is the effective part of the jaw. The gnathal lobe in *Anaspides* is differentiated into a thick basal molar process (*mol*) and a toothed distal incisor process (*inc*).

The musculature of the *Anaspides* mandibles is simple, but strongly developed (fig. 38 E). Each jaw has a small anterior dorsal muscle (*da*) and a large posterior dorsal muscle (*dp*), both arising on the tergum of the mandibular segment. Inasmuch as each mandible has a single dorsal point of articulation (*a*), and has little freedom of anterior and posterior movement, the dorsal muscles probably act as rotators of the jaw, but if they both pull together, they evidently may function also as adductors. The principal adductor muscles of the mandibles, however, are huge bundles of ventral fibers (*v*) filling the cavities of the mandibular bases and attached medially on a sheet of ligamentous tissue (*Lg*) suspended between the two jaws. These muscles of opposite sides pulling against each other effect a strong adductor action of the gnathal lobes. The intergnathal ligament when denuded of muscles (H) is seen to be a sheet of rather dense tissue expanded between the mandibles and connected posteriorly by two short arms with a second smaller ligament giving attachment to the maxillary adductors. The whole is hung from the back by three pairs of slender suspensory ligaments (*slgs*). The intergnathal ligament of lower Crustacea appears to be a structure of the same nature as the endosternum of the chelicerate arthropods, which gives attachment to the ventral muscles of the prosomatic appendages. Mandibles of the *Anaspides* type of structure appear to have no muscular mechanism of abduction: the jaws probably open by the elasticity of their basal connections.

The principal food of *Anaspides* is said by Smith (1909) and by Manton (1930) to be algal slime and organic detritus covering the rocks and plants on which the crustaceans live, but the latter eat also small animals such as the dead bodies of insect larvae or of their own species, and even small worms and tadpoles.

On the ventral surface of the maxillary region of the body (fig. 38 D) the sterna of the two maxillary segments are united in a single, median, deeply channeled sternal plate extending forward from the

138

separate maxilliped sternum (*VS*) to the mouth. On each side the maxillary sternum gives off two intersegmental arms; one is an *intermaxillary brachium* (*imB*) between the bases of the first and second maxillae (*1Mx, 2Mx*), the other a *postmaxillary brachium* (*pmB*) separating the second maxilla from the maxilliped (*Mxpd*). On its anterior end the maxillary sternal plate supports two large, flat, divergent lobes (*Pgn*) lying against the posterior surfaces of the mandibles. These lobes are the *paragnaths;* they are characteristic organs of Crustacea, but they are not segmental appendages serially homologous with the other mouth parts and the ambulatory limbs; they probably belong to the mandibular segment. The median groove of the maxillary sternum runs into the mouth between the paragnath bases.

The abdomen of *Anaspides* is superficially distinguishable from the thorax by the shape of its first five tergal plates (fig. 37 A), the lower ends of which are expanded into broad, rounded lobes overlapping from before backward. Each of the corresponding five segments (*XIII–XVII*) bears a pair of pleopods (*Plpds*). The last segment (figs. 37 A, 39 A, *XVIII*) is much longer than the others, tapers posteriorly, and has no division between the tergal and sternal surfaces; its appendages are the uropods (*Urpd*). The body ends with the broad, flattened telson (*Tel*), which bears the anus on its undersurface (fig. 39 B, *An*).

The first five pairs of pleopods are all essentially of the same structure in the female; in the male the first two pairs are modified in form for reproductive purposes, but those of the third, fourth, and fifth segments are like the female pleopods, and any pair of these (fig. 39 E) may be taken to show the typical pleopod structure. The sternum (*S*) of the body segment is a broad plate lying anterior to the bases of the appendages and united laterally with the lobes of the tergum (*T*); between the appendage bases it is produced into a small median point. Each pleopod has a single, large basal segment (*Prtpd*), broadly attached on the posterior margin of the sternum and supporting laterally a long, many-jointed, hairy flagellum (*Expd*), which clearly corresponds with the exopodite of a thoracic limb. Mesally, the basal segment bears a small, soft lobe (*Endpd*), which is regarded as a much reduced endopodite, though its simple structure in itself would scarcely suggest this interpretation. However, the two parts being appended from the basal segment indicates

139

that the latter is composed of both the coxopodite and the basipodite. A composite basal limb segment of this kind is termed a *protopodite* (*Prtpd*). On the fifth pleopods the endopodite lobe is either very small or absent.

The pleopods of the first and second abdominal segments of the male differ from those following and from the female pleopods in that each has a long, strong arm (fig. 39 F, G, *Gon*) from the inner end of the appendage base that projects forward beneath the thorax. The sternum of each of these segments is a small plate (S) lying behind the appendages, and produced on each side in a short post-pedal extension. The mesal arms of the first two male pleopods evidently are the endopodites of the appendages, but they serve during mating for the transfer of sperm from the male to the female, and therefore may be termed *gonapophyses*. Those of the first pair (F, *1Gon*) are relatively thick, widened distally, and their outer margins are folded ventrally and mesally to enclose deep grooves on the inner surfaces. The gonapophyses of the second pair (G, *2Gon*) are slender processes, each divided into a long basal segment and a short distal segment. In the functional position (H) the end of each second gonapophysis is held in the groove of the corresponding first gonapophysis.

The spermatozoa of *Anaspides* are enclosed in two horseshoe-shaped capsules, or spermatophores (fig. 39 I), formed in curved terminal parts of the genital ducts that open on the venter of the last thoracic segment (D, *Gprs*). The gonapophyses serve to introduce the spermatophores into the sperm receptacle on the sternum of the last thoracic segment of the female, but apparently the procedure has not been observed in *Anaspides*. Smith (1909), however, says that the spermatophores may sometimes be seen projecting from the receptacle of the female, and that they soon drop off after the spermatozoa have passed out of them. The openings of the oviducts, as already noted, are on the coxae of the fifth pair of legs of the female; the ova presumably flow posteriorly and are inseminated as they pass under the sperm receptacle. The female of *Anaspides* deposits her eggs on water plants, instead of carrying them on the pleopods as do many other Malacostraca.

The uropods, or pleopods of the sixth abdominal segment, consist each of two long, flat, hair-fringed lobes, an exopodite and an endopodite (fig. 39 C, *Urpd*), projecting posteriorly at the side of

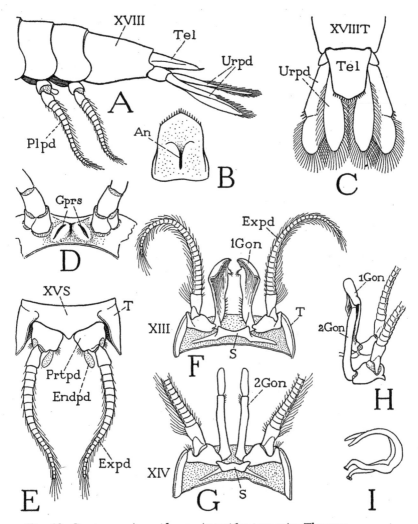

Fig. 39. Crustacea—Anaspidacea. *Anaspides tasmaniae* Thomson.

A, distal segments of abdomen, lateral. B, telson, ventral. C, end of abdomen, dorsal, with uropods. D, seventh thoracic segment (*XII*) of male, ventral, with gonopores. E, third abdominal segment of male, ventral, pleopods turned back. F, first abdominal segment of male, ventral, with first pleopods. G, second abdominal segment of male, ventral, with second pleopods. H, first and second left pleopods of male, ventral, with gonopophyses in functional position. I, two spermatophores (from Smith, 1909).

For explanation of lettering see pages 190–192.

the telson from a small basal segment on the posterior angle of the last abdominal segment. The bases of the lobes are overlapped mesally by the telson, and the inner lobes overlap the somewhat larger outer lobes. The telson and the uropods together compose the so-called "tail fan" of the crustacean, which, however, is much more fan-shaped in the decapods (fig. 40) than in *Anaspides*.

From observations on living specimens, Manton (1930) says that the entire body of *Anaspides* acts as a unit in locomotion to a much greater degree than is true with any other malacostracan, as might perhaps be inferred from its uniformity of structure. The first five pairs of pleopods are often used in conjunction with the legs during walking; and in swimming, the pleopods make the same movement that they do in walking. *Anaspides*, however, seldom swims freely except to go from one submerged rock or weed to another, but spends most of its time walking or half-swimming over the bottom in search of food and digging in the mud with its legs. The thoracic exopodites, Manton observes, are continually in motion, beating anteroposteriorly about 100 times a minute, and produce a backward flow of water along the sides of the body; the broad thin branchial epipodites beat in unison with the exopodites.

THE CRAYFISH, *CAMBARUS*

The crustaceans known as crayfishes are decapods of the group Astacura, which includes also the common lobster, *Homarus,* and a few species of other genera. Being widely distributed over the world, the fresh-water crayfishes have become favorite subjects for study in zoological courses. The genus *Astacus* of Europe and the western part of the United States is the usual crayfish of textbooks, but the crayfishes east of the Rocky Mountains in this country belong to several other genera, differing in certain respects from *Astacus,* and are grouped in the subfamily Cambarinae. Systematists recognize several genera and numerous species, but they are all so much alike in their general structure that we need not be particular as to the species, except with regard to the genital appendages of the male. The following descriptions and accompanying figures are based mostly on specimens of *Cambarus longulus* Girard, from St. Mary's River in Virginia.

The name "crayfish" does not mean anything in itself. It is generally supposed to be a phonetic corruption of the French name

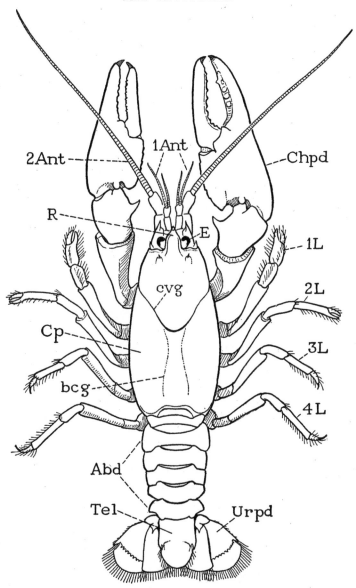

Fig. 40. Crustacea—Decapoda. *Cambarus longulus* Girard, male.
Abd, abdomen; *1Ant,* first antenna, antennule; *bcg,* branchio-cardiac
groove; *cvg,* "cervical" groove; *Chpd,* cheliped; *Cp,* carapace; *E,* compound eye; *1L–4L,* legs, second to fifth pereiopods; *R,* rostrum; *Tel,*
telson; *Urpd,* uropod.

for the animal, *écrevisse*, but Huxley (1880) suggested that it might as well have come from the Low Dutch name of *crevik*. The old English spelling was "crevis" or "crevise," with the *e* pronounced like long *a*. The "vis" later became "fish" because the animal lived in the water and therefore should be a fish. Inasmuch, then, as "crayfish" has a devious etymological history and no claim to zoological correctness, the common American version of "crawfish" should be equally acceptable.

General External Features

The crayfish (fig. 40) gives a good example of the type of crustacean structure in which the gnathal and thoracic regions of the animal are covered dorsally by an unsegmented shell, called the *carapace*. On referring back to *Anaspides* (fig. 37), it will be seen that the carapace of the crayfish is a product of the union of at least six more tergal plates with the three already united in *Anaspides*. The part of the animal covered by the carapace is the *gnathothorax*, but the dorsum of the last thoracic segment appears not to enter into the composition of the carapace. Anteriorly the carapace of the crayfish is marked by a "cervical" groove (fig. 40, *cvg*) that sets off the mandibular region, which terminates in an apical rostrum (*R*). The lower edges of the carapace come down to the bases of the legs, but between them and the legs on each side is a long, narrow opening that leads up into a spacious cavity, the *branchial chamber*, which contains the gills. The gills arise from the coxal segments of the limbs, and also from the articular membranes above them. In front of the carapace, but mostly concealed beneath its projecting edge and the rostrum, is a distinct head structure, the protocephalon, very similar to that of *Anaspides*, which carries the stalked eyes, the two pairs of antennae, and the labrum. Behind the carapace is the more slender *abdomen* (*Abd*), consisting of six segments freely movable on the thorax and on one another. The abdomen ends with a broad, flat telson (*Tel*) having the anus on its undersurface. The abdomen of the crayfish thus differs little from that of *Anaspides*.

On the underside of the body, beneath the anterior part of the carapace are the usual three pairs of gnathal appendages, namely, the mandibles, the first maxillae, and the second maxillae. The next three pairs of appendages, instead of only the first pair as in *Anaspides*, serve as accessory organs of feeding, and are designated the

144

first, second, and *third maxillipeds.* Since there are in all the same number of thoracic limbs in the crayfish as in *Anaspides,* the crayfish has only five pairs of *pereiopods.* The first pereiopods are specifically the *chelipeds* (fig. 40, *Chpd*) since they carry the large pincers, or *chelae,* of the crayfish; the next four pairs (*1L–4L*) are legs. The first and the second legs have small chelae, the last two pairs end with simple claws. Between the leg bases of opposite sides is seen the strongly sclerotized ventral wall of the thorax (fig. 43 A), in which all the sternal plates but the last are solidly united.

The six pairs of appendages of the abdomen resemble those of *Anaspides,* except that the first pair in the female crayfish are very small and unbranched. In the male the first two pairs are modified for reproductive purposes. The large lobes of the last pair, or uropods, form with the telson a broad tail fan. The female crayfish, unlike the female of *Anaspides,* does not deposit her eggs but carries them attached to the ventral pleopods.

Because of the structural differentiation of the several body regions of the crayfish, it is difficult to illustrate by any specific example the fundamental structure of a body segment. In the section on *Anaspides* it was observed that the back of each thoracic segment is covered by an arched tergal plate (fig. 41 A, *T*), but that a small laterotergal, or pleural, plate (*Pl*) intervenes on each side between the main dorsal plate and the base of the leg. The coxa of the leg is then articulated laterally (*a*) on the pleural plate, and mesally (*b*) on the sternum. A cross section of the thorax of a decapod (D) shows that the lateral articulations of the thoracic coxae are on the lower edges of plates (*Pl*) that form the inner walls of the gill chambers (*brC*). These plates, therefore, are the true lateral walls of the thorax of the crayfish, and correspond with the region of the laterotergal pleural plates of *Anaspides.* In *Anaspides* the gills project freely from the coxae of the legs (A, *Brn*); in the decapod (D) they are covered by long descending folds (*tf*) from the upper part of the carapace. A cross section of an abdominal segment (B) differs from that of the thorax in that the pleural areas (*Pl*) between the limb bases and the short tergal folds take a horizontal position in adaptation to the dorsoventral flattening of this part of the body. Again, in the maxillary region (C) the same parts (*Pl*) are also horizontal, forming strong lateral bridges between the bases of the second maxillae (*2Mx*) and the folds of the carapace (*tf*).

In the thorax of the crayfish the pleural and sternal regions of the united segments are joined by strong transverse bars between the bases of the limbs (fig. 41 F, *icx*). The segmental annuli of the abdomen are all separate from each other, and here it is seen that

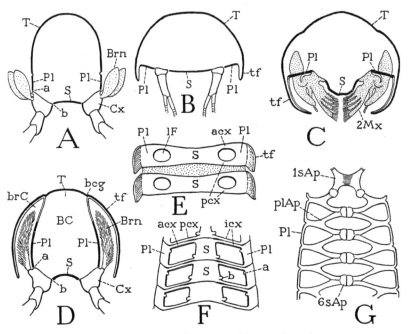

Fig. 41. Crustacea. Diagrams explanatory of decapod structure.

A, cross section of thoracic segment of *Anaspides;* pleura (*Pl*) are parts of dorsum carrying dorsal articulations (*a*) of coxae. B, cross section of astacuran abdomen, pleura horizontal and overlapped by tergal folds (*tf*). C, section of maxillary region; pleura form bridges between bases of maxillae and tergal folds. D, section of astacuran thorax, pleura vertical in lateral walls of body, long tergal folds (branchiostegites) extended downward to cover the gills. E, ventral surfaces of two consecutive abdominal segments, sternal and pleural regions connected by antecoxal (*acx*) and postcoxal (*pcx*) pleurosternal bridges. F, ventral surface of astacuran thorax, contiguous postcoxal and antecoxal bridges united, forming intercoxal brachia (*icx*). G, scheme of endoskeletal structure of the astacuran thorax, dorsal, composed of pleural (*plAp*) and sternal (*sAp*) apodemes arising from the intersegmental grooves.

For explanation of lettering see pages 190–192.

in each segment (E) the pleura and the sternum are united before and behind the limb bases by antecoxal and postcoxal pleurosternal connectives (*acx, pcx*). In the thorax (F) the union of the segments involves the pleurosternal connectives, so that the antecoxal connectives of one segment are united with the postcoxal connectives of the segment in front, and there is thus formed the series of *intercoxal brachia* (*icx*) separating the foramina of successive appendages. Each intercoxal brachium, however, is marked by a deep groove continuous on the one hand with the corresponding intersegmental groove between the united pleural plates, and on the other with that between the sterna. The endoskeleton of the decapod (G) is composed of pleural and sternal apodemes (*plAp, sAp*) inflected from the intersegmental grooves.

Carcinologists commonly describe a decapod as being divided into two tagmata, the first being the "cephalothorax," which is mostly covered by the carapace, the second the abdomen. However, inasmuch as there is a distinct head section anterior to the carapace, and the region of the carapace includes only the segments of the feeding organs and the pereiopods, the cephalothorax really includes a head tagma and a gnathothoracic tagma. In the following account of the trunk regions of the crayfish, therefore, the tagmata will be described as the *head* (*protocephalon*), the *gnathothorax*, and the *abdomen*. There being the same number of trunk segments in the decapod as in *Anaspides,* the segments may be similarly numbered, beginning with the segment of the second antennae as segment I.

The Head

The head of the crayfish is a typical protocephalon; it has essentially the same structure as that of *Anaspides,* but it is more strongly calcified and the epistomal region is more elaborately developed. Viewed anteriorly, with the appendages removed, the head skeleton (fig. 42 B) presents dorsally a broad surface perforated by the foramina of the eyestalks (*esF*), and reflected posteriorly into the underfolded margin of the carapace. From the groove between the head and the carapace arises a pair of flat, divergent apodemes (*Ap*), on which are attached broad muscles going to the dorsal wall of the proventricular "cardiac" sack of the stomodaeum. Below the ocular region mesally are the foramina of the first antennae (*1antF*), and laterally the huge foramina of the second antennae (*2antF*). Between the antennae is a narrow, vertical frontal ridge ending ventrally in a protruding lobe on the upper angle of the epistome (*Epst*). The epistome expands beneath the antennae and becomes

147

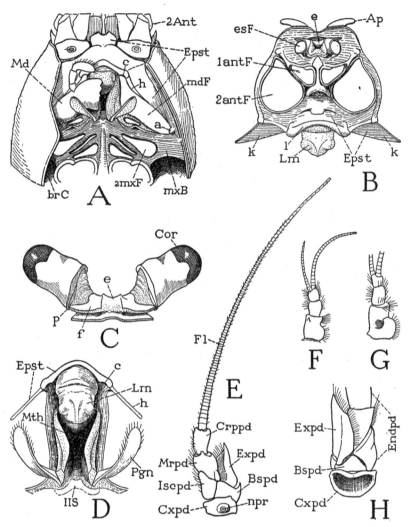

Fig. 42. Crustacea—Decapoda. *Cambarus longulus* Girard. The head and the protocephalic appendages.

A, head and anterior part of body, ventral, left mandible, maxillae, and maxillipeds removed, fold of carapace cut off on left side. B, skeleton of head, anterior. C, eyestalks and attachment on dorsal wall of head. D, mouth region, with labrum and edge of epistome, ventral. E, left second antenna, ventral. F, left first antenna, ventral. G, right first antenna, dorsal. H, base of left second antenna, dorsal.

For explanation of lettering see pages 190–192.

continuous laterally with narrow bars ascending along the sides of the antennal foramina that connect the epistome with the ocular region of the head. The median part of the epistomal margin carrying the labrum (*Lm*) is thickened to form a strong supralabral ridge (*l*), laterad of which the epistome is produced on each side into a triangular, winglike posterior extension (*k*) that unites laterally with the base of the inner lamella of the carapace fold.

In the front view of the head (fig. 42 B) the epistomal region is foreshortened because it turns posteriorly from beneath the antennae; its true form is better seen in the ventral view shown at A of the figure, which, however, is taken from another species. Though the epistome is inflected upon the ventral surface of the crayfish, it is a preoral structure since it carries the labrum, which projects over the mouth behind it (D, *Mth*). The epistome, therefore, is not a sternal plate either of the second antennal segment or of the mandibular segment, since these segments are primarily postoral. The epistome belongs to the protocephalon, and in most arthropods, including the amphipods and isopods among the Crustacea, it has a facial position (fig. 49 B). The strong development of the epistome in the decapods is correlated with the fact that the mandibles are articulated mesally on it at the ends of the supralabral ridge (fig. 42 A, *c*) and are hinged on the posterior margins (*h*) of the epistomal wings. These connections of the mandibles with the epistome have been secondarily evolved in the decapods; they are not present in *Anaspides* or the entomostracan crustaceans. The persisting primary articulations of the mandibles are those on the carapace (*a*).

The Gnathothorax

In a study of the gnathothoracic region of the body, it will be convenient to consider separately the carapace, the branchial chambers and the pleura, the mouth, the ventral skeleton, and the pleurosternal endoskeleton.

The Carapace— The carapace is composed of the united terga of at least the first ten body segments behind the head. The last component tergum probably pertains to the seventh thoracic segment, for, though there is no individual tergal plate corresponding with the eighth thoracic segment, this segment appears to make no contribution to the carapace, and its pleural and sternal parts are not united with those of the segments before it. The anterior part

149

of the carapace is set off by a U-shaped groove on the back (fig. 40, *cvg*) that runs downward and forward on the sides. This impression is commonly termed the "cervical groove," and the part of the animal before it is regarded as the "head," which is supposed to include the segments of the mandibles and the two pairs of maxillae. The dorsal muscles of the first and second maxillae, however, are shown by Schmidt (1915) in *Astacus* to have their origins on the carapace *behind* the groove, while the mandibular muscles arise in front of it. The part of the carapace before the groove in the crayfish, therefore, clearly corresponds with the first free tergal plate of *Anaspides* (fig. 37 A, D, *II*), which supports only the mandibles, the maxillary terga of *Anaspides* being combined with the tergum of the first thoracic segment. However, since the tergal region before the "cervical groove" gives attachment also to the dorsal muscles of the second antennae, it might be suspected of including the dorsal arc of the second antennal segment. Other intersegmental lines are entirely obliterated in the carapace of *Cambarus*, but the dorsum behind the "cervical groove" is marked by a pair of faintly impressed, sinuous, longitudinal lines called the *branchiocardiac grooves* (fig. 40, *bcg*). On the sides of the animal the carapace curves downward and ends with free margins close to the bases of the appendages. The entire margin on each side is fringed with fine, closely set hairs.

The Branchial Chambers and the Pleura— The open space on each side of the body between the edge of the carapace and the bases of the pereiopods leads upward, as has already been noted, into a large chamber that contains the gills (fig. 41 D, *brC*). The outer wall of each branchial chamber is soft and membranous, and is reflected upward directly from the edge of the hard outer wall of the carapace, on which it is closely adnate. When this inner wall of the carapace fold (*tf*) reaches the site of the branchiocardiac groove (*bcg*) on the dorsum, however, it becomes free and turns downward to unite with the upper edge of a long plate (*Pl*) that forms the lower part of the inner wall of the gill chamber and supports the appendages. The inner-wall plates of the gill chambers are thus seen to be the true lateral walls of the thorax covered by folds of the carapace that have grown down over the gills. They are commonly termed "epimera" by carcinologists, but we have already given reasons for calling the plates in question the *pleura*, because they so evidently correspond with the laterotergal, or pleural, plates

150

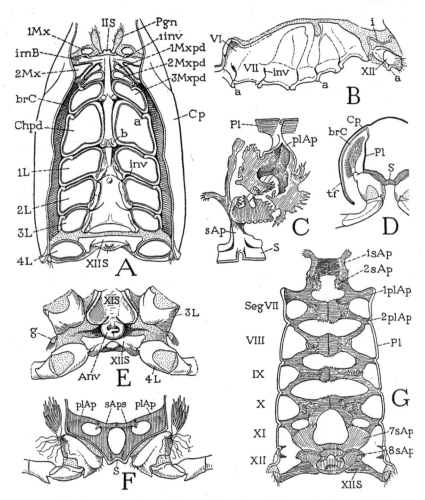

Fig. 43. Crustacea—Decapoda. *Cambarus longulus* Girard. The gnathothorax.
A, ventral view of gnathothoracic region of body of male, with appendages
removed. B, pleural wall of left branchial chamber. C, a right pair of pleural
and sternal apodemes, dried until parts completely separated, showing interlocking fimbriated margins, mesal view. D, section of first leg segment, anterior.
E, ventral surfaces of segments XI and XII of female, with annulus ventralis.
F, vertical section of lower part of segment of second legs of male, showing
relation of pleural and sternal apodemes, anterior. G, ventral gnathothoracic
endoskeleton, dorsal.

For explanation of lettering see pages 190–192.

of *Anaspides* (A, *Pl*) and with the sclerotic lateral areas known as the pleura in other arthropods.

The branchial chambers extend forward from the posterior end of the carapace into the segment of the second maxillipeds (fig. 43 A), which bear the first gills (fig. 44 G). When the interior of a branchial chamber is exposed, it is seen that the pleural plate on the inner wall (fig. 43 B) extends from the second maxilliped segment (*VI*) into the segment of the fourth pair of pereiopods and is marked by five intersegmental grooves showing that it is formed by the union of the pleura of six successive segments. The corresponding limbs articulate on small knobs (*a*) on the lower margins of their respective pleural areas. Behind this composite pleural plate is a second smaller plate (*XII*) in the pleural wall of the last thoracic segment, which supports the fifth pereiopod, but there are no gills on this segment.

The major pleural plate of the gill chamber (fig. 43 B) has a rounded upper margin with a notch behind the middle and is continuous with the membranous integument above it reflected from the inner wall of the carapace fold. The lower margin is somewhat thickened and scalloped over the limb bases. Deep slits in the lower ends of the intersegmental grooves (*inv*) mark the roots of the pleural apodemes of the thorax (G, *plAp*). Between the limb bases the pleuron is connected with the sternum by intercoxal pleurosternal brachia (A). Anteriorly the pleural plate (B) enlarges upward in the maxilliped region, creating here a pocketlike dorsal expansion of the gill chamber. From the anterior wall of the pocket the pleural sclerotization becomes horizontal and forms a strong bridge between the base of the second maxilla and the inner wall of the carapace fold (fig. 41 C, *Pl*). In the maxillary region is the pumping apparatus of the respiratory system that creates a water current through the gill chambers, but this structure and the gills themselves will be described in connection with the appendages and the general respiratory system.

The pleural plate of the last thoracic segment is a small, oval plaque (fig. 43 B, *XII*), to which the fourth leg is articulated (fig. 45 F). Its posterior setose part is somewhat exposed beyond the end of the carapace and is strongly connected with the sternum of the segment by a postcoxal pleurosternal arm. Above this pleural plate the membranous integument of the last thoracic segment contains a curious angulated bar (i) that connects the thoracic pleuron with the abdomen. The bar begins between the adjacent ends of the two pleural plates of the thorax as a small expansion, narrowly connected with each plate, and then proceeds upward and posteriorly over the second plate as a ridge on the *inner* surface of the membranous integument, which finally connects with the anterior margin of the first abdominal tergum.

The Mouth— The mouth of a crustacean lies immediately behind the cephalic epistome and between the bases of the mandibles. In the Malacostraca it is limited posteriorly by a small plate at the anterior end of the ventral skeleton that appears to be a remnant of the mandibular sternum. The segmental status of the mouth is uncertain since there is nothing in the adult animal that has been identified with the sternal arc of the postoral embryonic segment of the second antennae. For practical purposes the mouth must be between the jaws, and probably it has acquired this position secondarily, while the mandibular sternum has been either deleted or transposed to make way for the mouth.

The mouth of the crayfish (fig. 42 D, *Mth*) is a large, elongate, distensible opening above the gnathal lobes of the mandibles. The labrum (*Lm*) projects below its anterior end, and a ridge on the metastomal plate (*IIS*) guards its posterior end. Laterally the mouth is bounded by two thick integumental folds, from the posterior ends of which arise the long, flat paragnaths (*Pgn*). The mouth folds are separated from the bases of the mandibles by strips of flexible integument. The size and shape of the mouth vary with the separation or approximation of the lateral folds, but the oral aperture leads into a deep, funnel-shaped cavity, regarded as the oesophagus, at the inner end of which may be seen the smaller opening into the proventricular sac of the stomodaeum known as the "cardiac chamber of the stomach." A thick fold on the anterior wall of the oesophagus proceeds inward from the labrum.

The Ventral Skeleton— The entire undersurface of the gnathothoracic region of the crayfish contains a strongly developed ventral sclerotization composed of the segmental sterna and the pleurosternal connectives (fig. 43 A). At the anterior end of the ventral skeleton is the small metastomal plate (*IIS*), which is the only part that might be referred to the mandibular segment. From the metastomal plate a long median sternal bar extends posteriorly between the bases of

153

the second maxillae, the three pairs of maxillipeds, and the first
pair of pereiopods (*Chpd*) and gives off on each side a series of
brachia separating the foramina of the appendages. The small first
maxillae (*1Mx*) arise from the membranous areas behind the mandibles, but they are separated from the foramina of the second
maxillae by the wide first pair of sternal brachia (*imB*), which contain the pits (*1inv*) of the first sternal apodemes, showing that these
brachia include the postcoxal elements of the first maxillary segment. The second maxillae have a lateral position, but the intermaxillary brachia and the postmaxillary brachia unite laterad of
them in wide bridges (fig. 42 A, *mxB*) that join the inner lamellae
of the carapace folds and form the anterior limits of the branchial
chambers. These maxillary bridges, as already explained, are the
pleural plates of the maxillary segments (fig. 41 C, *Pl*); posteriorly
they are continuous with the pleura of the thoracic segments (D, *Pl*),
which abruptly assume a vertical position on the inner walls of the
gill chambers. The foramina of the three pairs of maxillipeds (fig.
43 A, *1Mxpd, 2Mxpd, 3Mxpd*) are crowded forward between the
widely separated foramina of the second maxillae (*2Mx*), so that
the outer ends of the coxae of the second maxillipeds, which carry
the first gills, come to be opposite the anterior ends of the branchial
chambers.

On the body segment of the first legs (fig. 43 A, *1L*) the sternal
sclerotization widens posteriorly and becomes successively broader
and deeply concave on the following two segments. The posterior
angles of each segmental area of the composite thoracic sternum are
produced into knobs that bear the sternal articulations of the coxae
(*b*). Since the lateral articulations (*a*) are anterior on the corresponding pleural areas, the axes of the coxae on this part of the body
are transversely oblique, almost at an angle of 45 degrees. Intersegmental pleurosternal brachia separate the foramina of the appendages back to the third pair of legs, but there are no pleurosternal
connections between the third and fourth legs.

The sternum of the last thoracic segment (fig. 43 A, *XIIS*) is an
entirely distinct plate between the bases of the fourth legs, separated
from the sternum in front of it by flexible integument. It is connected by postcoxal arms with the pleural plates of its segment (fig.
45 H), and the coxal axes of the last legs are directly transverse between the pleural and sternal articulations. This last sternal plate of

154

the thorax differs in shape in the male (fig. 43 A) and the female (E).

On the posterior part of the thoracic sternal region in the female of crayfishes belonging to the Cambarinae, but not in *Astacus*, there projects posteriorly from the sternum of the seventh segment a thick circular lobe partly overlapping the sternum of the eighth segment (fig. 43 E, *Anv*). This structure is known as the *annulus ventralis;* it contains a small sac, which is the sperm receptacle, or *spermatheca*, of the female. The ventral surface of the annulus is concave between two rounded marginal elevations, separated posteriorly by a suture-like groove. From one side a prominent ridge traverses the ventral concavity of the organ and ends below the marginal thickening of the other side, in some specimens going to the right, in others to the left. Just behind the ridge is a narrow slitlike aperture opening into a hard-walled spermathecal sac lying transversely within the annulus. An elaborate description of the annulus ventralis and sperm receptacles in several cambarine species is given by Andrews (1906, 1908).

During mating, the male crayfish introduces the spermatozoa into the spermatheca of the female by means of the genital processes on his first and second abdominal appendages, the structure of which will be described in the section on the appendages; but it is difficult to understand how the sperm is discharged by the female at the time the eggs are liberated. Andrews (1906), however, gives evidence that the sperm may be ejected by pressure resulting from a forcible retraction of the free last thoracic sternum against the annulus ventralis. The openings of the oviducts are on the mesal ends of the coxae of the third pereiopods of the female (fig. 45 I, *Gpr*), and beyond them the sternal concavity forms a channel widening posteriorly to the end of the seventh segment, where the annulus ventralis is situated. Andrews (1906) records observing the discharge of the eggs in two semiliquid streams that flow down the sternal channel, as the female lies on her back with the abdomen bent forward. Presumably the eggs are inseminated when they reach the annulus ventralis, and are then attached to the abdominal pleopods.

The Endoskeleton— The decapod crustaceans have an elaborately developed endoskeleton extending from the maxillary segment to the last thoracic segment. It is composed of intersegmental pleural and sternal apodemes, most of which in the Astacura unite over the

155

ventral nerve cord in a series of transverse bridges that support the alimentary canal and other viscera above them and give attachment on their undersurfaces to the ventral muscles of the appendages.

The endoskeleton of the crayfish, as seen from above (fig. 43 G), appears to consist of five horizontal median plates each supported on each side by a pair of convergent arms from the pleural wall of the gill chamber. In the seventh and eighth thoracic segments, however, the endoskeletal elements are entirely sternal (*7sAp, 8sAp*). When a specimen is thoroughly cleaned and dried, it is seen that each apparent median plate of the endoskeleton is divided along the middle and that the lateral arms are not extensions of the plates but are joined to them by interlocking fimbriations. Moreover, each half-plate is merely the expanded upper end of an intersegmental sternal apodeme, and the anterior and posterior arms attached on successive plates are seen to branch from a common stalk arising from an intersegmental groove of the pleuron. The anatomical components of the endoskeleton, therefore, from the second maxilliped segment to the segment of the third pereiopods, are a series of horizontal, Y-shaped intersegmental *pleural apodemes* along each side, and a double row of vertical, T-shaped *sternal apodemes*. The converging arms from successive pleural apodemes are united with the sternal apodemes between them by a mutual interlacing of their irregular, fimbriated opposing margins, and the plates of the two adjoining sternal apodemes are united with each other. At C of figure 43 is shown in mesal view the pleural and sternal apodemes of the right side arising on the intersegmental line between the fifth and sixth thoracic segments, after having been cleaned and dried until the associated parts have become entirely disconnected. In this condition the highly irregular shapes and frayed-out margins of the pleural arms and the sternal plates become very apparent. The long anterior arm of the sternal apodeme was not observed in the other segments.

The bases of the pleural and sternal apodemes arise not only from the intersegmental grooves of the pleura and sterna, but they invade the intercoxal pleurosternal brachia between them and thus become confluent with each other (fig. 43 F). In a cross section of the thorax (F) showing a pair of sternal apodemes and the corresponding anterior arms of the pleural apodemes behind them, the endoskeletal complex looks like a vertical pleurosternal plate perforated by a large median foramen and two smaller lateral foramina. The median bridge between the stalks of the sternal apodemes is termed the *mesophragm* of the endoskeleton, the lateral parts the *paraphragms*.

The anteriormost component of the ventral endoskeleton is the so-called "head apodeme." It consists of a pair of sternal apodemes arising from the intermaxillary sternal brachia, the roots of which are marked externally by conspicuous pits in front of the second maxillae (fig. 43 A, *1inv*). The two apodemal plates are firmly united by long, interlacing strands from their opposed margins, and thus form a broad median bridge (G, *1sAp*) supported on the lateral stalks. The posterior angles of the bridge are united with the anterior arms of the first pleural apodemes (*1plAp*), which arise between the second and third maxilliped segments, and also with a pair of small, irregular plates (*2sAp*) of the sternal apodemes of the intercoxal brachium between the first and second maxillipeds.

Between the segments of the last two pairs of legs (fig. 43 G, *XI*, *XII*) the only representative of the pleural apodemes is a small spur arising from the inner surface of the expanded end of the thoracico-abdominal connective rod inserted between the seventh and eighth thoracic pleura. Sternal apodemes, on the other hand, are present in both the seventh and eighth segments of the thorax. Those of the seventh segment (*7sAp*) are a pair of broad plates diverging forward and united by their anterior margins with the preceding endosternal plates and the posterior arms of the last pleural apodemes. The sternal apodemes of the eighth segment are crestlike lobes (*8sAp*) arising directly from the sternum and have no connection with the rest of the endoskeleton.

The Abdomen

The six-segmented abdomen of the crayfish tapers from the thorax to the telson; its dorsal surface is transversely rounded, the ventral surface somewhat concave. The segments are continuously calcified annuli, separated from each other, as is the first segment from the thorax, by flexible intersegmental conjunctivae. At its base the abdomen is freely movable on the thorax because it is yoked to the latter only by the pair of long slender bars, already described (fig. 45 F, *i*), that arise between the two pleural plates of the thorax and curve posteriorly in the membranous body wall over the second plates to attach on the anterior margin of the first abdominal tergum. The successive segmental rings of the abdomen, however, are firmly hinged on each other by paired articular knobs on the anterior tergal margins (fig. 46 B, *r*), which limit the movements of the segments on each other to motion in a vertical plane.

The arched tergal region of an abdominal segment (fig. 46 B, *T*) is produced on the sides in expansions that form free lateral lobes overhanging the undersurface of the segment. Anteriorly each tergal arch is extended into a smoothly convex, crescent-shaped area that glides under the tergum in front when the abdomen is straightened but becomes fully exposed when the abdomen is deflexed. The sternum of an abdominal segment is the narrow ventral part of the segmental annulus between the bases of the pleopods (B, C, *S*). Laterally, except on the last segment, the sternum expands and divides into antecoxal and postcoxal arms that unite laterad of the appendages with the pleural areas of the segment (*Pl*) mesad of the tergal lobes. On the second to the fifth segments the pleural areas are hardly to be distinguished from the overhanging lobes of the tergum, but on the first segment of the female (A) the small pleopods are relatively close together, and long pleural areas (*Pl*) intervene between them and the tergum. Because of the continuity of the sclerotization in the walls of the abdominal segments, however, there is only a topographical distinction between tergal, pleural, and sternal regions. The last abdominal sternum (E, *XVIIIS*) is a wider plate than the others, and it does not directly carry the uropods, which arise from a membranous integument behind the sternum and are pivoted on slender processes of the sternal margin. The terminal telson (D, *Tel*) is a flat lobe with a transverse line of flexion across its middle; the anus (E, *An*) is on the ventral surface of the proximal part.

The Appendages

In a restricted technical sense the "appendages" of an arthropod are the paired limbs that pertain to individual trunk segments, and presumably are serially homologous. The eyestalks and the paragnaths of the Crustacea are not generally regarded as true appendages according to this definition, and the status of the first antennae as segmental appendages is uncertain. Yet these organs are appendicular structures and as such will be included under the present heading.

The Eyestalks— Inasmuch as the head of the crayfish is buried beneath the front edge of the carapace and the rostrum, it is quite necessary that the eyes should be elevated on stalks, and, being on stalks, it is a further advantage to have the stalks movable. Crustaceans in which the head is fully exposed usually have sessile eyes. Eyestalks are present in most of the Malacostraca, but in the malacostracan amphipods and isopods, as in most other arthropods, the eyes are on the head surface.

The eyestalks of the crayfish (fig. 42 C) arise from the sides of a small median elevation, or *ocular lobe,* on the dorsal wall of the head (B, *e*). Each stalk has two parts generally termed segments, but the proximal "segment" is merely flexible on the supporting head lobe and has no muscles inserted on it; the large distal segment, however, is freely movable on the basal segment and is amply supplied with muscles arising in the latter. The proximal segment is funnel-shaped, expanding outward from its narrowed base, and is largely membranous except for a dorsal sclerotization of its basal part, which posteriorly is prolonged into an arm (C, *f*) on which the movable distal segment is pivoted (*p*). The large, fully sclerotized distal segment is somewhat bell-shaped, with its rounded outer end capped by the faceted cornea (*Cor*). It should be noted that the facets of the crustacean compound eye are square, not hexagonal as in insects. The eye-bearing segment can be turned in any direction on the supporting arm of the basal segment; in *Astacus*, Schmidt (1915) describes seven ocular muscles between the two segments of each eyestalk. In addition, the stalks may be moved by a pair of muscles that are inserted on the ocular lobe of the head and that arise on a long tendon from the epistome. The contraction of these muscles, according to Schmidt, produces an infolding of the flexible anterior wall of the lobe, which in turn somewhat elevates the attached stalks and directs them more anteriorly.

The First Antennae, or Antennules— The first antennae are relatively small appendages (fig. 42 F) arising close together on the head beneath the eyestalks and at the sides of the narrow frontal bar (B, *1antF*). Each antennule (F) consists of a basal stalk of three segments and a pair of slender, many-jointed distal flagella, of which the outer flagellum is the longer. The basal muscles of the antennules are said to arise on the rims of the antennular foramina

159

of the head, and they thus pertain to the protocephalon. The dorsal surface of the basal segments of each appendage is slightly concave and lies just beneath the corresponding eyestalk; in the depression is a crescentic slit, the outer lip of which is fringed with long hairs (G). The slit is the opening of a pocket, known as the *statocyst* (formerly thought to be an auditory organ and called the "otocyst"). The walls of the cyst are provided with flexible, innervated setae, and the cavity contains minute particles termed *statoliths,* which in most of the decapods are grains of sand introduced by the animal itself with its chelae just after moulting, but in some the grains are said to be crystals of calcium carbonate secreted within the cyst. By the secretion of glands beneath the bases of the setae the statoliths become attached to the setae, and their movements are supposed to orient the animal to gravity. According to Prentiss (1901), if iron filings are substituted for the statoliths in a shrimp, the animal will respond to the attraction of an electromagnet.

The Second Antennae— If the second antennae are appendages of a primarily postoral somite of the embryo, as they are said to be, they should belong to the series of body limbs, and, in fact, they seem to show a relationship to these appendages in their structure. The basal muscles of the second antennae arise on the carapace, except one that comes from the epistome. The shaft of each antenna (fig. 42 E) consists of five distinct segments, the last of which bears a single, long, multiarticulate flagellum, probably representing a sixth segment, since a pair of muscles is inserted on the base of the flagellum, while there are no muscles within the antennal flagellum itself. From the second segment beyond the base (E, *Bspd*) there arises laterally a broad lobe (*Expd*), which evidently represents the exopodite of a biramous limb, so that the rest of the shaft appears to be a four-segmented endopodite. The horizontal joint between the third and fourth segments allows a free up-and-down movement to the part of the antenna beyond it, but, since the axis of this joint is very oblique, when the antenna is flexed ventrally, the flagellum turns over to the opposite side of the body. The fifth segment flexes horizontally on the fourth, principally in a lateral direction, and the flagellum has a free movement on the fifth segment in a horizontal plane. On the ventral surface of the coxopodite is a conspicuous aperture, the *nephropore* (E, *npr*), which is the exit of the antennal excretory gland. On the dorsal surface of the antennal base (H) the coxal rim is very narrow, and the basipodite appears only as a triangular plate bearing the endopodite and the exopodite.

The Mandibles— The mandibles are the appendages next in order according to their position, but since they are covered from below by the several pairs of appendages immediately following them, which in turn underlap each other, the mouth parts of the crayfish cannot conveniently be studied in the order of their succession as here described. The student, therefore, is advised to turn first to the third maxillipeds and to work forward from them, then returning to the pereiopods.

The transversely elongate mandibles of the crayfish (fig. 42 A, *Md*) lie between the epistome (*Epst*) and the pleural bridges of the maxillary segment (*mxB*), where they extend obliquely forward and somewhat downward from the bases of the inner walls of the carapace folds to the mouth. They are strongly hinged on the lateral wings of the epistome (*h*) but are separated from the maxillary bridges by membranous areas that contain the first maxillae. The mandibles are closely embraced by the paragnaths behind them and are covered ventrally by the endites of the first and second maxillae and the first maxillipeds.

Each mandible (fig. 44 A) consists of a broad, quadrate basal part attached on the mandibular segment and of a large, free, bluntly toothed gnathal lobe (*gnL*) that projects below the mouth. At the base of the lobe arises anteriorly a three-segmented palpus (*Plp*). The posterior lateral angle of the mandible (*a*) is articulated to the inner wall of the carapace fold in the angle between the carapace and the maxillary bridge (fig. 42 A, *a*). The anterior mesal angle of the mandibular base is produced into a large process (fig. 44 A, *c*) by which the jaw articulates on the epistome at the side of the base of the labrum (fig. 42 A, D, *c*). The mandibular axis of movement, therefore, is oblique between these two points of articulation, and along this axis the anterior margin of the jaw is strongly hinged on the lateral wing of the epistome (*h*). The movement of the mandible is thus strictly limited to a partial rotation on the axis (fig. 44 A, *a–c*) between the carapace and the epistome; consequently the gnathal lobe turns only up or down on the radius *c–d* from the mesal articulation *c*. The two jaws, therefore, open and close from below like a pair of valves.

When the mandible is removed from its basal connections, it is seen that the body of the jaw has a deep inner cavity (fig. 44 B) continuous with the haemocoele of the body. The epistomal hinge (*h*) falls in line with the axis (*a–c*) between the articulations, but laterad of the axis the anterior margin of the jaw is produced dorsal to the hinge line in a large, triangular apodemal extension (A, B, *Ap*). The gnathal lobe has a smooth inner surface, but at its base are two thick prominences that serve for crushing rather than mastication.

The basal musculature of the mandible includes six muscles (fig. 44 B), three (*1, 2, 3*) attached on the anterior margin of the jaw, one (*4*) on the posterior margin, and two (*5, 6*) within the basal cavity. The first two anterior muscles (*1, 2*) arise laterally on the carapace and are inserted on the mandibular apodeme. These muscles, therefore, are *anterior dorsal adductors*, since they pull outward on the apodeme above the hinge line of the jaw. The third muscle of the anterior group (*3*) arises on the carapace dorsal to the mandible and is a small *dorsal abductor*. The posterior muscle (*4*) is the largest of all the jaw muscles; its fibers arise in a huge conical bundle on the dorsal wall of the carapace anterior to the "cervical groove" and are inserted on a long thick tendon attached to the posterior margin of the mandible close to the base of the gnathal lobe. This muscle is a powerful *posterior dorsal adductor* of the jaw. The two muscles inserted within the mandibular cavity arise on the endosternal "head apodeme" of the maxillary region. The larger of the two (*5*) is a thick bundle of fibers that spread out into almost the whole interior of the mandible and constitute a *ventral adductor* of the jaw. The other, smaller muscle from the endosternum (*6*) goes to the inner face of the mandibular apodeme above the hinge line of the jaw and is thus a *ventral abductor* directly opposed to the two dorsal adductors (*1, 2*). In a more generalized mandible, such as that of *Anaspides* (fig. 38 E), having only a dorsal point of articulation, the three anterior muscles of the decapod jaw are represented by the single dorsal promotor, or anterior rotator, and the ventral fibers of both jaws are united in a common intergnathal adductor. The dorsal adductor of *Cambarus* (fig. 44 B, *4*) is the dorsal remotor, or posterior rotator, of the *Anaspides* mandible.

The Paragnaths— The paragnaths are flat, elongate, somewhat spatulate lobes arising from the posterior ends of the lateral mouth

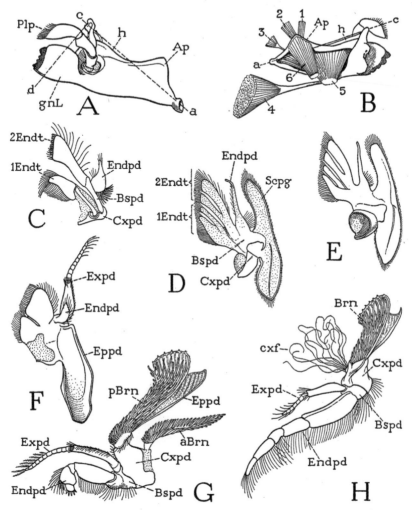

Fig. 44. Crustacea—Decapoda. *Cambarus longulus* Girard. The mouth parts and maxillipeds.

A, left mandible, ventral. B, left mandible and its muscles, dorsomesal. C, left first maxilla, ventral. D, left second maxilla, ventral. E, right second maxilla, dorsal. F, right first maxilliped, dorsal. G, left second maxilliped, lateral. H, left third maxilliped, lateral.

For explanation of lettering see pages 190–192.

folds (fig. 42 D, *Pgn*); they are closely applied to the undersurfaces of the mandibles. The paragnaths have no musculature and, according to Keim (1915), are innervated from branches of the mandibular nerves. Between their bases is a small metastomal plate (*IIS*), possibly a remnant of the mandibular sternum.

The First Maxillae— The very small first maxillae, or maxillulae, are situated in the membranous integument behind the mandibles at the sides of the mouth (fig. 43 A, *1Mx*) and have no connection with the ventral skeleton. Each appendage (fig. 44 C) appears at first inspection to be an assemblage of three thin, flat lobes with hair-fringed margins closely applied to the undersurface of the gnathal lobe of the corresponding mandible. In the isolated appendage, however, it is seen that the lobes are supported on a basal coxopodite (*Cxpd*). The first lobe (*1Endt*) is carried on an arm from the coxopodite and is evidently a coxal endite. The other two lobes are borne on a second arm (*Bspd*), which may be regarded as the basipodite carrying a basal endite (*2Endt*) and a much reduced endopodite (*Endpd*). On its anterior (dorsal) surface the endopodite is armed with a strong basal tooth.

The Second Maxillae— The maxillary appendages of the second pair (fig. 44 D), often called simply *the maxillae,* are somewhat larger than the first maxillae and have a more lateral position than any of the other mouth parts. They arise from large foramina of the ventral skeleton at the sides of the first maxillipeds (fig. 43 A, *2Mx*). Each appendage carries two long, bifid, densely fringed endite lobes (fig. 44 D, *Endt*), a slender, tapering endopodite (*Endpd*), and a long, flat, lateral lobe (*Scpg*), known as the *scaphognathite* presumably from its fancied resemblance to a boat. The segmentation of the maxilla is obscure, but the basal part (*Cxpd*) carrying the first pair of endites must represent the coxopodite. The distal pair of endites and the endopodite arise from a common basal part (*Bspd*), which therefore should be the basipodite. The endites project forward and mesally from the basal region of the maxilla and underlap the endites of the first maxilla.

The long scaphognathite of the second maxilla (fig. 44 D) is attached near its middle to the base of the appendage and extends horizontally forward and backward. Its more slender anterior lobe lies against the undersurface of the maxillary bridge of the ventral skeleton and beneath the lateral part of the mandible; the shorter and broader posterior lobe projects into the anterior part of the branchial chamber. The maxillary scaphognathites are vibratory organs that cause the forward flow of water through the gill chambers. Each organ in the European crayfish, *Astacus*, is shown by Schmidt (1915) to be provided with eight muscles, one arising in the basipodite of the maxilla, the other seven on the maxillary pleuron and on the maxillary apodeme of the endoskeleton. The seven body muscles enter the base of the maxilla through a cup-shaped opening on the dorsal side (fig. 44 E). The same musculature appears to be present in *Cambarus*. The lateral position of the second maxillae is correlated with the respiratory function of the scaphognathites, which will be described in the section on the respiratory system.

The First Maxillipeds— The first maxillipeds (fig. 44 F) somewhat resemble the second maxillae (D) in that each bears a pair of endite lobes and has a large coxal epipodite (F, *Eppd*), but they differ from the maxillae in having both an endopodite and an exopodite. The endopodite is a small, simple lobe (*Endpd*), but the exopodite (*Expd*) is a typically developed outer ramus consisting of a long basal segment and a terminal flagellum. The basal part of the maxilliped is evidently a combination of the coxopodite and the basipodite, that is, a protopodite, and the two endites apparently are to be referred one to the coxopodite, the other to the basipodite, as in the maxilla. The epipodite is of special interest; it is a long, broad, troughlike structure, concave dorsally, that extends posteriorly from the base of the appendage into the upper part of the anterior end of the gill chamber. The posterior lobe of the maxillary scaphognathite lies snugly in the dorsal concavity of the maxilliped epipodite (fig. 48 B), and the latter thus forms a conduit from the gill chamber by which water may be conducted into the pump chamber over the scaphognathite (A).

The Second Maxillipeds— The second maxillipeds (fig. 44 G) retain more of the form of legs than do any of the appendages before them. Functionally, however, they are a part of the feeding apparatus and project horizontally forward beneath the preceding mouth parts. Each appendage is a biramous limb with seven segments in the main shaft, though the basipodite and ischiopodite are united with each other. The large coxopodite (*Cxpd*) carries a bilamellate epipodite (*Eppd*) that bears the first coxal gill, or podobranchia (*pBrn*), and projects upward and posteriorly into the anterior end of the gill chamber beneath the epipodite of the first maxilliped. Associated with the coxa is also a single arthrobranchia (*aBrn*) arising on the articular membrane at the base of the coxa. The relatively short, five-segmented endopodite (*Endpd*) is armed distally with strong spines that probably serve for grasping food introduced between the third maxillipeds. The long, slender exopodite (*Expd*) is of the usual form, consisting of an elongate basal segment and a distal flagellum. Endites are absent.

The Third Maxillipeds— The maxillipeds of the third pair (fig. 44 H) are the most leglike in form of any of the appendages anterior to the chelipeds, but their principal function is concerned with feeding. Each appendage is a long, seven-segmented shaft, with a relatively small exopodite (*Expd*) arising from the basipodite (*Bspd*). The coxa carries a gill-bearing epipodite, and associated with each third maxilliped are two arthrobranchiae (not shown in the figure). On the anterior surface of the coxa arises a tuft of long, tangled filaments (*cxf*) such as occur also on the coxae of the first four pairs of pereiopods. The endopodites of the third maxillipeds project forward horizontally beneath the other mouth parts (fig. 45 A), and the inner surfaces of the long ischiopodites are armed with combs of strong teeth above fringes of long hairs directed mesally. The mouth of the crayfish lies above these combs of the third maxillipeds, and the food is introduced between them by the small chelae of the second and third pereiopods.

The First Pereiopods, or Chelipeds— The first pereiopods of the crayfish are the huge pincer-bearing chelipeds (fig. 40, *Chpd*) following the third maxillipeds (fig. 45 A). They have the same segmentation as the legs, except for a union of the ischiopodites with the basipodites. The forceps, or chela, of each appendage (B) consists of the greatly enlarged propodite (*Propd*), with a strong clawlike lateral extension that forms the *fixed finger,* and of the dactylopodite (*Dactpd*), which is articulated on the propodite at the base of the fixed finger and constitutes the *movable finger* or effective part of the instrument. The coxa of each cheliped bears a podobranchia attached on an epipodite, and associated with it are two arthrobranchiae. The part of the limb beyond the coxa turns up and down on a horizontal hinge with the coxa, but the meropodite moves laterally on the ischiopodite. The "knee" joint between the meropodite

166

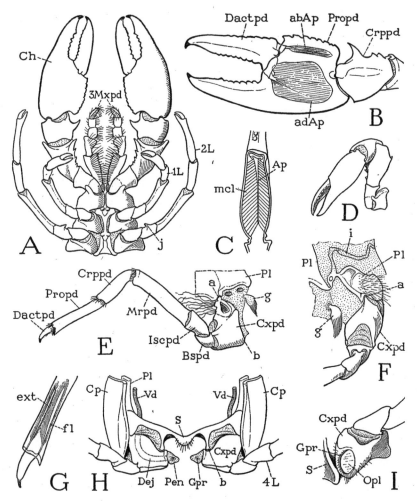

Fig. 45. Crustacea—Decapoda. *Cambarus longulus* Girard. The thoracic appendages.

A, third maxillipeds, chelipeds, and first two pairs of legs, male, ventral. B, left chela of female, with muscle apodemes. C, lengthwise section of base of chela, showing adductor apodeme and muscle fibers. D, left cheliped, flexed to full extent. E, third left leg and pleural attachment, posterior. F, base of fourth left leg and pleuron of last thoracic segment. G, dactylopodite and its muscles in propodite. H, ventral part of last thoracic segment of male, posterior, showing penes on coxopodites. I, base of left second leg of female, with genital aperture (*Gpr*) on coxopodite.

For explanation of lettering see pages 190–192.

and the carpopodite flexes as usual in a vertical plane, but at the next two joints the movement is horizontal.

When the chelipeds are extended anteriorly (figs. 40, 45 A), the broad chelae lie almost horizontally, or with their dorsal surfaces sloping downward to the sides, and the movable fingers are mesal. Inasmuch as the movable finger of the chela is the dactylopodite of the appendage, which ordinarily moves in a vertical plane, and since the fingers of the small chelae on the next two pairs of legs are dorsal (fig. 45 A), it is probable that the chelae of the chelipeds have become horizontal by a twisting of the appendages. The dactylopodite of the chela, therefore, is hinged on the "hand," or propodite, by dorsal and ventral articulations (B) so that it moves in a transverse plane; the "hand" is similarly movable on the carpopodite. When the upper or lower wall of the propodite is removed, the cavity of the segment is seen to be filled with a great mass of oblique muscle fibers attached on the dorsal and ventral walls. The fibers in the mesal part of the segment converge to their insertions on a slender apodeme (B, abAp) attached to the inner basal angle of the movable finger; the much more numerous fibers in the lateral part are inserted on both sides of a large, horizontal apodemal plate (adAp) attached on the outer basal angle of the finger. The two sets of fibers are respectively the abductor and adductor muscles of the dactylopodite. As seen in a vertical section of the propodite (C), the adductor fibers go at an angle of about 45 degrees from their origins to their insertions, and it might therefore seem that they can exert but a very limited pull on the base of the movable finger. It will be found, however, that in order to close the widely opened pincers, only a relatively small movement of the adductor apodeme is necessary to bring the points of the claws together.

The chelae of the chelipeds are the principal organs of the crayfish for grasping and holding prey, but when either appendage is strongly flexed at the knee joint and the proximal part is bent back on the coxa as far as it will go (fig. 45 D), the large chelae are still far from the mouth; however, they come within easy reach of the small pincers on the first two pairs of legs (A, 1L, 2L).

The Second and Third Pereiopods— These two pairs of appendages, which functionally are the first and second legs, have the form of simple ambulatory limbs, except that each bears a small chela at the end of the long slender propodite (fig. 45 A, 1L, 2L).

With the miniature chelae on these slender, flexible legs the crayfish is able to take food from the chelae of the chelipeds and insert it between the toothed inner edges of the meropodites of the third maxillipeds (A, *3Mxpd*). The food is then probably taken over by the strongly spined apical segments of the second maxillipeds (fig. 44 G), from which it must finally be transferred to the mandibles by the fringed endites of the first maxillipeds (F) and the two pairs of maxillae (C, D). All the appendages, therefore, in the series from the mandibles to the second legs, inclusive, are in one way or another implicated in the function of feeding.

The coxae of the third pereiopods, or second legs, of the female give exit to the genital ducts. Each genital aperture is a crescent-shaped slit (fig. 45 I, *Gpr*) on the mesal posterior part of the ventral surface of the coxa but is closed by a large operculum (*Opl*) that looks like an oval disc set into the coxal wall. The position of the two orifices is such that the discharged eggs are readily received into the channel of the sternum, where they can flow back to the annulus ventralis. In the male of *Cambarus* and some other genera a small, curved spine on the ventral side of the ischiopodite of each third pereiopod (fig. 45 A, *j*) serves as a clasping organ during mating, it being hooked over the joint between the second and third segments of the corresponding leg of the female.

The Fourth Pereiopods— This pair of appendages will serve best for a study of the structure of a typical ambulatory limb (fig. 45 E). The large coxopodite is articulated dorsolaterally (*a*) on the pleuron (*Pl*) and ventromesally (*b*) on the sternum of its segment. Since the pleural articulation is anterior and the sternal articulation is posterior, the axis of rotation of the coxa on the body is oblique in two directions. The movements of the coxa, therefore, turn the extended leg forward and mesally in promotion, and posteriorly and outward in remotion. Between the coxa and the basipodite the hinge is horizontal, and the leg is simply elevated or depressed at this joint. The hinge lines at the basi-ischial joint and at the ischiomeral joint are oblique in a vertical plane and permit only a limited sidewise movement. The rigidity of these two joints in the vertical plane allows the levator and depressor muscles of the basipodite to lift or lower the body on the distal part of the leg. The "knee" joint of the leg is the prominent bend between the meropodite and the carpopodite, and the movement of the carpopodite is restricted to

169

the vertical plane. The propodite again is movable transversely on the carpopodite, and finally the simple, clawlike dactylopodite turns up and down on the end of the propodite. The many-jointed arthropod limb, therefore, while it has no twisting motion such as that of the human arm, can perform almost any kind of movement by reason of the different actions at the joints between its seven segments. The leg muscles, which are fully described in *Astacus* by Schmidt (1915), are appropriate to the specific movements of the limb segments as determined by the intersegmental articulations. The muscles of the dactylopodite of the Crustacea arise entirely in the propodite (fig. 45 G).

The coxa of the fourth pereiopod, as the coxae of the preceding legs, carries on its upper end a gill-bearing epipodite (removed in fig. 45 E), and two arthrobranchiae arise from the articular membrane above it (also removed in the figure). On the anterior face of the coxa is a group of long slender filaments, such as those noted on the third maxillipeds, which are present also on all the legs but the last. It is probable that these filaments act as strainers guarding the entrances to the gill chambers between the mesal ends of the coxae. From the pleurocoxal membrane behind the fourth pereiopod arises a small pendent lobe bearing a brush of long hairs (E, F, g); the nature and function of this organ is not known.

The Fifth Pereiopods— The legs of the last pair are similar to the preceding legs, differing from them principally in that the clawlike dactylopodites are turned forward instead of downward (fig. 40). There are no gills associated with these legs. The coxopodite articulates laterally in a notch on the lower edge of the pleural plate of the last thoracic segment (fig. 45 F, *a*) and mesally on the sternum (H, *b*). The two articulations are approximately in the same transverse plane, so that the movements of these legs on the body are more directly anterior and posterior than are those of the other legs. In the female the mesal surfaces of the coxopodites of the last legs are evenly rounded; in the male they are produced into conical projections terminating in small, soft, protractile papillae (fig. 45 H, *Pen*) that contain the openings of the male genital ducts.

The Pleopods— Each of the abdominal segments carries a pair of appendages, but they are not all alike in either sex, and the first two pairs in the male are different from those of the female. Though the term *pleopod* signifies a "swimming leg," the abdominal ap-

170

pendages of the crayfish serve various purposes; the first five pairs in the female are used for carrying the eggs and the young, the first two pairs in the male are accessory genital organs for sperm transfer, while the last pair in each sex, distinguished as the *uropods,* are the principal swimming organs.

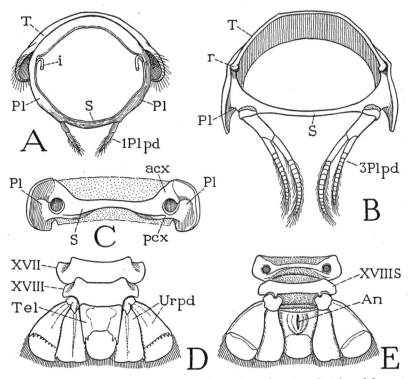

Fig. 46. Crustacea—Decapoda. *Cambarus longulus* Girard. The abdomen and the pleopods.

A, first abdominal segment of female, anterior. B, third abdominal segment of female, anterior. C, ventral surface of third abdominal segment of female, pleopods removed. D, last two abdominal segments, with telson and uropods, dorsal. E, same, ventral.

For explanation of lettering see pages 190–192.

The first pair of pleopods of the female (fig. 46 A, *1Plpd*) are small, simple appendages situated relatively close together on the undersurface of the abdomen. Each consists of two elongate basal segments and a short terminal flagellum. The following four pleopods of the female are much larger biramous appendages and arise near the lateral extremities of the ventral arc of the segment (B). The stalk of each of these pleopods consists of a long basipodite and a small incomplete basal ring apparently representing the coxopodite. The endopodite and the shorter exopodite have each an elongate basal segment and an articulated distal flagellum. The articles of the flagellum are not to be regarded as segments, since they are not interconnected by muscles, but, as shown by Schmidt (1915), a muscle enters the base of the flagellum and branches to each of the subdivisions. The same is true of the flagella of the maxillipeds, but not of the antennae.

The uropods (fig. 46 D, E) have the same form in both sexes. Each uropod is a biramous appendage in which the two rami are broad, flat, triangular lobes supported on a small basal segment, which is articulated on a pivotlike process of the posterior margin of the sixth abdominal sternum. The exopodite has a line of flexure across the middle armed dorsally (D) with small spines, and each lobe is fringed with long hairs. The uropods and the telson constitute the "tail fan," and are provided with numerous muscles, while the abdomen itself is filled with a great mass of muscles. In using the tail fan in swimming, the crayfish gives a powerful stroke of the abdomen downward and forward, by which it violently propels itself backward.

In the male crayfish the pleopods of the third, fourth, and fifth pairs and the uropods are like those of the female; the first two pairs, being modified and used for reproductive purposes, may be termed *gonopods*. Ordinarily the gonopods project horizontally forward in the sternal groove beneath the thorax, so that in describing them the anterior surfaces may be said to be dorsal, and the posterior surfaces ventral. Because those of the second pair are the less modified of the two, they will be described first.

By comparison with a pair of typical pleopods, the second male gonopods (fig. 47 B) are seen to be a simple modification of a biramous abdominal limb. Each has a strong, two-segmented basal stalk bearing an exopodite and an endopodite. The exopodite (*Expd*) is of the usual flagellar structure, but the much larger endopodite (*Endpd*) has a long, thick basal segment that is produced at its distal end into a small lobe (*n*) turned dorsally, and it bears a short terminal flagellum. The only special feature, then, of this gonopod is the lobe on the endopodite. The lobe stands vertically on the dorsal edge of the endopodite (F, *n*); it is triangular in shape, with the

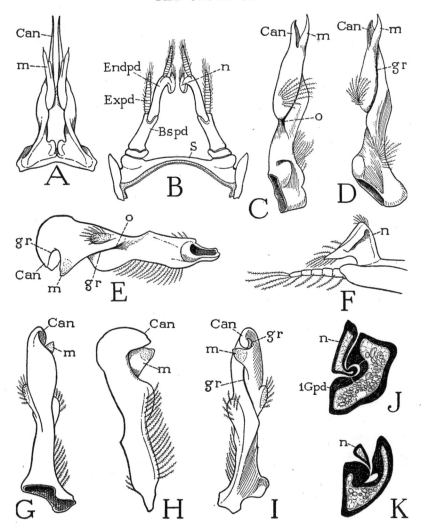

Fig. 47. Crustacea—Decapoda. The male gonopods.

A, *Orconectes virilis* (Hagen), first gonopods, ventral. B, same, second gonopods, ventral. C, *Orconectes limosus* (*Raf.*), first left gonopod, posteromesal (from Andrews, 1911). D, same, first left gonopod, ventral (from Andrews, 1911). E, *Cambarus longulus* Girard, first right gonopod, mesal. F, same, distal part of second left gonopod, mesal. G, same, first right gonopod, dorsal. H, same gonopod, lateral. I, same gonopod ventral. J, *Orconectes limosus* (Raf.), proximal section of first gonopod and lobe (F, *n*) of second gonopod (from Andrews, 1911). K, same, section through distal part of gonopod (from Andrews, 1911).

For explanation of lettering see pages 190–192.

anterior angle produced in a long, free point, and the oblique anterior margin is rolled laterally in a wide flange.

The first gonopods of a mature male are strong, unsegmented shafts arising close together on the ventral arc of the first abdominal segment (fig. 47 A), but their form is so different in different species that only a few individual examples can be given here. A relatively simple structure of the first gonopods is seen in *Orconectes limosus* (C, D), which has been fully described and illustrated by Andrews (1911). At about the middle of the ventral (posterior) surface of each appendage there begins on the mesal margin a deep groove (D, *gr*) that curves laterally, then runs forward, going beneath the base of a long tapering subapical lobe (*m*) that overhangs it laterally, and finally runs out on the sharp apex of the main shaft (*Can*), termed the *cannula* by Andrews. At its proximal end the groove is widely open by an aperture on the mesal surface that can be seen on turning the gonopod laterally (C, *o*).

The first gonopods of *Orconectes virilis* (fig. 47 A) have essentially the same structure as those of *O. limosus*, but the accessory lobes (*m*) are longer and the cannulae (*Can*) are drawn out into long, tapering processes. In *Cambarus longulus*, on the other hand (E, G, H, I), the distal parts of the appendages are short; the accessory lobe (*m*) is flat and triangular; the cannula (*Can*) is broad, rounded, and decurved. The seminal groove (*gr*) begins at about the middle of the mesal surface (E, *o*), curves laterally past the base of the accessory lobe (I, *m*), and then turns downward on the mesal surface of the broad cannula (E, I, *Can*). The second gonopods of *C. longulus* resemble those of *Orconectes virilis* (B).

During conjugation of the male and female crayfish, as described by Andrews (1911), the protruded genital papilla on either coxa of the last thoracic legs of the male (fig. 45 H, *Pen*) extends over the dorsal side of the corresponding first gonopod, and its decurved extremity is pressed into the aperture of the groove on the mesal surface. The triangular lobe of the second gonopod (fig. 47 B, F, *n*) on the same side is now pressed against the genital papilla and held there by the insertion of its marginal flange into the groove of the first gonopod (J, K) thus giving a watertight passage for the sperm from the papilla into the groove. Andrews notes that the lobe of the second gonopod may be moved back and forth in the groove of the first, but he doubts that this movement has anything to do

174

with the propulsion of sperm through the canal. It is the function of the first gonopods, however, to discharge the sperm from the apical cannulae into the aperture of the annulus ventralis of the female. In order to do this, the gonopods have to be lowered at an angle of about 45 degrees from the horizontal, and to hold them in this position, it is observed by Andrews, one or the other of the last thoracic legs is thrust transversely between the body and the gonopods.

The great variation in form of the first gonopods of cambarine crayfishes affords the best characters by which taxonomists are able to determine species. Numerous illustrations of their form and a review of their nomenclature for taxonomic purposes are given in several papers by Hobbs (1940, 1942, 1945).

The Respiratory System

The structures concerned with the exchange of respiratory gases between the water and the blood and the maintenance of a water current are all external in the crayfish. They include the gills, or *branchiae*, the *branchial chambers*, and the *respiratory pump*.

The gills of the crayfish are entirely enclosed in the branchial chambers beneath the descending lateral folds, or *branchiostegites*, of the carapace. In all there are 17 gills crowded into each chamber. Six of them in an outermost row arise on the coxae of the appendages, and hence are termed *podobranchiae;* the other eleven arise from the articular membranes above the coxae, and are distinguished as *arthrobranchiae*. In some decapods there are also *pleurobranchiae* arising from the pleural walls of the gill chambers. The podobranchiae of the crayfish pertain to the second and third maxillipeds and the first four pereiopods. The arthrobranchiae arise from the articular membranes of these same appendages, but there are two of them corresponding to each podobranchia, one anterior, the other posterior, except on the second maxilliped, which has only a posterior arthrobranchia. The fifth pereiopods carry no gills of either kind. The podobranchiae are larger than the arthrobranchiae and increase in size from before backward as the height of the branchial chamber increases; they slope first upward and posteriorly from the coxae, each partly overlapping the one behind, but the upper ends are curved forward, and the forward curvature becomes successively greater to the fifth gill.

175

Each gill superficially appears to be a mass of soft filaments directed away from the base of the organ (fig. 48 C). The branchial structure is more evident when most of the filaments are removed. It is then seen in a typical podobranchia, such as the one on the second leg (D), that the filaments arise entirely from a long, curved, axial shaft ending in a slender apical process, and that the branchial shaft is attached for much of its length to the outer side of a membranous, trough-shaped coxal epipodite (*Eppd*), which expands distally into two broad, thin, vertical lamellae with thickened margins and lengthwise corrugations. The structure of the epipodite with its corrugated lamellae is best seen when the gill is turned over, exposing the mesal surface (E), or in a dorsal view (F). A cross section (G) along the line *x–y* in F shows the trough shape of the epipodite (*Eppd*), which becomes accentuated distally, and the attachment of the tubular branchial shaft to the outer side of the epipodite trough. In the lobster, *Homarus*, the podobranchiae arise from the bases of the epipodites and project freely from them. By comparison with *Anaspides* (fig. 37) it will be seen that in the latter it is the epipodites themselves that serve as gills. The podobranchiae of the fourth pereiopods (fig. 48 H) in the crayfish have the same structure as those of the preceding segments, but the epipodite (*Eppd*) is much reduced in size, having an expanded base from which only a slender tapering arm runs out in conjunction with the proximal half of the branchial shaft. The arthrobranchiae are simple plumose gills (I) in which the axial shaft arises directly from the articular membrane above the coxa.

The branchial chambers have been sufficiently described in connection with the thoracic skeleton. By comparison with the thorax of *Anaspides* (fig. 41 A), in which the gills are completely exposed, a cross section of the decapod thorax (D) suggests that the branchiostegites are folds of the tergum (*tf*) that have grown down from the back to form protective covers for the gills. The outer walls of the branchial chambers are the inner lamellae of the tergal folds, the sclerotic inner walls are the united pleura of the segments involved.

Each branchial chamber extends from the second maxilliped segment to the end of the carapace. In the maxilliped region the sclerotic pleural wall of each chamber (fig. 41 D, *Pl*) turns abruptly downward and becomes horizontal as it continues forward in the maxillary

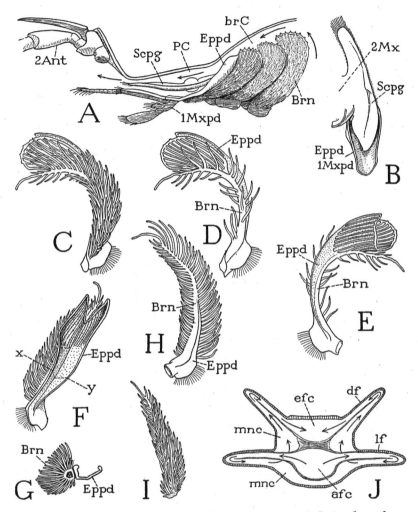

Fig. 48. Crustacea—Decapoda. The respiratory system. A–I, *Cambarus long-ulus* Girard.

A, diagram of the respiratory pumping apparatus. B, scaphognathite of right second maxilla in trough of epipodite of first maxilliped, dorsal. C, podobranchia of left second leg, lateral. D, same gill, partly denuded of filaments, exposing the supporting epipodite. E, same as D, mesal surface. F, podobranchia of second leg, dorsal. G, section of gill on line *x–y* in F. H, podobranchia of third leg. I, an arthrobranchia of second-leg segment, left, lateral. J, diagrammatic cross section of a gill of *Astacus*, showing internal circulatory system (from figures and descriptions by Bock, 1925).

For explanation of lettering see pages 190–192.

region (C, Pl), where it forms the broad pleural bridge connecting the maxillary foramen with the base of the inner lamella of the carapace fold (fig. 42 A, mxB). The branchial passage is thus continued anteriorly into a shallower chamber (fig. 48 A, PC), also covered laterally by the carapace fold but shut in ventrally by the large epipodite of the first maxilliped (Eppd) and the scaphognathite of the second maxilla (Scpg). This anterior part of the respiratory passage constitutes the branchial *pump chamber* (PC). The horizontal scaphognathite lies beneath the maxillary bridge, and its anterior end underlaps the body of the mandible.

We have already noted how the posterior lobe of the scaphognathite rests in the dorsal concavity of the large epipodite of the first maxilliped (fig. 48 B). Posteriorly the epipodite curves upward and forms a troughlike projection into the upper part of the branchial chamber above the first gill (A). The vibratory scaphognathites are the active agents of the respiratory pumping apparatus. If each scaphognathite vibrates in such a manner that its posterior lobe is depressed while the anterior lobe turns up and closes the exit from the pump chamber and then reverses itself on the next stroke, a posterior inhalent force will alternate with an anterior exhalent force. It is evident, then, that water entering the branchial chambers from below and behind the carapace will be drawn forward between and over the gills, finally to be collected in the epipodite troughs of the first maxillipeds and by them delivered into the pump chambers over the scaphognathites. Anteriorly the water is discharged from the pump chambers beneath the antennae.

The true respiration of the crayfish is the exchange of gases between the blood inside the gills and the water in the gill chambers, effected through the walls of the gill filaments. The circulation of the blood within the gills is more difficult to understand than the circulation of water in the branchial chambers, but a detailed account of the inner structure of the gills and a convincing explanation of the course of the blood through them have been given for *Astacus* by Bock (1925), who shows that the circulatory system within the gills is more complex than had previously been supposed.

The blood, after circulating through the body, collects in the ventral sinus of the thorax and abdomen. In the thorax it is conducted from the sinus into the gills through *afferent branchial veins;* after circulating through the gills the oxygenated blood is returned to the body by way of *efferent branchial veins*, from which it is finally conveyed to the heart in the *branchiocardiac veins*. Within the gill the blood flows distally through an *afferent canal* (fig. 48 J, *afc*) and returns through an *efferent canal* (*efc*). The afferent canal follows the outer side of the gill shaft, the efferent canal runs along the inner side; these two canals are separated by a common wall, or septum, of connective tissue. It is shown by Bock (1925), however, that a third canal, or sinus, which he calls the *mantle canal* (*mnc*), plays an essential part in the circulatory system of the gill. The mantle canal is a space of the gill lumen surrounding the afferent canal on three sides but interrupted where it comes against the walls of the efferent canal.

The lumen of each gill filament is divided into two channels by a lengthwise connective tissue partition, one accommodating the inflowing blood, the other the outflowing blood, the two channels being connected in the apex of the filament. In the lateral filaments of the gill the afferent channels come directly from the afferent canal of the shaft (fig. 48 J, as indicated by the arrows), but the efferent channel of the filament discharges into the outlying mantle canal (*mnc*). In the filaments adjoining the efferent canal of the shaft (*efc*), on the other hand, the afferent channels come from the mantle canal, and the efferent channels open directly into the efferent canal of the shaft. In order to get from the afferent canal into the efferent canal of the shaft, therefore, according to Bock's account, the blood has to circulate through two sets of filaments and is thus twice subjected to oxygenation.

AN ISOPOD, *LIGYDA*

The crustacean order Isopoda includes a large number of species, most of which are marine, but some live in fresh water, and others are terrestrial. The order includes eight suborders, but the most familiar isopods are those of the suborder Oniscoidea, known as woodlice, sowbugs, pillbugs, and slaters, which are mostly terrestrial, inhabiting moist places under logs and stones, though members of the family Ligiidae live fully exposed on the surfaces of rocks along the ocean shore. One of the ligiids, *Ligyda exotica* Roux, will be the principal isopod subject of the present chapter; its structure is fairly representative of the Oniscoidea, and in a broad way of that of the isopods in general, but some isopods are parasites on other crusta-

179

ceans and on fish and have undergone structural and developmental changes in adaptation to their parasitic habits.

Various anatomical features relate the Isopoda to the Amphipoda, the familiar members of the latter group being the "sandfleas" found in rubbish along ocean beaches. In particular, the head in these two orders of malacostracan Crustacea is a definite cephalic capsule, usually separated from the body by a short membranous neck. The head carries the eyes, both pairs of antennae, the mandibles, the first and second maxillae, and at least one pair of maxillipeds; it therefore includes the protocephalon, the three gnathal segments, and the first thoracic segment, all consolidated in a craniumlike tagma that often has a striking resemblance to the head of an orthopteroid insect. The body is divided into a thorax and an abdomen between the same two segments as in the other Malacostraca, but there is no thoracic carapace. The thorax is composed of seven free postcephalic segments, except in some isopods in which a second maxilliped segment enters into the composition of the head.

Carcinologists include the isopods and amphipods together with the Mysidacea, Cumacea, and Tanaidacea in a superorder Peracarida, characterized in part by the presence in the female of most species of large, thin mesal lobes of the coxae, termed *oostegites,* that come together beneath the thorax to form a brood chamber in which the eggs and young are carried. The head of the tanaidaceans resembles that of the isopods and amphipods, but in the other two peracaridan orders the head is a typical protocephalon, and the gnathal segments are combined with the anterior thoracic segments under a common carapace.

Ligyda exotica (fig. 49 A) is an isopod of the ocean shore, widely distributed in the tropical and warm temperate regions of both hemispheres. On the Atlantic coast of America it occurs from North Carolina to Brazil, and on the Pacific coast from California to Chile. It is often seen in great numbers on flat surfaces of rocks outcropping along the beach. When approached, the animals move swiftly away, but they do not take to the water. The body of *Ligyda exotica* is elongate oval, about an inch in length, and of a dull gray color. The head bears a pair of long second antennae, and between them a pair of diminutive antennules. Laterally on the top of the head are two large compound eyes. The back plates of the body segments are produced on each side into broad lobes over the bases of the ap-

180

pendages. Several pairs of anterior legs are directed forward, the others backward. From the rear end of the abdomen project two long, slender biramous uropods.

The Head and the Mouth Parts

The head of *Ligyda* is triangular in facial aspect (fig. 49 B), with the labrum (*Lm*) at the lower angle and the upper angles capped by the bulging compound eyes (*E*). The second antennae (removed in the figure) arise from large foramina (*2antF*) below the eyes; between them are the foramina of the minute first antennae. The mandibles (*Md*) depend from the sides of the head and close behind the labrum. Two lines crossing the face above the antennae are mere external wrinkles, but a deep groove (*es*) below them, between the bases of the mandibles, forms an internal shelflike ridge and sets off the region of the epistome (*Epst*) as a specific area of the facial surface of the head bearing the labrum. The epistome of the isopod is thus an exact replica of the clypeus of an insect (fig. 78 A, *Clp*); on its basal angles are the anterior articulations of the mandibles (fig. 49 B, *c*) just as in the insects.

On the back of the head (fig. 49 C) is the large, quadrate neck foramen (*For*), below which are suspended the maxillipeds (*Mxpd*). The foramen is margined laterally by narrow rims, which are confluent dorsally in a wider flange (*VT*) distinctly separated by a groove from the dorsal wall of the cranium before it. Ventrally the foramen is closed by the neck membrane, which, however, contains a crossbar from the posterior end of the maxilliped sternum (*VS*). Since the dorsal flange over the foramen, the marginal rims on the sides, and the sternal bar below appear to be parts of a circle on the posterior part of the head carrying the maxillipeds, the inference can hardly be avoided that they together represent the reduced annulus of the maxilliped segment united dorsally and laterally with the maxillary part of the cranium. On each side of the neck foramen an apodeme (*tAp*) is inflected from the groove in front of the marginal rim, which evidently is an ingrowth between the tergal regions of the maxillary and the maxilliped segments.

The ventral wall of the head, when exposed by removal of the mouth parts (fig. 49 H), is seen to have a continuous median sclerotization, from which three pairs of lateral arms are given off. The first arms are *intermaxillary brachia* (*imB*) since they go be-

181

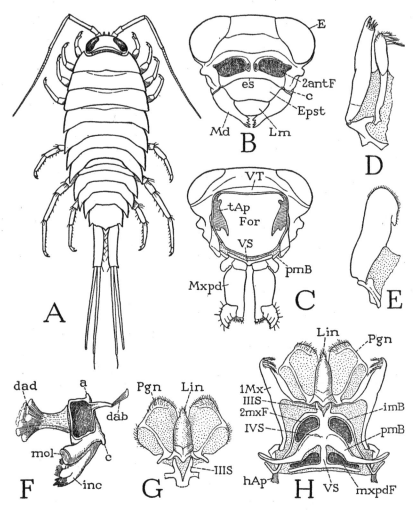

Fig. 49. Crustacea—Isopoda. *Ligyda exotica* Roux.

A, entire animal, male, length of body 24 mm. B, head, anterior, antennae removed. C, head and maxillipeds, posterior. D, right first maxilla, ventral. E, right second maxilla, ventral. F, left mandible, anterodorsal. G, hypopharynx, ventral. H, ventral head skeleton with hypopharynx and mesal lobes of first maxillae attached, second maxillae and maxillipeds removed.

For explanation of lettering see pages 190–192.

tween the bases of the first and the second maxillae; the second are *postmaxillary brachia* (*pmB*) between the second maxillae and the maxillipeds; the third are *postmaxilliped brachia* extending transversely behind the bases of the maxillipeds. The ventral skeleton of the head, therefore, includes the united sternal regions of the first maxillary segment (*IIIS*), the second maxillary segment (*IVS*), and the maxilliped segment (*VS*). The slender intermaxillary brachia (*imB*) extend laterally and posteriorly, and each ends with a loop in a pocket of the head wall above the outer end of the much thicker postmaxillary brachium of the same side. From the angle of the loop is given off a large apodeme (*hAp*, only the base shown in the figure) that extends forward in the head cavity. The pair of apodemes arising thus posteriorly in the head of *Ligyda* corresponds with the first sternal apodemes (the "head apodemes") of *Cambarus*, though in the decapods the apodemes arise anteriorly from points near the mesal ends of the intermaxillary brachia (fig. 43 A, *1inv*).

Anterior to the sternal skeleton arise the paragnaths (fig. 49 H, *Pgn*), a pair of large, flat, soft lobes, and between them an elongate, conical median lobe (*Lin*), which may be termed the *lingua*. The three lobes have a common skeletal base (G), which gives off a strengthening arm into each paragnath and is itself supported on the arms of the bifurcate first maxillary sternum (G, H, *IIIS*). The paragnaths and the lingua together are highly suggestive of the three-lobed hypopharynx of certain lower insects (fig. 75 E); in entomological terminology the lateral lobes are the *superlinguae*.

The mandibles of *Ligyda* are strong biting and masticatory organs. Each jaw (fig. 49 F) has a boxlike base, from which projects a large gnathal lobe differentiated into a toothed incisor process (*inc*) and a molar process (*md*) with a flat, oval mesal surface. The molar processes of the opposite jaws are opposed to each other when the jaws are closed; the incisor processes come together behind the labrum. Each mandible is hinged to the edge of the cranium by the lateral margin of its base and is doubly articulated by an anterior condyle (F, *c*) and a posterior condyle (*a*) at opposite ends of the hinge. The anterior condyle articulates on the basal angle of the epistome (B, *c*), the posterior condyle on the subgenal margin of the cranium. The jaw is activated by at least two muscles; one is a large dorsal adductor (F, *dad*) attached by a wide thick tendon on the mesal margin of the mandibular base, the other is a slender dorsal abductor (*dab*) attached on a lever arm of the lateral margin of the mandible near the posterior articular condyle. A small ventral adductor from the head apodeme inserted in the cavity of the mandible is perhaps present, but, if so, the writer has not with certainty identified it in preserved specimens, though such a muscle is present in the amphipods.

The isopod jaws, hanging downward from the head and swinging transversely against each other, are clearly more effective as biting organs than are the mandibles of the decapods. The change in the mandibular mechanism is dependent on the entire reconstruction of the head. The close parallelism in the structure of the head and mandibles between the isopods and the pterygote insects with biting and chewing jaws might suggest that the insects originated from isopods, but, as will be shown later, the mandibles of the more primitive insects have a quite different structure and mechanism.

The maxillae of *Ligyda* are much simplified appendages. Each first maxilla (fig. 49 D) consists of two elongate lobes, united proximally by membrane and arising from a basal sclerite that articulates on the intermaxillary sternal brachium just mesad of the base of the head apodeme (H). The appendages have wide membranous connections with the head, and the mesal lobes are attached by slender rods to the arms of the first maxillary sternum (*IIIS*). From a study of *Ligyda* alone it is impossible to identify the parts of the first maxillae, but the lobes are usually regarded as endites of the base. The second maxillae (E) are still more simplified, each consisting of a single broad lobe supported on a basal sclerite that articulates with the postmaxillary brachium.

The maxillipeds of *Ligyda* are elongate flattened appendages (fig. 49 C, *Mxpd*). They are attached to the posterior part of the ventral wall of the head (H, *mxpdF*) behind the postmaxillary brachia (*pmB*) at the sides of the narrow maxilliped sternum (*VS*), the posterior end of which is produced into a pair of postmaxilliped brachia that form a transverse bar (C, H, *VS*) behind the bases of the appendages. Each maxilliped (C) consists of a small basal segment bearing an epipodite and of an elongate distal segment with a movable subapical lobe.

The Thorax and the Legs

The seven segments of the thorax of *Ligyda* are of similar shape (fig. 49 A); all are free and flexible on each other. The tergal plates are produced on the sides into broad folds (fig. 50 D, *tf*) projecting over the leg bases. The ventral surfaces of the segments between the legs are membranous except for the presence of weakly developed transverse sternal bars (D, E, G, S). In the female the apertures of the genital ducts are on the venter of the fifth segment (G) behind the narrow sternum; in the male the ducts open through a pair of penes on the seventh segment (H, *Pen*).

The legs have each six free segments (fig. 50 A, B), but the long basal segment by which each leg is suspended from beneath the tergal fold is the basipodite (*Bspd*) and not the coxopodite, as at first sight it might appear to be. The coxopodite is reduced to a narrow ring completely fused to the base of the tergal fold above it (E, F, H, *Cxpd*), but it bears a large condyle on which the basipodite is articulated. The segments of the functional part of the leg (A, B), therefore, are the basipodite, the ischiopodite, the meropodite, the carpopodite, the propodite, and the dactylopodite. The dactylopodite (C) bears an apical claw (*Dac*), which has been regarded as the dactylopodite itself, but no muscles are attached on this claw, which is merely flexible on the dactylopodite. The basal muscles of the leg arising on the tergum are inserted on the basipodite, and presumably are the primarily tergocoxal muscles that have been transferred to the basipodite with the suppression of the coxopodite as a movable segment of the limb. In the genus *Asellus* of the isopod suborder Asellota, the coxae, though small, are free and slightly movable segments.

The lateral folds, or lobes, of the thoracic tergal plates in some of the isopods are set off from the median parts of the terga by grooves, and in some forms they are freely flexible on the back plates. This condition has given rise to the idea that the folds are primarily platelike expansions of the coxae themselves articulated on the terga, and that in those forms, such as *Ligyda*, in which the lobes are continuous with the terga, they have secondarily united with the latter. This interpretation is deduced by Calman (1909) from a study of the genus *Idotea*, in which he says it can be shown in a series of species that the coxae expand to form plates that

185

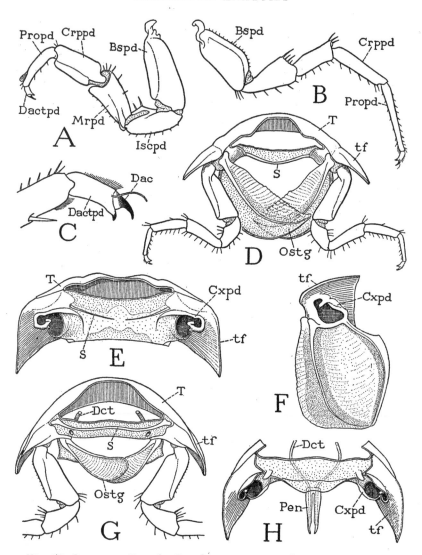

Fig. 50. Crustacea—Isopoda. *Ligyda exotica* Roux. The thorax and the legs.
A, first left leg of male, lateral. B, sixth left leg of male, lateral. C, dactylopodite and dactyl of sixth leg. D, second thoracic segment of female, posterior.
E, fifth thoracic segment of female, ventral, appendages removed. F, left
coxopodite and oostegite of fourth thoracic segment of female, ventral. G, fifth
thoracic segment of female, showing genital ducts and openings, anterior. H,
seventh thoracic segment of male, ventral, with genital ducts and penes.

For explanation of lettering see pages 190–192.

186

eventually replace the primary folds of the terga. In *Idotea baltica* (*Pallas*), however, the coxae are distinct rings on the undersurfaces of the tergal folds, just as they are in *Ligyda*, and yet the folds are separated by dorsal grooves from the median tergal plates, on which they are slightly flexible. The muscles of the basipodite always arise on the tergal region mesad of the groove setting off the lateral fold, not on the fold itself, as they might be supposed to do if the fold is a part of the coxopodite. Inasmuch as the abdominal terga are produced into lateral lobes like those of the thorax (fig. 49 A), it seems rational to suppose that the thoracic lobes are primarily tergal and that, with the fusion of the coxae on their undersurfaces, the lobes in some cases have become flexible on the back to allow more freedom of movement to the leg. A separation of the lobes from the back plates is even more pronounced in some of the amphipods, in which the lobes unquestionably *appear* to be "coxal plates."

In a mature female the lower surface of the thorax, from the first to the fifth segment inclusive, is concealed by a strongly convex covering of large, thin, semitransparent, underlapping lobes arising from the mesal margins of the coxae on each side (fig. 50 D, *Ostg*). These lobes are the oostegites, so named because they form a deep protective pouch under the body in which the eggs and young are carried. The oostegites underlap each other from behind forward, and those on the right underlap those on the left. On the second, third, and fourth segments the oostegites are large, broad lobes (F), each strengthened by a rib in the anterior part and a thickened posterior margin; they are supported by the long basipodites of the legs (D). The first and fifth pairs are shorter and narrower than the others and lie more transversely (G), so that they close the two ends of the baglike brood pouch.

The Abdomen and the Pleopods

The abdomen of *Ligyda* is somewhat ovate in outline as seen from above (fig. 49 A) with the larger end forward. The first five tergal plates resemble the thoracic terga, except that the first and second are smaller than the others. The sixth tergum is a quadrate plate produced into an obtuse angle forming the apical point of the body between the bases of the uropods.

The undersurface of the abdomen is covered by two rows of large, flat, soft lobes, which are the exopodites of the first five pairs

187

of pleopods, underlapping each other from before backward like a double series of scales. The lobes enclose above them a branchial chamber beneath the lower surface of the abdomen, which contains the gills borne on the bases of the pleopods. The true ventral wall of the abdomen (fig. 51 A) can be seen only by removal of the pleopods. Each segmental area of the venter except the last is mostly membranous but is bordered anteriorly by a marginal sternal bar, the outer ends of which curve posteriorly and then mesally around the bases of the pleopods and enclose keyholelike foramina from which the pleopods arise. On the third, fourth, and fifth segments the middle of each sternal bar is produced into a large, tapering process directed posteriorly. The sternum of the sixth segment is a broad plate (*XVIIIS*) between the lateral lobes of the tergum, and the appendages of this segment, the uropods (*Urpd*), project posteriorly from it. The telson of *Ligyda* is represented only by a pair of lobes enclosing the anus (*An*) on the underside of the projecting end of the sixth tergum.

The pleopods of *Ligyda* and other isopods have little resemblance to those of *Anaspides* or *Cambarus*, and they carry the gills. The first five pairs arise anteriorly from the ventral areas of their respective segments, and all but the first have long transverse connections with the body (fig. 51 A), extending from the median sternal processes to the bases of the lateral tergal lobes. The first pleopods (fig. 51 C) differ from the following appendages, but they are alike in the two sexes. Each has a thick, transversely elongate basal part, or protopodite (*Prtpd*), which bears on its lateral end a small epipodite (*Eppd*) and, arising from its ventral surface, a large flat lobe regarded as the exopodite; an endopodite is absent. The ventral surface and the outer end of the protopodite is traversed by a deep cleft, the inner end of which turns forward within the protopodite, and on its anterior lamella is attached a transverse row of short muscle fibers, which evidently serve to open the cleft. The function of this structure is unexplained. The second pleopods are different in the two sexes. In the male each appendage (D) has a large, flat exopodite (*Expd*) like that of the first pleopods, and also an epipodite (*Eppd*), but in addition to these parts there arises from the protopodite a long, sclerotic, two-segmented, elbowed endopodite (*Endpd*), which presumably has some copulatory function. The much simpler corresponding pleopod of the female (F) bears, in the position of

188

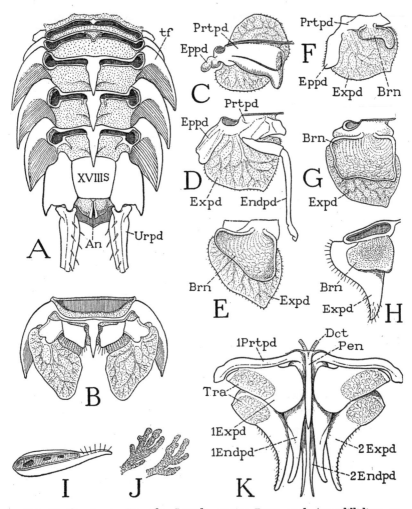

Fig. 51. Crustacea—Isopoda. *Ligyda exotica* Roux, and *Armadillidium* sp. The abdomen and pleopods.

A, *Ligyda exotica* Roux, abdomen of male, ventral, pleopods removed. B, same, fifth abdominal segment of male, ventral. C, same, first left pleopod of male, dorsal. D, same, second left pleopod of male, dorsal. E, same, fourth left pleopod of male, dorsal. F, same, second left pleopod of female, dorsal. G, same, third left pleopod of female, dorsal. H, *Armadillidium* sp., third left pleopod of male, dorsal. I, same, posterior edge of exopodite of first right pleopod, showing marginal depression with three "spiracles." J, same, branched lobes of tracheal organ. K, same, first and second pleopods and penis of male, ventral, showing respiratory organs (*Tra*) in exopodites.

For explanation of lettering see pages 190–192.

the male endopodite, a small, dorsal gill lobe (*Brn*), which is re-garded as the endopodite of the appendage, though the interpretation may be questioned. The next three pairs of pleopods, of which the third in the female (G) or the fourth in the male (E) may be taken as an example, are again alike in both sexes and resemble the second pleopod of the female (F), except that the gill lobe (*Brn*) is much larger and there is no epipodite on the protopodite. Since *Ligyda exotica* is often found on rocky beaches well back from the "spray zone," it is not clear how its gills function as respiratory organs in dry air. The uropods of *Ligyda* are simple biramous appendages (fig. 49 A), each consisting, as already noted, of an elongate basal stalk and two slender rami.

The common terrestrial inland isopods, such as *Porcellio* and *Armadillidium*, have, in addition to gills, respiratory organs for air breathing contained in the first two pairs of pleopods. These organs appear as conspicuous "white bodies" within the exopodites of the first and second pleopods in both sexes (fig. 51 K, *Tra*). When dissected from a fresh specimen, the "bodies" are found to be masses of hollow, branching lobes having a soft, granular texture (J). If these organs are truly respiratory in function, as they are supposed to be, the branched lobes are of the nature of primitive tracheae; they open through apertures in a troughlike depression on the posterior margin of the exopodite (I). In the male of *Armadillidium* each pleopod of the first two pairs (K) has a long endopodite. The next three pairs of pleopods are alike in the two sexes, and resemble the corresponding pleopods of *Ligyda;* each has a flat gill lobe (H, *Brn*) arising dorsally from the protopodite. The underlapping exopodites, as in *Ligyda*, enclose the gills in a branchial chamber beneath the abdomen, but the apertures of the tracheal organs are fully exposed. The gills of these terrestrial isopods are supposed to function in particularly moist places, or when water enters the gill chamber.

Explanation of Lettering on Figures 37–51

a, pleural articulation of coxa.
abAp, abductor apodeme.
Abd, abdomen.
aBrn, arthrobranchia.
acx, antecoxal pleurosternal bridge.
adAp, adductor apodeme.

afc, afferent canal of gill.
An, anus.
Ant, antenna; *1Ant*, first antenna, or antennule, *2Ant*, second antenna.
antF, antennal foramen.

Anv, annulus ventralis.
Ap, apodeme.

b, sternal articulation of coxa.
BC, body cavity, haemocoele.
bcg, branchiocardiac groove.
brC, branchial chamber.
Brn, branchia, gill.
Bspd, basipodite.

c, epistomal articulation of mandible.
Can, cannula.
Ch, chela.
Chpd, cheliped or cheliped foramen.
Cor, cornea.
Cp, carapace.
Crppd, carpopodite.
cvg, "cervical" groove.
cxf, coxal filaments.
Cxpd, Cx, coxopodite.

da, anterior dorsal muscle.
dab, dorsal abductor muscle.
Dac, dactyl, claw of dactylopodite.
Dactpd, dactylopodite.
dad, dorsal adductor muscle.
Dct, duct.
Dej, ductus ejaculatorius.
df, dorsal filament of gill.
dp, posterior dorsal muscle.

e, ocular plate of protocephalon.
E, compound eye.
efc, efferent canal of gill.
Endpd, endopodite.
Endt, endite.
Eppd, epipodite.
Epst, epistome.
es, epistomal sulcus.
esF, foramen of eyestalk.
Expd, exopodite.
ext, extensor muscle.

f, basal sclerite of eyestalk.

fl, flexor muscle.
Fl, flagellum.
For, neck foramen of head.

g, brushlike appendage of thorax of *Cambarus.*
gnL, gnathal lobe of mandible.
gr, seminal groove.
Gon, gonapophysis.
Gpd, gonopod.
Gpr, gonopore.

h, epistomal hinge of mandible.
hAp, head apodeme (first ventral apodeme).

i, connecting bar between thoracic pleuron and abdomen.
icx, intercoxal pleurosternal brachium.
IIS, metastomal plate, perhaps remnant of mandibular sternum.
imB, intermaxillary brachium.
inc, incisor process of mandible.
inv, point of invagination.
Iscpd, ischiopodite.
I–XVIII, enumeration of segments, beginning with second antennal.

j, spur on second leg of male *Cambarus.*

k, postantennal wing of epistome.

l, marginal ridge of epistome.
L, leg.
lf, lateral filament of gill.
lF, limb foramen.
Lg, intergnathal ligament.
Lin, lingua.
Lm, labrum.
ltg, laterotergite.

m, accessory lobe of first gonopod.
mcl, muscle.
Md, mandible.
mdB, base of mandible.

mdF, mandibular foramen.

mnc, mantle canal of gill.

mol, molar process or area of mandible.

Mrpd, meropodite.

Mth, mouth.

Mx, maxilla; *1Mx*, *2Mx*, first and second maxillae.

mxB, maxillary bridge (maxillary pleuron).

mxF, maxillary foramen.

Mxpd, maxilliped.

mxpdF, maxilliped foramen.

n, lobe of second gonopod.

npr, nephropore.

o, orifice of seminal groove.

Opl, operculum.

Ostg, oostegite.

p, pivot of distal segment of eyestalk.

pBrn, podobranchia.

PC, pump chamber of respiratory system.

pcx, postcoxal pleurosternal bridge.

Pen, penis.

Pgn, paragnath.

Pl, pleuron.

plAp, pleural apodeme.

Plp, palp.

Plpd, pleopod.

pmB, postmaxillary brachium.

Propd, propodite.

Prpds, pereiopods.

Prtc, protocephalon.

Prtpd, protopodite.

r, articular knob of tergum.

R, rostrum.

S, sternum.

sAp, sternal apodeme.

Scpg, scaphognathite.

Seg, body segment.

slgs, suspensory ligaments.

T, tergum.

tAp, tergal apodeme

Tel, telson.

tf, tergal fold, branchiostegite of decapod thorax.

Tra, tracheal respiratory organ.

Urpd, uropod.

v, ventral muscles of mandible.

Vd, vas deferens.

vmdmcls, ventral mandibular muscles.

VS, sternum of maxilliped segment.

VT, tergum of maxilliped segment.

* VII *

THE CHILOPODA

THE chilopods are the true centipedes with one pair of legs on each body segment. They, together with the diplopods, pauropods, and symphylans, are distinguishable from the other mandibulate arthropods by the division of the trunk into only two tagmata, a head and a body, wherefore these four orders have been classed together as members of one group, termed the Myriapoda because of their large number of legs as compared with the insects. The chilopods, however, are opisthogoneate, the genital opening being at the posterior end of the body, while the others are progoneate, having the genital opening near the anterior end of the body. Furthermore, it is not certain that the segmental composition of the head in the diplopods and pauropods is the same as that of the chilopods and symphylans; the chilopod head carries two pairs of maxillary appendages behind the mandibles, while in the diplopods and pauropods there is only one postmandibular head appendage. On the other hand, the symphylans resemble the chilopods in having two maxillary appendages on the head, and in this respect these two groups are like the insects. But again, the chilopods, the diplopods, and the pauropods have certain head characters not present in the insects, and the mandibles of the diplopods and symphylans have a structure peculiar to these two groups. However, all the myriapodous forms and the insects have one common feature by which they differ from the other arthropods, which is that the pretarsal segment, or dactylopodite, of the legs has only *one* muscle, which is the usual depressor, or flexor, with its fibers arising in segments proximal to the tarsus. Altogether, then, the inconsistently

193

distributed characters of the seemingly related myriapods and insects make a difficult problem for the phylogenists.

A feature peculiar to the chilopods, diplopods, and symphylans, though not present in all of them, is a pair of sensory organs known as the *organs of Tömösvary*, situated on the sides of the head a short distance behind the bases of the antennae. The essential part of each of these organs is a group of sensory cells in the epidermis innervated from the optic lobe of the brain. In some forms the sense cells lie at the bottom of a deep open pit, in others they are contained in a small, circular cavity of the cuticle with a central aperture, in others the cuticular cavity is horseshoe-shaped with a median cleft to the exterior, again the cavity is an oval groove without an external opening, and, finally, the sensory elements may be covered merely by an undifferentiated cuticle. The structure of the organs of Tömösvary in the chilopods and diplopods is described by Hennings (1904, 1906) and by Pflugfelder (1933), that of the symphylan *Hanseniella* by Tiegs (1940). The function of the organs is not known.

The chilopods are divided taxonomically into four orders, the Scutigeromorpha, Lithobiomorpha, Scolopendromorpha, and Geophilomorpha. The first two are anamorphic, in that the body segmentation is completed after hatching; the other two are epimorphic, their segmentation being complete when they leave the egg. The chilopod structure will be best understood by a study of a representative of each of the four orders.

SCUTIGERA

Scutigera coleoptrata (L.) is the common "house centipede" (fig. 52), so called because it is more frequently found in houses or other buildings than out of doors. *Scutigera* is the typical genus of the subclass Scutigeromorpha, which includes about 14 genera in the single family Scutigeridae. The members of this group differ in one remarkable respect from the other centipedes in that their breathing orifices are on the middle of the back, where each opens into a bilobed, lunglike tracheal respiratory organ, while in the other chilopod groups the spiracles are on the sides of the animal and open into tracheal tubes that branch throughout the body. For this reason the scutigerids are sometimes classed as Notostigmata and the rest of the centipedes as Pleurostigmata. The pleurostigmatic centipedes

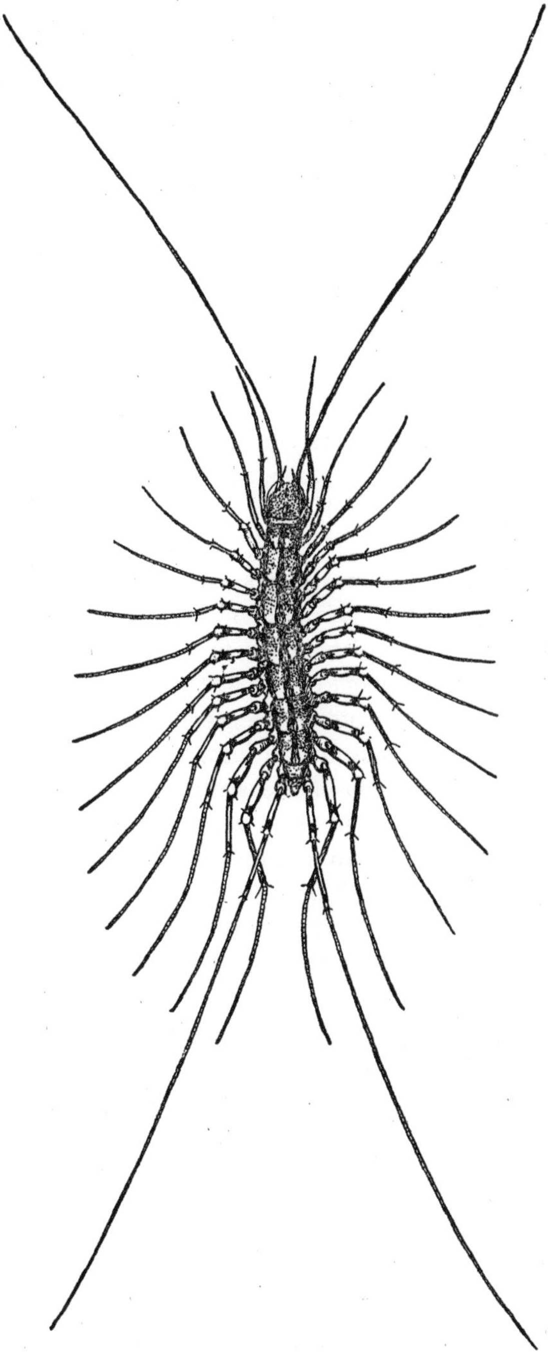

Fig. 52. Chilopoda. *Scutigera coleoptrata* (L.). The house centipede.

195

live mostly on the ground under stones and logs, or beneath loose bark, but such confined places would not be suitable to the scutigerids with their spiracles on the back. The Scutigeromorpha, therefore, live in more open places, and *Scutigera coleoptrata*, when it invades human dwellings, never seeks to hide itself beneath any loose object but depends on its speed to escape capture, in which it is usually successful.

The house centipede (fig. 52) is easily recognized by its long slim legs crowded along the sides of the body, its much longer, threadlike antennae, and its equally long and slender hind legs. The head (fig. 54 A, B) carries dorsally the eyes and the antennae; on its underside are the mandibles, a pair of first maxillae, and a pair of second maxillae. Immediately behind the head is a much reduced first body segment, which has a very small tergal plate (A, *1T*) but carries the huge poison claws, or maxillipeds (*Mxpd*). Following the poison claws are 15 pairs of legs, between which on the ventral surface are 15 simple sternal plates overlapping from before backward (fig. 53 E). On the corresponding part of the dorsum, however, there are only 8 tergal plates. The first of these terga (fig. 54 A, *2T*) pertains to the segment of the first pair of legs (*1L*); the next tergum covers the segments of the second and third legs, and the third those of the fourth and fifth legs, but the longer fifth tergum extends over the segments of the sixth, seventh, and eighth pairs of legs; the next three terga again correspond each with two pairs of legs, while the short ninth tergum (fig. 53 A, *9T*) belongs only to the segment of the fifteenth pair of legs (*15L*).

Behind the last leg-bearing segment is a well-developed legless segment, which is the seventeenth body segment (fig. 53 A, B, C, *17*), covered by the tenth tergal plate (A, *10T*). In the female of *Scutigera* (B, C) this segment carries ventrally a forceps with two-segmented prongs, or gonopods (*Gpd*), between which is the genital aperture (C, *Vul*). In the male (D) the corresponding segment is smaller than that of the female, but on its posterior ventral margin it bears a pair of small, setigerous, styluslike gonopods (*1Gpd*). A second pair of similar gonopods (*2Gpd*) is situated on a transverse fold behind the bases of the first pair, which fact suggests that in the male there are two segments in the genital region, though the second is much reduced. There is, in fact, reason to believe that in both sexes of all the chilopods there were primarily two appendage-bearing segments between the last leg-bearing segment of the adult and the apical segment, or telson (*Tel*), since it has been shown by Heymons (1901) that in the embryo of *Scolopendra* (fig. 60 H) there are two small segments in this region, each bearing rudiments of a pair of gonopods. Inasmuch as the genital aperture lies behind

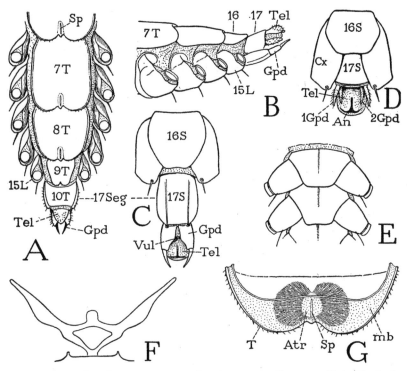

Fig. 53. Chilopoda—Scutigeromorpha. *Scutigera coleoptrata* (L.). The body segments.

A, posterior part of body of female, dorsal; numerals indicate body segments. B, same segments of female, lateral. C, end segments of female, ventral. D, end segments of male, ventral. E, two consecutive sternal plates, with coxae. F, endosternal arms of sternum. G, pair of tracheal lungs in posterior end of tergum, seen from below through inflected membrane (*mb*).

For explanation of lettering see pages 223–224.

the second of these segments, when two are present, this segment is called the *genital segment* (*gSeg*), and the one preceding, the *pregenital segment* (*pgSeg*). Behind the genital segment is the anus-bearing telson (*Tel*).

The eight tergal plates of the 15 leg-bearing segments of *Scutigera*

197

have rounded, medially indented posterior margins that widely overlap the tergum following (fig. 53 A). Just before the notch on each of these terga but the last is a short, median slit (*Sp*), which is the aperture, or *spiracle*, of a respiratory organ (G) that can be faintly seen through the tergal wall. The external aperture (G, *Sp*) opens into a thin-walled median sac, or *atrium* (*Atr*), from each side of which are given off numerous fine, dichotomously branching tracheae, massed in two lateral bodies with such an even contour as to suggest that the whole is covered by a thin tunic; under the microscope, however, no such covering is to be seen. The respiratory organ is virtually a pair of tracheal lungs with a common opening; it is contained mostly in the posterior fold of the tergum closed below by the membranous inflection (*mb*) from the tergal margin.

The ventral body muscles of *Scutigera* are not attached on the sternal plates but on an endosternal structure (fig. 53 F) composed of two lateral bars that diverge upward and laterally from a common median support on the posterior end of the sternum to points close to the edges of the tergum. The two bars are connected by a bridge supporting a median membranous lobe on which muscles are attached. This "endoskeletal" structure is not a cuticular ingrowth or apodeme; it slowly dissolves in caustic, and is a tissue entirely similar to that of the intergnathal ligament of the head, on which the ventral muscles of the mandibles and maxillae are attached (fig. 55 B, *Lg*).

The head of *Scutigera* (fig. 54 A) differs in several respects from that of the other chilopods. Instead of being flattened, the cranium is strongly convex dorsally; its lateral margins slope downward from the back of the neck to the labrum (*Lm*). The large eyes are situated laterally behind the middle of the head, and in front of them arise the antennae, separated by a long, steeply declivous epistomal surface (*Epst*) from the marginal labrum (*Lm*). On the chilopod head there is none of the familiar lines, or "sutures," of an insect's head. In *Scutigera coleoptrata* (B) a line marking an internal ridge of the cranial wall runs forward and mesally from the inner angle of each eye and is joined to its mate by a reversed U-shaped groove, from which a faint median ridge runs back between the eyes.

The large eyes of *Scutigera* have superficially the appearance of many-faceted compound eyes such as those of the crustaceans and insects. Internally, however, the eye does not have the structure of

198

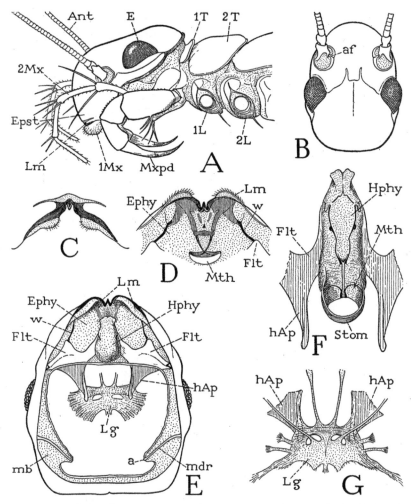

Fig. 54. Chilopoda—Scutigeromorpha. *Scutigera coleoptrata* (L.). The head.
A, head and anterior body segments, left side. B, head, dorsal. C, labrum,
anterior. D, labrum and preoral epipharyngeal surface, ventral. E, head,
ventral, mouth parts removed, showing epipharyngeal surface (*Ephy*), hypopharynx (*Hphy*), fultural sclerites (*Flt*) with head apodemes (*hAp*) supporting
intergnathal ligament, and articular rods of mandibles (*mdr*). F, hypopharynx
and adjoining parts of fulturae. G, intergnathal ligament, muscles removed,
supported on ends of head apodemes, dorsal.

For explanation of lettering see pages 223–224.

a true compound eye. Examples of the structure of myriapod eyes will be given in the next chapter.

Each long, threadlike antenna of *Scutigera* is supported on a short basal stalk, or *scape,* of two small segments set in a membranous depression, or *socket,* of the head wall and pivoted on a short process, the *antennifer* (fig. 54 B, *af*), projecting from the mesal rim of the socket. The antenna as a whole is thus free to swing about in all directions, and it is provided with basal muscles arising on the head wall. The distal part of the antenna, or *flagellum,* is divided into a great number of short annulations that give it flexibility, but at two points it has true joints that divide it into three segments movable by muscles. As shown by Imms (1939), a dorsal muscle arising in the scape and two ventral muscles arising in the first flagellar segment are inserted on the basal ring of the second segment of the flagellum, three corresponding muscles arise in the base of this segment and are inserted on the base of the third flagellar segment, and, finally, a similar set of muscles traverses the last segment, becoming extremely attenuated toward the extremity.

The labrum of *Scutigera* is not a free lobe of the head, and is only partly set off from the epistomal region above it by a transverse membranous area (fig. 54 A, *Lm*). It is marked by two dark bands on the lower margin of the cranium (C) converging to a median notch in which is a strong tooth. On each side of the notch the labrum bears an elongate, soft, setigerous marginal lobe.

The ventral surface of the head is entirely concealed by the mouth parts and the poison claws. Though the poison claws are the appendages of the first body segment, they are functionally associated with the feeding organs and are hence called maxillipeds. In order to conserve material, the appendages, beginning with the maxillipeds, may be studied first, since their successive removal will expose the underside of the head.

The poison claws, or maxillipeds, of *Scutigera* are seven-segmented, leglike appendages (fig. 55 H, *Mxpd*) turned forward at the sides of the head (fig. 54 A), with their distal, clawlike segments curved downward and posteriorly in the manner of a rattlesnake's fangs. The large flattened coxae (fig. 55 H) arise close together beneath the posterior part of the head, and each is produced forward in a broad endite (I, *cxnd*) armed with four strong spines. The telopodite of each appendage is attached laterally to the coxa

200

by a small first trochanter (*1Tr*), which is followed by a long second trochanter, or prefemur (*2Tr*), a short femur, a tibia, and the long, curved claw. The claw of the poison fang of *Scutigera* is indistinctly divided into a tarsus and a pretarsus (*Tar, Ptar*), but in other chilopods these two segments are united. At the tip of the pretarsus is the opening of the duct (*Dct*) of the poison gland (*Gld*), which latter is a long cylindrical sac extending back to the middle of the second trochanter. Though the pretarsal claw is not freely flexible on the tarsus, it has a strong flexor musculature; attached on its base is a long tendon that traverses the tarsus and tibia and gives insertion to five bundles of muscle fibers arising in the femur and the second trochanter (*flptar*). The pretarsus of the maxillipeds, in common with that of the legs in all the centipedes, has no extensor muscle. Each of the other segments of the maxilliped is likewise provided with a flexor muscle, though only that of the tarsus (*fltar*) is shown in the figure. The poison claws of the centipede are the organs by which the animal captures and kills its prey.

The second maxillae arise in front of the maxilliped coxae and project forward from beneath the expanded basal angles of the first maxillae. Each second maxillary appendage of *Scutigera* (fig. 55 G) is a slender, six-segmented limb; in no other arthropod does the second maxilla so much resemble a leg. The coxa (*Cx*) is a large plate with a mesal arm behind the base of the first maxilla and a broad lateral arm extended upward and posteriorly in the membranous lateral wall of the head (fig. 54 A), but with no articulation on the cranium. The five-segmented telopodite constitutes a spiny palpus projecting forward beyond the front of the head.

The first maxillae (fig. 55 F) are thick, soft, triangular lobes showing evidence of not more than three segments. They lie close together against the undersurface of the head, but are not united except as they are connected by two transverse sternal bars of the head wall between their bases. The broad basal parts of the appendages may be supposed to be the coxae, and the tapering distal parts the telopodites, which end with seta-covered lobes that project beneath the labrum (fig. 54 A, *1Mx*). Mesally, from the base of each appendage, there project two small endite lobes. The two closely adjacent first maxillae (fig. 55 F) form the functional underlip of the chilopods, which thus differ from the insects, in which the underlip, or labium, is composed of the united second maxillae. In

201

the mesal surface of each first maxilla of *Scutigera* is a deep pouch from which issues a great mass of long, delicate, filamentous setae. Within the inner part of the pouch, intermingled with the bases of the hairs, are innumerable minute spindle-shaped rods characterized by a fine, spiral surface thickening running left to right from the base almost to the transparent tapering tip. Electron microscope studies of these setae have been made by Richards and Korda (1947), but their structure is visible under an ordinary microscope with a strong light. The pouches are eversible, and the brushes of setae are thought to be cleaning organs; the rods, however, are deeply seated in the pouches and are readily detached.

The mandibles will be fully exposed on removal of the maxillae. They lie longitudinally (fig. 55 B) against the lateral parts of the ventral head wall and are connected with the cranial margins by deeply infolded membranes. Each mandible (A) is elongate, weakly sclerotized, widened at its anterior end, tapering posteriorly, and curved mesally. The broad, free anterior part is the gnathal lobe of the jaw (*gnL*) and is applied against the side of the hypopharynx. The tapering posterior part, or mandibular base (*mdB*), is open mesally into the head cavity and gives attachment to the mandibular muscles (B); its narrow posterior end is connected with the cranial margin behind the eye by a delicate rod (A, *mdr*) in the supporting infolded membrane. The attachment point of this rod to the mandible (*a*) represents the usual posterior cranial articulation of the jaw, which in the chilopods is intermediated by the rod, thus allowing the mandible a free movement as far as the suspensory membrane will permit. On the laterodorsal surface of the gnathal lobe is an anterior articular process (*c*), but it merely hooks over a ridge on the epipharyngeal surface of the head (fig. 54 D, E, *w*) and is hence only a point of loose contact and not a fixed articulation. A line between the two articular points (fig. 55 A, *a*, *c*), however, probably gives the mandible an axis for lengthwise rotation.

The walls of the broad gnathal lobe of each mandible contain two principal plates, one ventromesal (fig. 55 D), the other dorsolateral (C), separated dorsally and ventrally by wide membranous areas. Each plate is set off by a nonsclerotized incision or break from the mandibular base, so that the gnathal lobe appears to be flexible on the latter. The ventromesal plate bears distally three large, strong, loosely attached, tricuspid teeth, and on its mesal margin proximal

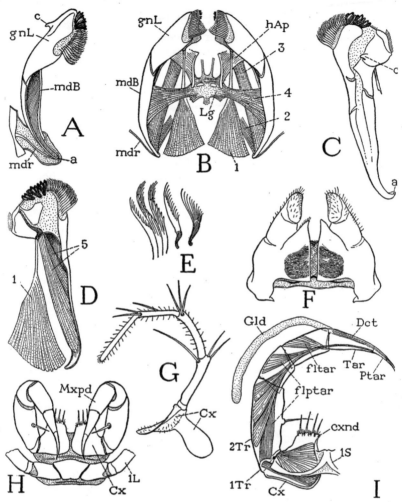

Fig. 55. Chilopoda—Scutigeromorpha. *Scutigera coleoptrata* (L.). The mouth parts.

A, right mandible, mesoventral, showing posterior articular rod (*mdr*) in membrane inflected from edge of cranium. B, mandibles and their muscles, ventral, with intergnathal ligament supported on head apodemes. C, left mandible, dorsal, showing anterior articular process (*c*). D, left mandible, mesal, with muscles of gnathal lobe. E, brushlike setae of mandible. F, first maxillae, ventral. G, left second maxilla, ventral. H, macillipeds and first leg segment, ventral. I, left maxilliped, dorsal, poison gland displaced, showing muscle of pretarsus.

For explanation of lettering see pages 223–224.

to the teeth a broad fringe of thick, brushlike setae, several varieties of which are shown separately at E of the figure. The dorsolateral plate carries the anterior articular process (C, c).

In the normal position of the mandibles (fig. 55 B) the teeth of the distal margins are dorsal, toward the mouth, and the brushes guard the mouth entrance ventrally. The chilopods are reluctant to demonstrate the action of their mouth parts under a microscope, and the only movements of the mandibles to be seen in a live specimen are those of protraction and retraction. Protraction brings the toothed margins of the jaws together, and, as the mandibles are thrust forward, the maxillary lobes slide backward against their undersurfaces.

The musculature of the mandibles (fig. 55 B) consists of extrinsic muscles arising on the head wall, or on the ventral head apodemes (to be described presently, fig. 54 E, hAp), and intrinsic muscles contained within the mandibles themselves. The extrinsic muscles of each jaw include two dorsal muscles and two ventral muscles. The first dorsal muscle (fig. 55 B, 1) is a large conical bundle of fibers arising posteriorly on the cranium and inserted by a strong tendon on the base of the dorsal plate of the gnathal lobe (D, 1). This muscle is evidently the retractor of the mandible, but it must also somewhat flex the gnathal lobe on the mandibular base. The second dorsal muscle (B, 2) arises on the cranium and is inserted dorsally on the posterior end of the jaw. The ventral musculature includes two bundles of fibers. One is a long muscle (B, 3) arising anteriorly on the head apodeme (hAp) of the same side and inserted posteriorly on the concave inner side of the mandible; it is the principal protractor of the jaw. The other ventral muscle (4) arises medially on an intergnathal ligament (Lg) supported on the head apodemes, and its spreading fibers are inserted within the cavity of the mandible; the two corresponding muscles from the opposite jaws are the mandibular adductors. The intrinsic musculature of the mandible of Scutigera includes two muscles (D, 5) arising within the cavity of the jaw, the longer one attached ventrally on the base of the gnathal lobe, the shorter one dorsally. These two muscles evidently can serve only to flex the gnathal lobe on the base of the jaw.

The undersurface of the head will be now fully exposed with the removal of the feeding appendages. It is then seen that a long, epipharyngeal surface extends back from the labrum to the mouth (fig. 54 D, E, *Ephy*), though the mouth is normally concealed above the base of a large median lobe of the ventral head wall, known as the *hypopharynx* (E, *Hphy*). Most of the epipharyngeal area is membranous, but from weak lateral sclerotizations just behind the labrum two sclerotic rods converge posteriorly (D) to a crossbar, from which slender tapering arms continue to the mouth, where they appear to cross each other and end in the dorsal wall of the stomodaeum. Laterally two other bars (D, E, *w*) diverge posteriorly to the anterior angles of two large sternal plates of the ventral head wall (*Flt*). It is in the angles between these bars and the plates that the anterior articular processes of the mandibles are held.

The hypopharynx of *Scutigera* is a large, soft, elongate median lobe of the ventral head wall (fig. 54 E, *Hphy*), arising just in front of the bases of the first maxillae and projecting forward between the gnathal lobes of the mandibles well beyond the mouth, which latter is thus hidden in a ventral view of the head. The name "hypopharynx" given to this suboral lobe of the head is, of course, a misnomer, since the organ in question is entirely outside the mouth and has no anatomical relation to the pharynx. The term *metastome* would be more appropriate, but "hypopharynx" is current in myriapod and insect nomenclature. The size and shape of the hypopharynx will appear different in different specimens; in some the organ is swelled out into a large vesicle, in others it is variously contracted. The dorsal surface (fig. 54 F) is much longer than the ventral surface, since it extends posteriorly to the mouth (*Mth*); it is conspicuously marked by a pair of lateral rodlike thickenings, beginning anteriorly as S-shaped loops on the sides and converging posteriorly to be continued by parallel median extensions to the mouth. Here the two rods give off upcurved ribs into the lateral mouth walls and then run on into a pair of thicker bars in the floor of the stomodaeum (*Stom*).

At the sides of the hypopharynx (fig. 54 E) are the two ventral plates of the head wall above-mentioned. Each plate (*Flt*) expands laterally and is attached by its posterior lateral angle to the margin of the cranium; mesally it is extended forward as a tapering arm against the side of the hypopharynx, and posteriorly it gives off a large apodeme (*hAp*) into the head cavity. These plates are called by German writers the *kommandibulares Gerüst*, but since they are more intimately associated with the hypopharynx than with the

205

mandibles, the writer (1951) has termed them *hypopharyngeal fulturae* (i.e., hypopharyngeal supports). Literally, however, the plates in question are *premandibular sternal sclerites of the head;* they are characteristic features of the Chilopoda, and are present in some Diplopoda, but no corresponding plates are known in the crustaceans or insects.

The apodemes of the fultural plates have been known as "hypopharyngeal apophyses"; Fahlander (1938) calls them "mandibular apodemes," but inasmuch as they pertain to the sternal plates and not to the hypopharynx, or the mandibles, they may be better termed the *ventral head apodemes* of the chilopods, though, because of their premandibular origin, they clearly have no relation to the usual "head apodemes" of Crustacea that arise between the two maxillary segments. The inner ends of the apodemes in *Scutigera* support a broad sheet of soft tissue (fig. 54 E, *Lg*) that in a freshly dissected specimen appears to fray out on each side into the adductor muscle fibers of the mandibles and the two pairs of maxillae. When the muscles are removed, however, there is left a thin ligamentous plate of definite shape (*G*), supported from below on the posterior ends of the head apodemes (*hAp*) and from above by slender suspensory arms attached dorsally on the cranium; lateral and posterior arms give attachment to muscles. The structure is very similar to that described by Fahlander (1938, fig. 6 A) for *Thereuonema*. The apodemes are simply imbedded in the tissue of the ligament and are readily detached. The ligament has a partly fibrous structure and slowly dissolves in caustic, as do the ventral "endoskeletal" arms of the body segments. The ligament is quite comparable to the endosternum of *Limulus* and the arachnids, but it can have no relation to the central plate of the insect tentorium, which is a sclerotized chitinous tissue.

The legs of *Scutigera* (fig. 56 F), in spite of their great length, have each only seven true segments. The basal part of most of the legs is a large, complex structure on the side of the supporting body segment between the tergum and the sternum (A, B). The basal segment of the telopodite is a very small, ringlike first trochanter (A, *1Tr*), which is followed by a large second trochanter, or prefemur (F, *2Tr*). The two trochanters are but weakly attached to each other, so that the legs of *Scutigera* readily break off between these two segments. The second trochanter is followed by an elongate femur (*Fm*) and a still longer tibia (*Tb*). The extremely long, slender distal part of the leg with numerous short annulations is the tarsus (*Tar*); the leg ends with a small simple claw, which is the pretarsus (*Ptar*). The tarsus is divided by a joint beyond the middle

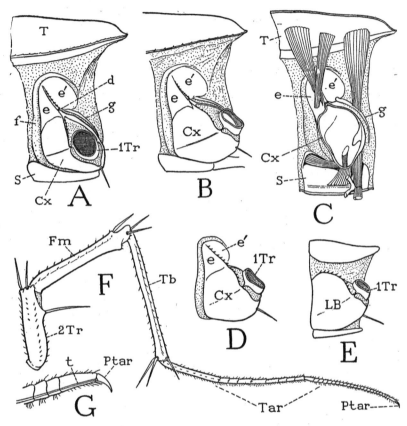

Fig. 56. Chilopoda—Scutigeromorpha. *Scutigera coleoptrata* (L.). The legs. A, left side of segment from middle of body, showing leg base composed of coxa (*Cx*) and subcoxal sclerites (*e, e′, f, g*). B, left side of eleventh leg-bearing segment. C, mesal view of right half of a body segment, with basal muscles of leg. D, leg base of thirteenth leg-bearing segment, subcoxal sclerites united with coxa. E, leg base of last leg-bearing segment, subcoxal sclerites confluent with coxa. F, a left leg, except coxa and first trochanter. G, pretarsus and distal part of tarsus, showing flexor tendon (*t*) of pretarsus.

For explanation of lettering see pages 223–224.

into two principal subsegments, which are merely flexible on each other, there being no interconnecting muscles. Though no muscles are present within the tarsus, a slender tendon (G, *t*), attached on the ventral angle of the base of the pretarsal claw, traverses the tarsus and the tibia and gives attachment to the flexor muscles of the pretarsus in the proximal segments of the leg. The pretarsal musculature of the legs is thus seen to be the same as that of the maxillipeds (fig. 55 I). In none of the chilopods, or other myriapods, or in the insects does the pretarsus have an extensor, or levator, muscle. The extraordinary length of the hind legs of *Scutigera* (fig. 52) is due mostly to the extreme length of the slender, threadlike tarsi.

The nature of the basal parts of the chilopod legs is difficult to understand. The typical structure is best developed on the middle segments of the body (fig. 56 A, B), and the principal element in the complex appears to be the coxa (Cx), since it supports the telopodite and retains the small first trochanter $(A, 1Tr)$ when the leg breaks off. The coxa is articulated (d) to a plate above it divided into two parts (e, e') by a spiny median ridge. Curving over this plate and running down the anterior side of the coxa is a long integumental fold (f), and from the posterior part of the plate a narrow sclerite (g) goes downward behind the coxa. These peripheral sclerites of the leg base have been interpreted as pleurites by Verhoeff (1906), who terms the supracoxal plate (e, e') the "katopleure," the fold above and before it (f) the "anopleure," and the postcoxal strip (g) the "coxopleure." In some of the other chilopods, however, there are distinct pleural sclerites between the leg bases and the terga (figs. 60 E, 61 B, *pl*). There is also the theory that the peripheral sclerites of the leg base represent a primitive "subcoxal segment" of the limb, an idea for which there is no convincing evidence, but the parts in question may be termed "subcoxal sclerites" merely to denote their position relative to the apparent coxa.

An examination of the basal musculature of the leg (fig. 56 C) shows that the anterior and posterior dorsal muscles are attached, respectively, on the "katopleure" and the "coxopleure," which fact would suggest that these sclerites are differentiations of the coxa itself. Moreover, on the more posterior body segments the subcoxal elements become united with the coxa (D), and, finally, on the last leg-bearing segment (E) there is no distinction of coxal and subcoxal parts; the base of the leg is here an undivided plate (LB), which has all the appearance of a simple coxa implanted between the tergum and the sternum, and the dorsal muscles of the leg are attached on it. The nature of the "subcoxal" sclerites will be further discussed in connection with the other chilopods to be described.

LITHOBIUS

The genus *Lithobius,* of which there are many closely related species, is representative of the chilopod subclass Lithobiomorpha, the members of which are anamorphic in their postembryonic development as are the Scutigeromorpha, although structurally in some respects the two groups are quite different. In their general appearance the Lithobiomorpha more resemble the epimorphic Scolopendromorpha, particularly in that the head and body are flattened, the legs relatively short, the antennae placed at the anterior end of the head, and the spiracles situated along the sides of the body.

Fig. 57. Chilopoda—Lithobiomorpha. *Lithobius* sp.

Lithobius (fig. 57), in common with *Scutigera,* has 15 pairs of legs, but in *Lithobius* there is a tergal plate for each pair of legs, though the terga are not all of the same size. Immediately behind the head is a short tergum (fig. 59 A, *1T*) of the maxilliped segment; the next is a long tergum (*2T*) belonging to the first pair of legs; then follows an alternation of short and long terga back to the eighth and ninth segments, in both of which the terga are long (fig. 57). With segment 10 the short and long succession begins again and continues to segment 15, but the tergum of segment 16 is also long, though shorter than the one preceding. On the undersurface of the body there is no sternal plate of the maxilliped segment (fig. 59 I), but there are 15 sterna of approximately equal size (fig. 58 F) for the 15 leg-bearing segments. Between the segmental sternal plates (*S*) are small, paired intersternites (*Is*) that do not represent segments. The spiracles lie in the membranous pleural areas of the segments above the leg bases and somewhat behind them (fig. 58 C, *Sp*). In the species of *Lithobius* here illustrated spiracles are present on segments 4, 6, 9, 11, 13, and 15, which are all segments having long terga. On segment 2 is a small papilla in the spiracle position, but it has no opening, though in some species there is a functional spiracle on this segment. The last spiracle, on segment 15, is very small. Species of *Lithobius* differ in the number of spiracles, those on segments 4 and 11 being said to be the only ones that are never absent. As in *Scutigera,* there are in *Lithobius* no pleural sclerites above the leg bases.

Following the last leg-bearing segment is a small pregenital segment. In the male this segment has a well-developed tergal plate, but normally it is entirely concealed by retraction beneath the preceding tergum; on the undersurface, however, the sternum is fully exposed (fig. 58 G, *17S*) and bears on its posterior margin a pair of vestigial gonopods (*Gpd*). Between the gonopods projects the tip of a large intromittent organ (*Pen*), or penis (I), which is ordinarily retracted above the sternum. The body ends with a simple, anus-bearing telson (G, *Tel*). In the female the tergum of segment 17 is reduced to lateral sclerotizations of the dorsum (J, *17T*), but the sternum is well developed (H, J, *17S*) and bears a pair of three-segmented gonopods (*Gpd*), one of which is shown separately at E. Between the bases of the gonopods and behind the supporting sternum is the female genital opening.

The legs of *Lithobius* (figs. 57, 58 A) are much shorter than those of *Scutigera,* owing principally to the great length of the tarsi in *Scutigera,* but they have the same segmentation (fig. 58 C), and the tarsi are each divided into two subsegments (B, C, *Tar*). As in *Scutigera,* the legs of *Lithobius* break off between the two trochanters. The simple, clawlike pretarsus (*Ptar*) is provided with a long, slender tendon (B, *t*) attached ventrally on its base, which gives insertion to large flexor muscles (*flptar*) arising in the tibia and the femur. The leg base of *Lithobius* (C) has the same structure as in *Scutigera,* except that the peripheral subcoxal sclerites (*scx*) are somewhat less developed and the anterior sclerite is continued ventrally between the sternum and the coxa (C, F). The coxa has an articular point (C, *i*) on the ventral subcoxal arc, as well as a dorsal articulation on the sclerite above it. On the last leg-bearing segment, however, as also in *Scutigera,* the long hind leg is supported on a simple basal plate (J, *LB*), in which the subcoxal sclerites of the preceding segments appear to be incorporated. The same condition will be

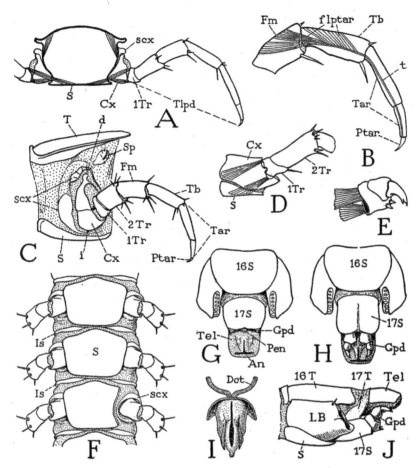

Fig. 58. Chilopoda—Lithobiomorpha. *Lithobius* sp. The body segments and the legs.

A, cross section of body segment with legs attached. B, segmentation of leg beyond second trochanter. C, left side of a body segment with leg, showing subcoxal sclerites around base of coxa. D, base of last leg, left side, with simple coxa. E, gonopod of female. F, undersurface of three consecutive body segments, part of third sternum cut off to show a subcoxal fold beneath coxa. G, terminal body segments of male, ventral. H, same segments of female, ventral. I, penis and ducts, ventral. J, terminal body segments of female, lateral.

For explanation of lettering see pages 223–224.

again encountered in the Scolopendromorpha and the Geophilomorpha. It is a question, therefore, whether the subcoxal sclerites of the chilopods are pleural elements that have been united with the coxae on the posterior segments, or whether they are coxal derivatives that have become detached from the coxae on the more anterior segments. At present there is no satisfactory answer to the question.

It is in the structure of the head that the Lithobiomorpha differ conspicuously from the Scutigeromorpha. The head of *Lithobius* is flattened (figs. 57; 59 A), the eyes have a forward position on the sides of the head, and the relatively short, thick antennae arise directly from the anterior end of the head. To accommodate this position of the antennae, the epistomal region and the labrum have been inflected upon the undersurface of the head in *Lithobius* (fig. 59 B), where together they conceal the epipharyngeal surface, and the broad labrum (*Lm*) underlaps the anterior ends of the mandibles (*Md*). The undersurface of the lithobiid head, therefore, resembles the undersurface of the head of a decapod crustacean (fig. 42 A), but the modification must be secondary in each case, since the head structure of *Scutigera* undoubtedly is more primitive than that of *Lithobius*. Whatever its position, the epistome is still to be regarded as the anterior part of the true dorsal surface of the head. The labrum of *Lithobius* is entirely separated from the epistomal margin.

The antennae of *Lithobius* are much shorter and relatively thicker than those of *Scutigera,* and are divided into many small sections, but all of the sections have been shown by Imms (1939) to be true antennal segments, since each one is individually provided with muscles arising in the segment proximal to it. In the species here described there are about 20 segments in each antenna, but in other species the number may be smaller or much greater.

Inasmuch as the mouth parts of the centipede underlap each other from behind forward, they may be more conveniently studied in reverse order, beginning with the poison claws, which, though they are the appendages of the first body segment, are a part of the feeding apparatus.

The poison claws, or maxillipeds, of *Lithobius* (fig. 59 I, *Mxpd*) lie horizontally against the underside of the head, with the fangs turned mesally. The large, flat coxae (*Cx*) are weakly joined in a long median suture, forming a broad plate with a toothed margin

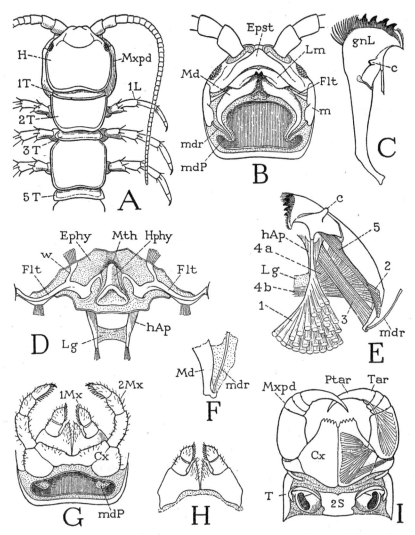

Fig. 59. Chilopoda—Lithobiomorpha. *Lithobius* sp. The head and the mouth parts.

A, head and first five body segments, dorsal. B, head, ventral, mandibles in place, but both pairs of maxillae removed, exposing mandibular pouches (*mdP*). C, left mandible, laterodorsal. D, circumoral region of ventral wall of head, showing hypopharynx, fultural sclerites, and the head apodemes. E, right mandible and its muscles, dorsal. F, posterior end of mandible with articular rod. G, first and second maxillae, and rear end of head, ventral. H, first maxillae, ventral. I, maxillipeds and first leg-bearing segment, ventral.

For explanation of lettering see pages 223–224.

covering the maxillae from below. The telopodites have each only four distinct segments, but it is to be observed that the tendon of the flexor muscle of the claw is attached, not on the base of the claw, but to the inner side well within the base, which fact is good evidence that the claw, or poison fang of *Lithobius,* consists of the pretarsus (*Ptar*) united with the tarsus (*Tar*). It is only in the Scutigeromorpha among the chilopods that the pretarsus is separated from the tarsus. The two pairs of maxillae (fig. 59 G) arise close together from wide bases in front of the maxillipeds and extend forward against the undersurface of the head. The coxae of the second maxillae (G, *Cx*) are broadly united with each other behind the bases of the first maxillae; the palpuslike telopodites consist each of four segments, including a terminal claw, and reach forward beneath the epistome and the bases of the antennae. The first maxillae (H) are relatively simple appendages, each having a large basal segment produced mesally in a triangular endite and supporting laterally a short, two-segmented telopodite. The bases are closely approximated but are not united, and the two appendages together form a triangular underlip that projects forward beneath the mandibles and the labrum.

The mandibles of *Lithobius* (fig. 59 B, *Md*) are in general form similar to those of *Scutigera,* but their long, tapering, incurved posterior ends are contained in pockets (*mdP*) of the membranous ventral wall of the head, inflected at the sides of the first maxillae and extended posteriorly above the bases of the second maxillae (G, *mdP*). An articular rod (B, *mdr*) connects the end of each mandible with a marginal plate of the cranium (*m*). Anteriorly an articular process on the dorsal side of the mandible (E, *c*) catches loosely in the angle between the corresponding fultural sclerite (D, *Flt*) and its epipharyngeal branch (*w*). The entire mandible is weak and flexible, and a break in the sclerotization at the base of the gnathal lobe (C, E) would appear to make the latter flexible on the basal part of the jaw.

The mandibular musculature of *Lithobius* (fig. 59 E) is essentially the same as that of *Scutigera.* A large dorsal muscle of the gnathal lobe (*1*) arises with widely spreading fibers on the cranial wall and is evidently a flexor of the lobe as well as a retractor of the mandible as a whole. A second smaller dorsal muscle (*2*) is attached proximally on the base of the jaw. A long protractor (*3*) goes from the base of the head apodeme (*hAp*) to the posterior end of the mandible. The adductor fibers corresponding with muscle 4 of *Scutigera* (fig. 55 B) are differentiated into two groups in *Lithobius* (fig. 59 E), a distal group (*4a*) attached on the head apodeme and a proximal group (*4b*) attached on the interapodemal ligament (*Lg*). Finally, there is an intramandibular flexor of the gnathal lobe (5).

On removal of the mandibles the circumoral region of the ventral head wall will be exposed (fig. 59 D). Medially is the hypopharynx (*Hphy*), which projects forward below the mouth (*Mth*), and on each side of it is a long, transverse premandibular fultural sclerite (*Flt*) attached laterally to the marginal plate of the cranium (B, *m*). Anteriorly each fultura gives off a short branch (D, *w*) into the membranous epipharyngeal wall (*Ephy*), its mesal end is produced into a recurved arm imbedded in the lateral wall of the hypopharynx, and from its posterior margin a long, tapering apodemal process (*hAp*) extends posteriorly into the head cavity. The two apodemes are connected by a broad transverse ligament (*Lg*) below the pharynx. The relation of the mandibular muscles to the head apodemes and their connecting ligament was noted in the last paragraph.

OTOCRYPTOPS

The Scolopendromorpha, of which *Otocryptops* is an example, are similar in their general appearance to the Lithobiomorpha but are to be distinguished superficially from the latter by their greater number of legs, some having 21 pairs and others 23 pairs. Correspondingly, they have 25 or 29 body segments, including the maxilliped segment, two segments in the genital region, and the telson. The typical genus of the order, *Scolopendra*, has 21 pairs of legs; *Otocryptops*, described here (fig. 60 A), is one having 23 pairs.

Otocryptops sexspinosa (fig. 60 A) is a common centipede associated with *Lithobius* in the eastern part of the United States. The back plates of the body are more uniform in size than are those of *Lithobius*, but they show a tendency to alternation of longer and shorter plates, especially on the anterior segments. The large first tergum possibly includes the tergum of the maxilliped segment, since this segment has no separate tergal plate of its own. On the undersurface of the body most of the segmental sterna (fig. 60 E, S) are separated by narrow intersternites (*Is*). The legs, except the last two pairs, are seven-segmented, and associated anteriorly and dor-

215

sally with the narrow coxa (Cx) is a large subcoxal sclerotization (scx). Between the latter and the tergum are several small sclerites (pl) that may be regarded as *pleurites* (pl), though some writers term them *paratergites*. Spiracles, associated with the pleurites, are present on segments 4, 6, 9, 11, 13, 15, 17, 19, 21, and 23. As in the other chilopods, the coxal and subcoxal sclerotizations appear to be confluent on the last leg-bearing segment (F), in which the leg is supported on a large lateral plate (LB) completely occupying the space between the tergum and the sternum. By comparison with the preceding segment (23) it would appear that the principal part of this plate on segment 24 is derived from the subcoxal element.

The legs of the next to the last pair (on body segment 23, fig. 60 F) differ from the others in having eight segments instead of seven, including the coxa (Cx) and a small first trochanter united with the long second trochanter. The legs of the last pair (A) have the same structure, except that there is no first trochanter (F). The unusual segmentation of these two appendages may be supposed to be explained on the assumption that the last two apparent segments before the pretarsus are subdivisions of the tarsus, since the tarsus is thus divided in the embryo of all the legs of *Scolopendra* (fig. 60 H), as it is in *Scutigera* and *Lithobius* (figs. 56 F, 58 B). An examination of the musculature of either of the last two legs of *Otocryptops*, however, shows that while no muscles have their origins in the penultimate segment, the antepenultimate segment not only gives origin to a long series of fibers attached on the tendon of the pretarsal claw but contains also a levator and depressor muscle of the penultimate segment. This same type of segmentation and musculature is seen also in the last leg of a geophilid (fig. 61 G). Attems (1926c) suggests that the simplest way to explain this apparent anomalous condition is to assume that some of the pretarsal muscles have shifted their origins back to the proximal subsegment of the tarsus. The explanation, however, is too simple; such a transposition of the pretarsal muscles is not known in any of the other myriapods or in the insects. The musculature of a leg of *Scolopendra* (fig. 61 H), which does have two tarsal subsegments, shows that the last group of pretarsal fibers still takes its origin in the tibia (Tb) just as in any seven-segmented leg (F). It is evident, therefore, that in the eight-segmented leg of *Otocryptops* or the geophilid (G) the antepenultimate segment is the tibia (Tb); the presence of levator

216

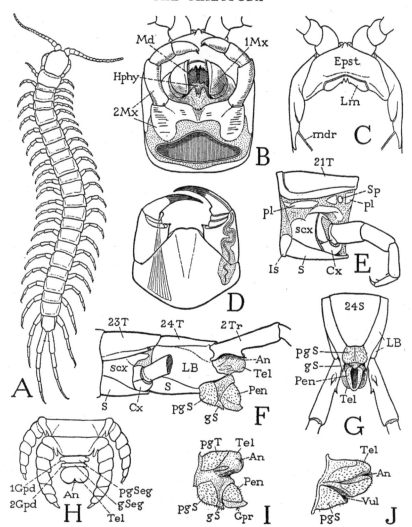

Fig. 60. Chilopoda—Scolopendromorpha. *Otocryptops sexspinosa* (Say), except H.

A, entire centipede. B, head and mouth parts, ventral. C, anterior part of head, ventral, mouth parts removed. D, maxillipeds (poison claws), ventral. E, middle segment of body and leg, left side, showing pleural sclerites (*pl*) above subcoxal sclerites of leg base. F, terminal body segments of male, left side. G, same, ventral. H, terminal part of body of embryo of *Scolopendra*, dorsal (from Heymons, 1901). I, genitoanal region of male, left side. J, same of female.

For explanation of lettering see pages 223–224.

and depressor muscles attached on the base of the penultimate segment, moreover, is confirmatory evidence that this segment alone is the tarsus (*Tar*). Finally, it is to be noted that the distribution of the pretarsal muscles is the same in all the legs (fig. 61 F, G, H) regardless of the number of segments or the division of the tarsus. The extra segment of the eight-segmented leg of *Otocryptops*, therefore, must be in the trochanteral region. The free six-segmented telopodite of the last leg clearly lacks the first trochanter (fig. 60 F), but otherwise the segmentation of this leg corresponds with that of the eight-segmented leg preceding.

The genital and anal region of *Otocryptops* projects beyond the twenty-fourth segment beneath the bases of the hind legs (fig. 60 F). In the male the pregenital segment has a distinct tergum and sternum (I, *pgT*, *pgS*), and beyond the latter there projects a small genital sternum (*gS*), from above which is protruded a large intromittent organ (*Pen*), open on the ventral surface (G). The structure is simpler in the female (J): the genital passage here opens (*Vul*) above the single sternum (*pgS*), which evidently is that of the pregenital segment. As already noted, there are in the embryo of *Scolopendra* (H), according to Heymons (1901), two small but distinct segments in the genital region between the segment of the last legs and the telson. The telson of the adult is a short simple lobe in each sex (F, I, J, *Tel*).

The head and mouth parts of *Otocryptops* are not essentially different from those of *Lithobius*. The Cryptopidae have no eyes, though eyes are present in *Scolopendridae*. On the underside of the head the epistome (fig. 60 C, *Epst*) covers a large area behind the bases of the antennae and carries the narrow labrum (*Lm*) on its posterior margin. The rest of the ventral surface of the head (B) is occupied by the mandibles (*Md*), the hypopharynx (*Hphy*), and the two pairs of maxillae (*1Mx*, *2Mx*). Below these parts lie the maxillipeds (D), the coxae of which are completely united ventrally in a large basal plate supporting the movable, four-segmented telopodites; the free distal margin of the coxal plate presents a pair of slightly rounded, sharp-edged lobes, but it is not toothed or spined. The claws are articulated laterally on the large basal segments of the maxilliped telopodites, the two intervening segments in each appendage being very small and incomplete laterally. The long poison gland of each claw extends into the base of the coxa.

218

The claw itself is seen to be composed of the united tarsus and pretarsus, since, as in *Lithobius,* the tendon of the flexor muscle is attached well within its base.

The second maxillae (fig. 60 B, *2Mx*) will be fully exposed on removal of the maxillipeds. The large coxal segments are connected by a median bridge, the telopodites are four-segmented, including a small, somewhat subapical claw on the end of the third segment. The first maxillae (*1Mx*) likewise have a common base, from which two soft, setigerous, median endites project forward between the bases of the thick, ventrally extended, two-segmented telopodites. The apical segments of the latter have broad, concave mesal surfaces and are ordinarily closed against each other like a pair of valves beneath the mandibles, but when the animal is experimentally stimulated in the mouth region, they move apart and the palpuslike telopodites of the second maxillae are then convulsively and repeatedly thrust between them. The movement is evidently a feeding reflex and suggests that the space between the first maxillary lobes is the outer entrance to the mouth, and that the food is pushed into it to the mandibles by the second maxillary palpi. No movement of the mandibles was observed.

The mandibles of *Otocryptops* are similar in form and structure to those of *Lithobius,* except that they are abruptly angulated instead of curved. The gnathal lobes close against the hypopharynx between them and lie against a pair of premandibular sternal plates of the head wall above them, which plates support the hypopharynx medially and give off posteriorly into the head cavity a pair of apodemes. Each mandible has a loose anterior articulation with the corresponding fultural plate, and posteriorly it is connected with the cranial margin by a delicate rod (fig. 60 C, *mdr*) in the membrane that suspends the mandible from the head. The mandibular musculature is the same as that of *Lithobius.*

A GEOPHILID

The geophilomorph chilopods are those slender centipedes with numerous body segments and short legs (fig. 61 A). In the number of segments and in details of general structure the Geophilomorpha are much less standardized than are the members of the other orders. The number of body segments in different species is said to vary from 35 to 181 and is not constant even within a single species. Cor-

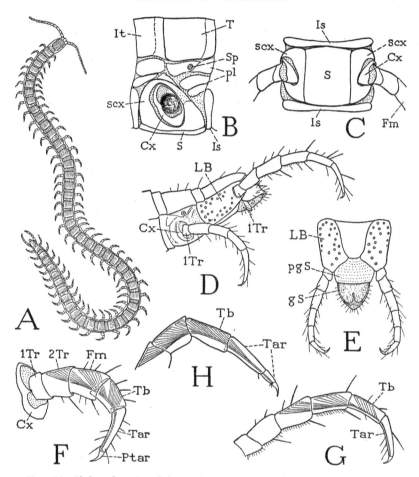

Fig. 61. Chilopoda—Geophilomorpha.

A, a geophilid, unidentified species. B, *Strigamia bothriopus* Wood, a middle segment of body and leg base, left side. C, same, ventral surface of a body segment. D, same species as A, posterior part of body and legs. E, same, end of body and last legs, ventral. F, same, typical leg segmentation. G, same, last leg, except coxa. H, *Scolopendra* sp., distal part of last leg, showing subdivision of tarsus, and pretarsal muscle.

For explanation of lettering see pages 223–224.

responding to each pair of legs are a main tergal and a sternal plate, but these plates are separated by narrower intertergites and intersternites throughout the length of the body. The order is divided into ten families, but since space cannot be devoted here to a description of the characters of the various forms, the following brief account is based on a species of the family Geophilidae, from which the order gets its name.

A typical geophilid is shown at A of figure 61. The specimen is 40 mm. long and has 50 body segments between the head and the genital region. The first dorsal plate behind the head (fig. 62 A, E, *1*) is the tergum of the maxilliped segment; the following terga belong to the 49 leg-bearing segments. The first of these (*2*) is a single plate, but the others are accompanied by intervening intertergites (fig. 61 B, *It*), which belong to the anterior parts of their respective segments; the last intertergite lies before the main tergal plate of the last leg-bearing segment (D). On the undersurface of the body (C) the main sternal plates (*S*) lie between the large leg bases and are separated by narrow intersternites (*Is*), which are more strictly intersegmental than are the intertergites.

The legs, except those of the last pair, are all alike, and each has the usual seven segments (fig. 61 F). The coxa (*Cx*) is a small ring at the base of the limb (B, C) and is almost surrounded by a large subcoxal sclerotization (*scx*), which is continuous ventrally (C) in a wide arc between the coxa and the sternum. The peripheral sclerotization of the leg base in the geophilid thus gives the best example among the arthropods of what might be regarded as a subcoxal segment of the leg, but it seems hardly probable that a primitive basal limb segment should be retained best developed in an otherwise highly specialized order. Intervening between the tergum and the subcoxal ring of the leg base on the side of the body segment (B) are a number of small sclerites, or thickened areas of the integument (*pl*), which may be regarded as true pleurites, or laterotergites. A spiracle (*Sp*) is present in this region on each leg-bearing segment but the first and the last.

On the last leg-bearing segment (fig. 61 D) the coxal and subcoxal areas of the limb base appear to be blended in a single large plate (*LB*) occupying the entire space between the tergum and the sternum. In the geophilids this plate is characteristically dotted with the dark-rimmed orifices of dermal glands. The free part of the terminal legs is somewhat longer than that of the other legs and has seven segments instead of six, and yet the small basal segment appears to be the first trochanter (*1Tr*) by comparison with the preceding legs. As already pointed out in the discussion of the similarly eight-segmented legs of *Otocryptops*, the extra segment is not a

221

subdivision of the tarsus, because the antepenultimate segment (G, *Tb*) contains a branch of the flexor muscle of the pretarsus and clearly corresponds with the tibia of the preceding legs (F, *Tb*). The tarsus of the last leg (G, *Tar*), furthermore, is identical with the tarsus of the other geophilid legs (F, *Tar*), and there is no suggestion of a tarsal subdivision such as that present in the legs of *Scolopendra*

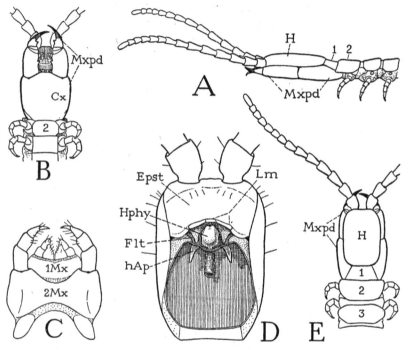

Fig. 62. Chilopoda—Geophilomorpha. Unidentified species shown at A of figure 61.

A, head and anterior part of body, lateral. B, same, ventral. C, first and second maxillae, ventral. D, head, ventral, mouth parts removed. E, head and anterior part of body, dorsal.

For explanation of lettering see pages 223–224.

(H). The eight-segmented leg, therefore, must have an extra segment interpolated in its proximal part, if the supporting plate (*LB*) includes the coxa as it appears to do.

The relatively small, slender head of the geophilid (fig. 62 A, *H*) rests on the large maxillipeds (*Mxpd*). Eyes are absent, and the antennae, each of 14 distinct segments, arise from the anterior end of the cranium, as in *Lithobius* and *Otocryptops*. On the undersurface of the head (D), behind the antennal bases, is a large epistomal region (*Epst*), which bears on its posterior margin, the narrow labrum (*Lm*). The geophilid labrum (*Lm*) differs from the labrum of the other orders in that it lacks a median tooth and is divided into a short median part, armed with a row of teeth, and two longer lateral parts. Premandibular sclerites of the ventral head wall (*Flt*) are present as in the other chilopods; their lateral arms are attached to the cranial margins but are mostly concealed by the underlapping edges of the epistome, their mesal arms clasp the sides of the hypopharynx (*Hphy*), and each sclerite gives off a short apodeme (*hAp*) extending posteriorly into the head. The head apodemes of the geophilomorphs, however, have no connecting ligament.

The true ventral surface of the head is almost entirely covered by the huge maxillipeds (fig. 62 B), the coxae of which are completely united in a large shield-shaped plate (*Cx*). The telopodites are four-segmented, and the points of the fangs project beyond the basal segments of the antennae. In each appendage the fang is articulated directly on the basal segment of the telopodite, the small second and third segments being incomplete laterally; the insertion of the long flexor muscle within the base of the fang shows that the latter is a composite segment formed by the union of the tarsus with the pretarsus, as in *Lithobius* and *Otocryptops*. The two maxillary appendages (fig. 62 C) are closely associated; in each pair the coxae are entirely united in a wide basal plate. The mandibles of the Geophilomorpha are highly variable in form; some of them resemble the mandibles of other chilopods, but in the geophilid here described each mandible has a wide, triangular anterior expansion and a narrow curved posterior extension. The adductor muscles are all attached directly on the head apodemes.

Explanation of Lettering on Figures 53–62

a, posterior articulation of mandible.
af, antennifer.
An, anus.
Ant, antenna.
Atr, atrium of spiracle.

c, anterior, epipharyngeal articulation of mandible.
Cx, coxa.

cxnd, coxal endite.

d, articulation between coxa and subcoxal sclerites.
Dct, duct.

e, e', dorsal subcoxal sclerites, "katopleure."
E, eye.

223

Ephy, epipharynx.
Epst, epistome.

f, precoxal fold, "anopleure."
flptar, flexor muscle of pretarsus.
Flt, fultura of hypopharynx, premandibular sternal sclerite of head.
fltar, flexor muscle of tarsus.
Fm, femur.

g, postcoxal sclerite, "coxopleure."
Gld, gland.
gnL, gnathal lobe of mandible.
Gpd, gonopod; *1Gpd*, *2Gpd*, first and second gonopods.
Gpr, gonopore.
gS, sternum of genital segment.
gSeg, genital segment.

H, head.
hAp, head apodeme.
Hphy, hypopharynx.

i, ventral articulation of coxa.
Is, intersternite.
It, intertergite.

L, leg.
LB, limb base, union of coxa and subcoxal sclerites.
Lg, intergnathal ligament of head apodemes.
Lm, labrum.

m, marginal sclerite of cranium.
mb, membrane.

Md, mandible.
mdB, base of mandible.
mdP, mandibular pouch.
mdr, articular rod of mandible.
Mth, mouth.
1Mx, first maxilla.
2Mx, second maxilla.
Mxpd, maxilliped, poison claw, first leg.

Pen, penis.
pgS, sternum of pregenital segment.
pgSeg, pregenital segment.
pgT, tergum of pregenital segment.
pl, pleurites.
Ptar, pretarsus, dactylopodite.

S, sternum.
scx, subcoxal sclerites.
Seg, body segment.
Sp, spiracle.
Stom, stomodaeum.

t, tendon.
T, tergum.
Tar, tarsus.
Tb, tibia.
Tel, telson.
Tlpd, telopodite, part of leg beyond coxa.
Tr, trochanter; *1Tr*, first trochanter, *2Tr*, second trochanter or prefemur.

Vul, vulva.

w, epipharyngeal arm of fultural sclerite of hypopharynx.

224

* VIII *

THE DIPLOPODA

THE diplopods are distinguished from all other arthropods by the fact that most of their body segments are "double" and, as implied in the group name, carry each two pairs of legs. The double nature of the segments is evident not only from their two pairs of legs but also in the presence of two pairs of ganglionic centers in the ventral nerve cord, two pairs of lateral pouches of the male gonads, and two pairs of cardiac ostia and alary muscles. The double segments are termed *diplosomites.* Since it would seem most probable that the diplopods have been derived from ancestors having single segments, it may be supposed that the diplosomites represent two primitive segments united; in their ontogenetic development, as shown by Pflugfelder (1932), the double segments are formed from two consecutive sets of somite rudiments that are not differentiated into separate segments. It is not clear what advantage the diplopods have gained from having two pairs of legs on each functional segment, or, otherwise stated, in having the segments united in pairs. By comparison with the agile chilopods, they are sluggish creatures, their gait is a leisurely crawl, and they make no effort to escape capture. Their curious organization, however, gives the diplopods a distinctive place among the arthropods.

In certain features of the head and the mandibles the diplopods are evidently related to the chilopods, while, on the other hand, there is little doubt that they are allied to the pauropods and the symphylids. The four groups, therefore, have generally been classed together as the Myriapoda. The Diplopoda, Pauropoda, and Symphyla, however, differ from the Chilopoda in that the genital openings are near the anterior end of the body, instead of at the posterior end. They are hence distinguished as the Progoneata, or progoneate myriapods, in contrast to the opisthogoneate chilopods. It is possible, however, that the progoneate condition is not primary, since Tiegs (1940) has shown that there is evidence in the ontogenetic development of the symphylids of a primitive posterior connection of the genital ducts with the exterior. The definitive anterior exit ducts, he says (1947), in both Pauropoda and Symphyla are not persisting coelomoducts but ectodermal ingrowths.

The class Diplopoda includes a large number of species, which taxonomists divide into three principal groups: the Pselaphognatha, the Opisthandria, and the Proterandria. The Pselaphognatha are minute forms comprised in the family Polyxenidae, having not more than 13 body segments, and are recognizable by the presence of large tufts of hair on the back and sides of the body. The Opisthandria are of larger size, and the males are characterized by having the last pair or the last two pairs of legs modified as clasping organs by which the male in copulation holds the female, while he places the sperm in her genital openings by means of his mandibles. The Proterandria include most of the more common types of diplopods; their distinguishing feature is the modification in the male of one pair or both pairs of legs, usually on the seventh body segment, to serve as the functional intromittent organs, termed *gonopods*, by which the sperm is transferred from the male genital openings on the third body segment to the receptacles of the female on the corresponding segment. The Proterandria are divided into four orders: the Polydesmoidea, the Nematophora, the Juliformia, and the Colobognatha. Representatives of the Polydesmoidea and the Juliformia will be described here as examples of the more familiar types of diplopods. For a full account of the structure and classification of the diplopods the student is referred to the comprehensive works of Silvestri (1903), Verhoeff (1926–1931), and Attems (1926b).

The diplopods feed principally on decaying vegetable matter, particularly on the dead leaves of certain kinds of trees and shrubs; fresh leaves and fruit they seldom eat, and carnivorous habits have been observed only in members of one family, which are said to attack earthworms and phalangids. The alimentary canal is a wide, simple tube extending straight through the body, and has a single pair of excretory Malpighian vessels arising from the anterior end

226

of the proctodaeum. Salivary glands of the anterior part of the head open into the preoral cavity through the epipharyngeal wall, and glands in the anterior region of the body discharge on the base of the inner wall of the gnathochilarium.

A distinctive feature of the diplopods is the nature of the respiratory system. The breathing organs are bundles of tracheae given off from hollow sternal apodemes that are open to the exterior by apertures called spiracles, closely associated with the bases of the legs. Descriptions of the apodemes and tracheae of various diplopod genera will be found in papers by Effenberger (1907), Ziegler (1907), Krug (1907), Wernitzsch (1910), Reinecke (1910), Voges (1916), Ripper (1931), and Seifert (1932). In general, the tracheae are fine, unbranched tubes, with spiral taenidia, and do not unite with each other, but in the Polyxenidae and Glomeridae they are branched. The apodemes, which are variously developed, serve also for the attachment of the leg muscles, and usually the apodemes of the first three body segments have no tracheae. It seems probable that the apodemes were developed originally for muscle attachments; the spiracles of the diplopods evidently can have no relation to the lateral spiracles of the chilopods and insects.

In their mode of development the diplopods are anamorphic, as are the Scutigeromorpha and Lithobiomorpha among the chilopods. The young of most species leave the egg with only three pairs of legs, which are the first three pairs of the adult, and have not more than seven body segments. The definitive segmentation is acquired gradually at the subsequent moults, accompanied by an increase in the number of legs. The new segments are formed in a zone of growth at the posterior end of the body, which remains active throughout life. Inasmuch as adult diplopods of different groups differ much in the number of body segments, the anamorphic progression of segment formation is not the same in all of them. A detailed account of the manner of segment generation in the zone of growth is given by Pflugfelder (1932). The postembryonic development of *Julus* is described by Krug (1907), that of several species of Polydesmoidea by Miley (1927), Pflugfelder (1932), and Seifert (1932), and a general account of diplopod anamorphosis is given by Attems (1926b). Each successive stage of growth is accompanied by a moult, preliminary to which the young diplopod encloses itself in an earthen nest. At each moult the head capsule of the exuviae is

227

disarticulated from the body, and ecdysis is accomplished by the animal's crawling out of the old cuticle of the trunk. Miley (1927) observes that in *Euryurus* the shed cuticle is "split down the mid-ventral line and also somewhat laterally, just above the articulation of the legs on one side or on both sides."

The Head and the Mouth Parts

The diplopod head (fig. 63) is strongly convex above (D) and flattened beneath. Eyes are present in some forms (C, D) and absent in others (A). The antennae arise well back from the anterior margin, so that before them is a wide epistomal area (A, D, *Epst*), but the latter is in no way specifically separated from the rest of the cranial wall. With the margin of the epistome is united the narrow, dentate labrum (A, C, *Lm*). On the sides of the head are the large, convex bases of the mandibles (A, B, C, D, *mdB*), and the ventral surface of the head is covered by a broad flat appendage known as the *gnathochilarium* (B, D, *Gnch*). The antennae are seven-segmented appendages (A, *Ant*), as are the legs, and each segment is individually musculated. The eyes, when present, have superficially the appearance of compound eyes (C, D), but they are merely groups of simple eyes like those of the chilopods, as seen at F, G, and H on figure 69.

Postantennal *organs of Tömösvary* are present in most of the diplopods, except Juliformia and Colobognatha. In some forms the sensory cells lie beneath open pits or grooves; in others they are entirely covered by the cuticle. In the polydesmoids each organ is marked externally by an oval groove of the cuticle with a thickened central disc (fig. 63 A, *OT*); the sense cells lie beneath the groove and have no communication with the exterior.

Beneath the epistome (fig. 63 F, *Epst*) is a long palatal, or epipharyngeal, surface (*Ephy*) which extends posteriorly from the labrum (*Lm*) to the mouth (*Mth*) and forms the roof of a large preoral food cavity (*PrC*) closed ventrally by the gnathochilarium. At the base of the inner surface of the gnathochilarium is a group of flattened suboral lobes that constitute the hypopharynx (*Hphy*), which in the diplopods does not have the form of a projecting tongue-like organ, such as that of the chilopods and insects. The preoral cavity is occupied mostly by the gnathal lobes of the mandibles, which close against the hypopharynx. The epipharyngeal surface usually presents a transverse ridge behind the labrum (E) and a low median elevation.

The segmental composition of the diplopod head is a subject on

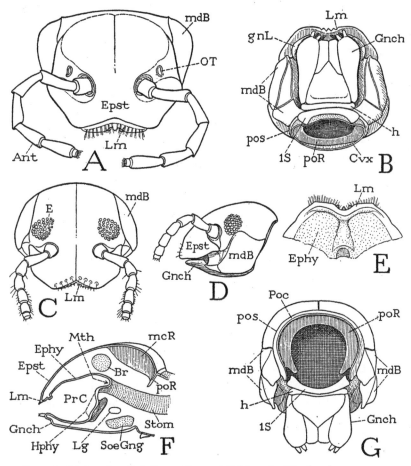

Fig. 63. Diplopoda—Polydesmoidea and Juliformia. The head.
A, *Apheloria coriacea* (Koch), head, anterior. B, *Habrostrepus* sp., head, ventral. C, *Arctobolus marginatus* (Say), head, anterior. D, same, head, lateral. E, *Fontaria virginiensis* (Drury), preoral epipharyngeal surface and labrum, ventral. F, same, diagrammatic longitudinal section of head. G, *Habrostrepus* sp., head, posterior.
For explanation of lettering see pages 248–249.

which there is difference of opinion. The principal question is whether or not it contains a second maxillary segment. The anatomy of the adult head throws little light on the matter. On the back of the head is a narrow postoccipital arch (fig. 63 G, *Poc*), set off by a postoccipital sulcus (*pos*), which forms a large internal postoccipital ridge (*poR*) for the attachment of muscles from the body. The pattern of the structure is thus almost identical with that of the back of an orthopteroid insect head (fig. 78 B), which presumably contains the second maxillary segment. In the diplopod, however, the base of the mandible (fig. 63 B, D, G, *mdB*) comes back to the postoccipital sulcus, apparently leaving no room for more than one maxillary segment. The subject will be further discussed in connection with the gnathochilarium.

The mouth parts of the diplopod include the mandibles, the gnathochilarium, and the hypopharynx. The diplopod mandible (fig. 64 A, B) is the most remarkable type of jaw found among the arthropods. It consists of a large, convex basal part (*mdB*) and of a movable, independently musculated gnathal lobe (*gnL*) articulated on the base. As already noted, the basal part is set on the side of the head (fig. 63 D, *mdB*) between the margin of the cranium and the gnathochilarium. It consists usually of two distinct plates, one anterior, the other posterior, and the former may have secondary divisions. The mandibular base looks so much like a part of the head wall that it has been regarded as being the "pleuron," or "pleurites," of the head; German writers commonly call it the *Backen,* or "cheek." That the part in question, though little movable, is the true base of the mandible, however, is evident from the insertion, within its cavity (fig. 64 A), of huge muscles that clearly represent the ventral adductors of the mandibles of other arthropods. One or two groups of these muscles (D, *4b*) go into each jaw from opposite ends of a thick, intergnathal ligament (*Lg*), and another posterior group (G, H, *4a*) arises on the head apodeme (*hAp*) of the same side. A muscle inserted distally on the dorsal margin of the mandibular base arises posteriorly on the cranium.

The gnathal lobes of the mandibles (fig. 64 A, *gnL*) are articulated on the distal ends of the basal plates, from which they turn mesally within the preoral cavity, where they are mostly concealed from below (fig. 63 B, *gnL*) by the gnathochilarium. Each lobe is strongly but variously toothed at its apex (fig. 64 A, B, C) and has a rasping surface on the basal part of its mesal margin. The two opposed, independently movable lobes constitute the functional jaws of the diplopod. Each is provided with a flexor muscle of great size (B, *1*)

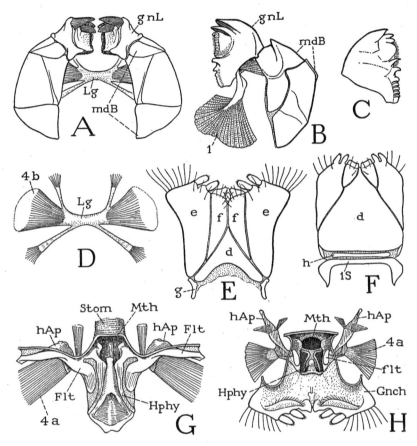

Fig. 64. Diplopoda—Polydesmoidea and Juliformia. The mouth parts.
A, *Apheloria coriacea* (Koch), mandibles, ventral. B, *Arctobolus marginatus* (Say), left mandible, with muscles of gnathal lobe, ventral. C, same, gnathal lobe of mandible, dorsal. D, *Fontaria virginiensis* (Drury), intermandibular ligament and muscles. E, *Apheloria coriacea* (Koch), gnathochilarium. F, *Arctobolus marginatus* (Say), gnathochilarium and first segmental sternum. G, same, hypopharynx and futural sclerites (*Flt*) bearing head apodemes (*hAp*), dorsal. H, *Apheloria coriacea* (Koch), dorsal surface of gnathochilarium, with hypopharynx, showing reduced futurae (*flt*) supporting long head apodemes.
For explanation of lettering see pages 248–249.

arising dorsally on the cranium and inserted by a thick, rigid tendon on its mesal basal angle. In some forms at least, a second, smaller muscle of the lobe arises in the distal part of the mandibular base. It is thus interesting to note that the musculature of the freely movable gnathal lobe of the diplopod mandible corresponds exactly

231

with that of the flexible lobe of the chilopod mandible (fig. 59 E), suggesting that the diplopod jaw could have been developed from a mandible like that of the chilopods.

In describing the act of feeding by *Strongylosoma pallipes* Oliv., Seifert (1932) says the contraction of the muscles of the *Backen* (mandibular bases) pulls the latter mesally, and with them the gnathal lobes, which work independently by their own muscles. When this diplopod feeds on dry leaves, the food is first flooded with a secretion from the mouth that softens the surface between the leaf veins so that it is easily scraped or pulled off. The first three pairs of legs may serve as accessory feeding organs, small pieces of leaves being held by them and finally thrust between the mandibles and the gnathochilarium by the first pair.

The gnathochilarium is the functional underlip of the diplopod. It is attached by its base to the posterior part of the undersurface of the head (fig. 63 B, *Gnch*). The sclerotic ventral surface of the organ in most of the diplopods is divided by grooves into several characteristic parts (fig. 64 E, F), including a triangular median plate (d) known as the *mentum,* two lateral lobes (E, *e, e*) called the *stipites,* bearing each a pair of apical papillae, and a pair of median distal plates (f, f) termed the *laminae linguales.* In some species there is a narrow transverse sclerite, the *prebasilare* (F, h) intervening between the mentum and the first body sternum ($1S$), and at the sides, proximal to the stipites, are two small sclerites termed the cardines (E, g). The nomenclature of the parts of the gnathochilarium is based on the commonly held opinion that the organ is a composite of the usual first maxillae and the labium, or second maxillae. Students of the development of the gnathochilarium, however, are not in agreement as to its composition: some claim that it is formed by the union of both pairs of maxillae, others contend that it contains only the appendages and a sternal plate of the first maxillary segment, while still others assert that it is formed of the second maxillae alone, the first maxillae being suppressed during development. According to Silvestri (1903), the lines in the outer wall of the typical gnathochilarium are differentiated secondarily during embryonic development, and in the pselaphognathid *Polyxenus* the gnathochilarium is shown by Reinecke (1910) to be a simple structure without subdivisions.

The gnathochilarium is supported at its base on a ventral plate,

232

commonly called the "hypostome," which is the sternum of the first body segment (fig. 64 F, *1S*). This segment, however, has been shown by Silvestri (1903, 1950) to be the second maxillary, or labial, segment of the chilopods and insects, and to have no appendages in the diplopods. The gnathochilarium, therefore, according to Silvestri, is a product of the union of the first maxillae, together with a sternal plate of their segment, and the diplopod head does not contain a second maxillary segment. We shall later see that, according to Tiegs (1947), the same is true of the pauropod head.

The hypopharynx of other arthropods is represented in the diplopods by the assemblage of flat lobes or folds of the integument situated in a depressed suboral area at the base of the inner wall of the gnathochilarium (fig. 64 G, H, *Hphy*). Typically there are a median lobe and two lateral lobes, which assume different shapes and proportions in different species. A supporting apparatus of the hypopharynx, consisting of a pair of transverse fultural sclerites attached laterally on the cranial margins, is fully developed in some of the diplopods, as in the spiroboloid shown at G (*Flt*). In the polydesmoid (H), however, the fulturae are reduced to small plates (*flt*) at the sides of the hypopharynx. In either case the sclerites support a pair of head apodemes (*hAp*) that give attachment to the second adductor muscles of the mandibles (*4a*). In the spiroboloid (G) the apodemes are short and thick; in the polydesmoid (H) they are elongate arms projecting posteriorly and dorsally in the head, with their apices attached to the central discs of the organs of Tömösvary (fig. 63 A, *OT*). Midway on each of these apodemes are attached two small muscles arising on the dorsal wall of the cranium, so that in this case the apodemes would appear to act as levers for lifting the hypopharynx. The presence of fultural sclerites of the hypopharynx bearing the head apodemes in Chilopoda, Diplopoda, and Pauropoda (see Tiegs, 1947) is a strong point in evidence of a relationship between these three myriapodous groups.

The Body Segments and Legs of a Polydesmoid Diplopod

The body of a polydesmoid diplopod (fig. 65 A) consists of 20 segments, as indicated by the number of back plates. The first segment behind the head and the last two segments are legless. The second, third, and fourth segments have each one pair of legs, the others bear each two pairs, including in the male the appendages of the first pair on the seventh segment, which are modified for reproductive purposes and become the gonopods. The consecutive segments are freely movable on each other, since they are united by wide, conjunctive membranes, but the conjunctives are ordinarily not visible, since normally the anterior part of each segment is deeply inserted into the posterior part of the segment preceding. Inasmuch as the double segment is the characteristic feature of the diplopod, it will be appropriate to understand first the structure of a diplosomite, preferably one from the middle part of the body.

Each diplosomite of *Apheloria coriacea* is a continuously sclerotized annulus with no sutural divisions between the tergal, pleural, and sternal areas (fig. 65 D, E). The segment, however, is demarked into an anterior ring and a posterior ring by an encircling groove a little before the middle. The anterior ring is termed the *prozonite* (*Prz*), the posterior ring the *metazonite* (*Mtz*). The dorsal surface of the metazonite is extended on each side in a broad lateral lobe (B, D, E); the long lateral surfaces of the segment between the tergal lobes and the bases of the legs (E, *Pl*) may be regarded as the pleura, or pleural areas, and the ventral wall carrying the legs the sternum (*S*). Since both pairs of legs are attached on the metazonite (D, C), it is evident that the two annular parts of the diplosomite do not represent the component segments. The prozonite has a smooth polished surface and fits into the metazonite of the preceding segment, thus permitting the characteristic ventral flexion or curvature of the body. Near the margin of each tergal lobe on most of the segments is a small pore (B, D, E, *rp*), which is the outlet of a *repugnatorial gland;* the pores are present on segments 5, 7, 9, 10, 12, 13, and 15 to 19 (figs. 65 A, 67 A). When the animal is irritated, the glands discharge a liquid with an acrid odor, said to contain free hydrocyanic acid, which is deadly to most other small arthropods in confined spaces.

On the undersurface of a diplosomite (fig. 65 C) it is seen that the two legs on each side arise close together and that those of the second pair are somewhat nearer the mid-line. In front of the outer angle of each leg base is a conspicuous oval pit (*aSp, pSp*). These pits lead into hollow cuticular ingrowths of the body wall (F, *aAp, pAp*) that serve both as apodemes for muscle attachments and as air passages for tracheae opening from them. The external

234

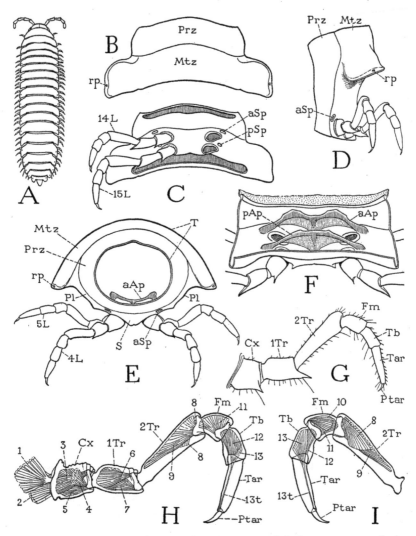

Fig. 65. Diplopoda—Polydesmoidea. Structure of body segments and the legs.

A, a typical polydesmoid diplopod. B, *Apheloria coriacea* (Koch), tergum of tenth segment. C, same, undersurface of tenth segment. D, same, lateral view of a diplosomite. E, same, fifth segment, anterior. F, same, inner surface of ventral wall of a diplosomite, showing ventral apodemes. G, same, a left leg, anterior. H, *Euryurus* sp., left leg and musculature, anterior. I, same, distal part of leg, posterior.

For explanation of lettering see pages 248–249.

openings of the apodemes, therefore, are the breathing apertures, or spiracles, of the diplopod; they are not simple holes, each orifice being guarded by a grating of branched processes that admits air but keeps out solid particles.

An examination of the internal surface of the ventral wall of a diplosomite of *Apheloria* (fig. 65 F), when the muscles and other soft tissues have been removed by caustic, shows that it is crossed by two apodemal arches (*aAp, pAp*) between the opposite pairs of external pits. Each apodeme consists of a thick, arcuate anterior bar and a thin posterior lamella strengthened medially by a thickening from the marginal bar. The two apodemes are inclined forward, so that the posterior one partially overlaps the one in front. The lateral parts of the apodemal arms are hollow air chambers, from which numerous delicate tracheae arise in bundles (fig. 66 D, *Tra*). Since the apodemes give attachment to leg muscles and are present in the first three body segments, which have no tracheae, and inasmuch as many other arthropods have sternal apodemes for muscle attachment, it seems probable that the respiratory function of the diplopod apodemes is secondary.

The legs of *Apheloria*, as in most of the diplopods, are uniformly seven-segmented (fig. 65 G), and the tarsi are undivided, but the identity of the segments has been differently interpreted by different writers. Silvestri (1903) regarded the basal segment as a subcoxa, the second segment as the coxa, and the third as a trochanter. On the other hand, according to Attems (1926b), the first segment is the coxa, the second the prefemur, or trochanter, the third the femur, and the fourth a postfemur. Hansen (1930) calls the fourth segment the tibia and the fifth the cotibia. However, if we simply follow the usual nomenclature of the arthropod leg segments, beginning with the coxa, the segmentation of the diplopod leg will be as given at G and H of figure 65. The long third segment in the leg of *Apheloria* (G) or *Euryurus* (H) thus becomes the second trochanter, though it resembles the femur of an insect leg, but in other diplopods the leg segments may be subequal in length, and in the arthropods generally the relative length of a leg segment is no criterion of its identity. Between the coxa and the first trochanter there is sometimes present a small ring, but no muscles arise within it or are inserted on it, so that there is nothing to suggest that the ring represents a true segment.

236

The musculature of the diplopod leg is fairly simple. The limb as a whole turns forward and backward on the base of the coxa, and Silvestri (1903) has given a comparative account of the coxal muscles, all of which arise on the corresponding sternum and the sternal apodemes, but in some forms some of the muscles cross over from the apodeme of one side to the coxa of the opposite side. Each of the first four segments in *Euryurus* (fig. 65 H, I) following the coxa has levator and depressor muscles, but the long tarsus has only a depressor, or flexor (*12*). The simple, clawlike pretarsus (*Ptar*) has a single muscle, the usual flexor (*13*), arising in the tibia, with its fibers inserted on a long tendon (*13t*) traversing the tarsus from its attachment on the base of the pretarsus.

If we now turn to the anterior part of the body of *Apheloria*, it will be seen that several of the segments following the head differ from the diplosomites and, in various ways, from one another. The first back plate immediately behind the head (fig. 66 A) is known as the collar, or *collum* (*Col*). On the ventral surface below the collum is a small but well-developed sternal plate (F, *1S*) bearing a pair of lateral apodemes (*Ap*); its extended anterior angles are connected with the posterior margin of the head by two crescent-shaped cervical sclerites (*cvpl*). The collum and this associated sternal plate will be here treated as representing a legless first segment of the body.

The first pair of legs (fig. 66 G, *1L*) are suspended from the anterior margin of a second small sternal plate (*2S*) lying free in the ventral membranous integument behind the first sternum. This plate bears a pair of typical sternal apodemes (*Ap*) connected by a transverse bridge, but the apodemal pits lie posterior to the leg bases instead of anterior to them as in the diplosomites. Corresponding to this first leg-bearing sternum is a small tergal plate (B, *2T*) behind the collum, which has large lateral lobes, from which descend normal pleura (C, *Pl*), but, instead of joining the sternum, the pleura are connected with each other ventrally by a narrow bridge *behind* the sternum.

The third body segment (fig. 66 H) carries the second pair of legs and is the segment of the genital openings in both sexes. The sternum is a distinct plate lying between the lower ends of the prozonite of the back plate, but the lower ends of the pleural areas of the metazonite are united in a narrow bridge behind the sternum. The sternum bears a pair of long tapering apodemal arms (*Ap*) connected by a transverse bar with a triangular median extension. There are no tracheae that have their origins in the first three segments, these segments and the head being tracheated from the fourth segment. The second pair of legs (H, *2L*) are carried by the sternum of the third segment. In the male of *Apheloria* the coxa of each of these legs bears mesally a small genital papilla, or penis (*Pen*), through which a vas deferens (*Vd*) opens to the exterior after turning downward anterior to the apodemes.

In the female the venter of the third segment contains a pair of deep pockets lying behind and somewhat to the sides of the coxal bases. These pockets are closed posteriorly by a narrow poststernal bridge of the pleura; they contain the external female genitalia, which are known as the *vulvae*. When a vulva is protracted, or removed from its pouch (fig. 66 I), it is seen to consist of a short, thick, cylindrical basal stalk bearing a pair of soft, apical, valvelike lobes (*vlv*). Covering the lateral ends of these two transversely elongate vulval lobes is a smaller opercular lobe (*opl*) arising from the side of the stalk. Beneath the operculum is the large opening of the oviduct; it will be freely exposed when the operculum, which covers it like a lid, is lifted. A fourth, larger sclerotized lobe projects mesally from a fold of the pouch wall around the base of the vulva and forms a hoodlike covering over the vulval lobes when the latter are retracted. At the bottom of the depression between the vulval lobes (J) is a median groove that opens into a thick-walled, serpentine sperm receptacle.

The fourth body segment (fig. 66 D) carries the third legs and is the last segment with a single pair of appendages; otherwise it has the typical segmental structure. The tergum (*4T*) is produced into lateral lobes, below which the pleura converge ventrally and are continuous with the sternum; the coxal foramina are entirely enclosed in the ventral sclerotization (E). The apodemal pits of the fourth segment are the first pair of spiracles (*Sp*) and are particularly large; the tracheae (D, *Tra*) arising from the hollow lateral parts of the apodemes are distributed to the first four body segments and the head.

The next two segments, the fifth and the sixth, are typical diplosomites, each bearing two pairs of legs and containing two separate sternal apodemes.

238

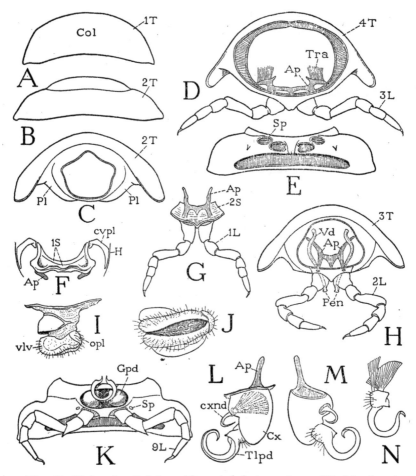

Fig. 66. Diplopoda—Polydesmoidea. *Apheloria coriacea* (Koch). Anterior body segments and gonopods.

A, first tergum, collum. B, second tergum. C, second segment, anterior, sternum (G, 2S) removed, pleura united ventrally behind sternum. D, fourth segment, posterior. E, same segment, ventral. F, first sternum, ventral. G, second sternum with first legs, anterior. H, third segment of male, with second legs, showing genital outlets on coxae, anterior. I, right vulva of third segment of female, posterior. J, same, ventral. K, seventh segment of male, ventral, with gonopods and legs. L, left gonopod, anterior. M, same gonopod, posterior. N, telopodite of gonopod, detached with muscles.

For explanation of lettering see pages 248–249.

The seventh body segment is of particular interest in the male because it is on this segment in most of the proterandrious diplopods that one pair or both pairs of legs are modified in structure to serve for transfer of the sperm from the genital openings on the third segment of the male to the sperm receptacles of the female. The modified appendages, as already noted, are known as *gonopods*. In *Apheloria*, as in other Polydesmoidea, only the first pair of appendages on the seventh segment are gonopods (fig. 66 K, *Gpd*), the second pair being ordinary seven-segmented legs (*9L*). In the female the same segment has the usual two pairs of legs. The gonopods of different species present a great variety of structure; their characters are much used by taxonomists for species identification.

The gonopods of *Apheloria coriacea* are sunken into a deep, transversely oval depression before the legs on the venter of the seventh segment (fig. 66 K, *Gpd*). Each gonopod (L) has a large, thick, conical basal segment (*Cx*), probably the coxa, from which an apodemal arm (*Ap*) projects into the body cavity. Distally, on its mesal surface, the coxa bears a long, sickle-shaped process, regarded as the telopodite (*Tlpd*), which tapers to a sharp, recurved point. Proximal to the base of the telopodite is a coxal endite in the form of a strongly bent hook (*cxnd*) set into an oval membranous area of the coxal wall (M). The telopodite is movable by antagonistic muscles (N) arising on the posterior wall of the coxa; the hook is provided with a single large fan-shaped muscle (L) arising on the anterior coxal wall. The mesal surface of the telopodite is almost flat, but it is traversed by a minute canal extending to the tip from a depression on its base. In the natural position (K) the telopodites of the opposite gonopods curve mesally and forward with their ends together or overlapping. Between them they enclose a cuplike space, the diameter of which is just sufficient to receive the two penes of the second legs (H, *Pen*), so that the apertures of the penes can be applied to the cavities at the inner ends of the telopodite canals, which are the sperm conduits of the intromittent apparatus. Copulation and the process of sperm transfer in two other polydesmoids, *Strongylosoma pallipes* and *Polydesmus edentulus,* are fully described by Seifert (1932). By bending the anterior part of the body downward, the male diplopod brings the penes on the second legs into contact with the gonopods on the seventh segment, and the sperm is discharged into the cavities on the bases of the telopodites, where it is mixed with a secretion from glands in the gonopods. During copulation, then, the sperm is discharged through the sperm canals of the gonopods into the sperm receptacles in the vulvae of the female, probably being activated by the copious secretion of the vulval lobes. That the telopodites and the coxal hooks also play an active role in copulation is indicated by their strong musculature.

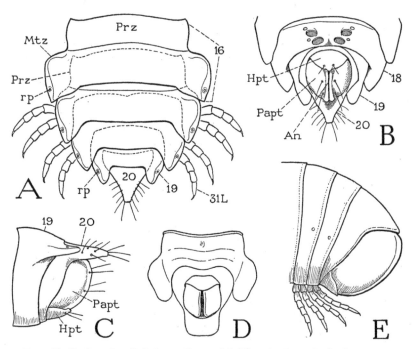

Fig. 67. Diplopoda—Polydesmoidea and Juliformia. Posterior body segments. A, *Apheloria coriacea* (Koch), last five body segments, dorsal, with overlapped prozonites indicated by dotted lines. B, same, terminal segments, ventral. C, same, terminal segments, lateral. D, *Euryurus* sp., anal and preanal segments, ventral. E, *Arctobolus marginatus* (Say), terminal segments, lateral.

For explanation of lettering see pages 248–249.

According to Seifert (1932), the females of *Strongylosoma pallipes* lay their eggs in small round chambers constructed in the earth, the walls of which are glazed, apparently with a secretion from the anal glands. In each chamber 40 to 50 round, ivory-white eggs are deposited, after which the nest is carefully closed. Two or three nests

241

may be made in succession by the same female to accommodate eggs ripening later in the ovaries. Other polydesmoids, however, may construct nests on objects above ground.

The next 11 segments of *Apheloria* are ordinary diplosomites and are essentially all alike in structure, but those beyond the sixteenth segment become successively narrower (fig. 67 A), and the body somewhat abruptly tapers posteriorly to a point. Segments 19 and 20 are legless (B, C). Segment 19 is a simple ring with relatively large tergal lobes, in which are the apertures (A, *rp*) of the last pair of repugnatorial glands. Segment 20 is produced dorsally into a tri-angular median lobe that forms the apical point of the body; its side margins bear a pair of valvelike lateral anal lobes, or *paraprocts* (B, C, *Papt*), and from its sternal arc projects a rounded subanal lobe, or *hypoproct* (*Hpt*). The anus lies in the vertical cleft between the lateral lobes. The twentieth segment, together with the anal lobes, probably represents the telson, since in the diplopods gener-ally the zone of growth lies in the preceding segment.

The Body Segments of a Juliform Diplopod

The order Juliformia includes the cylindrical diplopods with long flexible bodies, numerous segments, and relatively small legs; when disturbed they coil themselves ventrally. Most of them have eyes (fig. 63 C, D), and there is a narrow sclerite, the "prebasilare," be-tween the gnathochilarium and the first body sternum (fig. 64 F, *h*). On the seventh segment of the male are usually two pairs of gono-pods, but if only one pair is present there are no legs. The order includes three suborders: the Juloidea, the Spiroboloidea, and the Spirostreptomorpha. The following description is based mostly on the spiroboloid *Arctobolus marginatus* (Say), common in the east-ern and southern parts of the United States.

The body of *Arctobolus marginatus* consists of about 50 segments. The segments beyond the second are short rings (fig. 68 A), some-what wider above than below, and are but little inserted into each other; they are not differentiated into a prozonite and metazonite as in the polydesmoids, and the animal coils itself by a telescoping of the ventral parts of the segments rather than a separation of the dorsal arcs. The spiroboloids differ from other diplopods in that the fourth and fifth pairs of legs, as well as those preceding, are borne each on a separate segment (A, *4L*, *5L*). The first segment

242

having two pairs of legs is the sixth, which carries the sixth and seventh legs. The next segment is that of the gonopods in the male and has two pairs of ordinary legs in the female.

The first body tergum is the collum (fig. 68 A, *Col*), a large, hoodlike plate extending over the back of the head as far as the eyes. The sternum of the collum segment (fig. 64 F, *1S*) is a smaller and simpler plate than that of *Apheloria* and is more closely associated with the base of the gnathochilarium and the back of the head. The first legs hang from beneath the lower angles of the collum (fig. 68 A), but they are attached on a V-shaped second sternal plate (B, *2S*) suspended in the membranous ventral integument below the collum, and for this ventral plate there is no corresponding tergal plate. The legs of the second pair (C, *2L*) hang from a similar third sternum (*3S*), but this sternum is attached to the anterior margin of the second tergal plate (*2T*), which is expanded ventrally on the sides of the body (A), and the pleural areas of the segment are inflected mesally as a pair of flat triangular lobes (C, *Pl*), the points of which approach each other behind the coxae but do not unite. It is evident, therefore, that in this diplopod the second segment is incomplete in that it has no tergum, and that the actual second tergum (A, *2T*) pertains to the third segment. The first two pairs of legs are closely associated with the head.

In the male of *Arctobolus* the vasa deferentia come together and open into a depression behind the coxal bases of the second pair of legs, but there are no penes. In some of the Juliformia, as in *Habrostrepus* (fig. 68 D), a pair of penes (*Pen*) arises posterior to the bases of the coxae, and in others, as in *Parajulus* (E), there may be only a single median penis. The first and second legs of *Parajulus* (*1L, 2L*) and related forms are modified in a very unusual manner.

The third, fourth, and fifth segments of *Arctobolus* are simple rings (fig. 68 A), each bearing a single pair of ordinary walking legs. The coxae of these legs (F) are produced into apical lobes. The sixth segment is the first diplosomite with two pairs of legs (A), and the seventh is the segment of the gonopods in the male. It will be noted that there are the same number of tergal plates (six) anterior to the gonopod segment in the spiroboloid as there are in the polydesmoid *Apheloria*, and also the same number of legs (seven), but that in the polydesmoid the fourth and fifth legs are carried by a single segment. On the other hand, the polydesmoid has a complete

243

segment for the first legs, which segment in the spiroboloid is not represented by a tergal plate. While the number of apparent segments in the two forms is thus the same, the segments are not entirely identical.

The gonopods of *Arctobolus marginatus* are ordinarily concealed in a deep cavity on the venter of the seventh segment between a posterior sternal bridge and a transverse fold of the integument behind the preceding segment. When they are protracted, or removed (fig. 68 H), it is seen that an anterior V-shaped sternal plate (S) lies in front of the gonopods and is closely associated with them. The first gonopods (*1Gpd*) are attached to the ends of the sternal arms and clasp the bases of the second gonopods (*2Gpd*). Each first gonopod (I) is two-segmented; the basal segment, or coxa (*Cx*), is produced into a broad endite (*cxnd*); the distal segment, or telopodite (*Tlpd*), is an arm of irregular shape muscularly movable on the coxa. The second gonopod (J, K) is larger than the first, but its basal part is normally concealed in the sternal cavity. The coxa (*Cx*) runs out dorsally into an apodemal extension (*Ap*) and bears ventrally the telopodite (*Tlpd*), which is a large, thick arm with a bifid tip, turned mesally from its coxal articulation, strongly musculated, and deeply grooved on its dorsal surface for the reception of sperm from the third body segment. In the natural position (H) the telopodites of the second gonopods are clasped between the two branches of the first gonopods.

The gonopods are not in all proterandrious diplopods limited to the seventh segment. In the Nematophora the second legs of the sixth segment may be gonopods, or also the first or the second legs of the eighth segment. In the Colobognatha the gonopods are the second appendages of the seventh segment and the first appendages of the eighth segment, and in this group the gonopods resemble ordinary legs; in the genus *Brachycyba*, for example, each gonopod is merely a diminutive six-segmented leg.

The body segments following the segment of the gonopods in *Arctobolus* are all typical diplosomites and are almost circular in transverse outline (fig. 68 G). The segmental sterna are distinct plates (S) extending anterior to the bases of the first pair of legs; the coxae of the second pair are set into notches of the posterior sternal margin. Associated with the base of each leg is a spiracle. The spiracular apodemes in *Arctobolus*, however, though they give

244

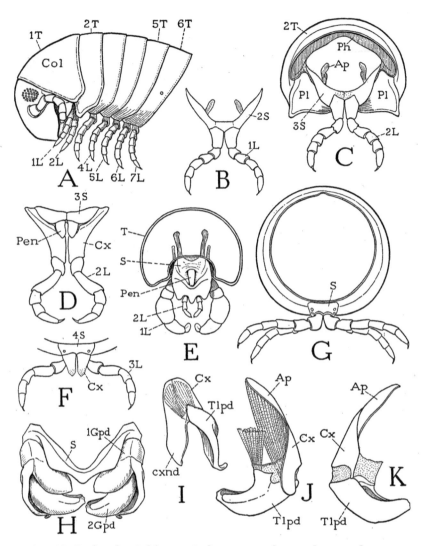

Fig. 68. Diplopoda—Juliformia. Body segments, legs, and gonopods.

A, *Arctobolus marginatus* (Say), head and anterior six body segments, lateral. B, same, first legs of male, anterior. C, same, second segment of male, anterior. D, *Habrostrepus* sp., male, second legs and supporting third sternum, paired penes arising at bases of coxae, posterior. E, *Parajulus* sp., male, first legs and segment of second legs with single median penis, posterior. F, *Arctobolus marginatus* (Say), sternum of fourth segment, with third legs, anterior. G, same, a diplosegment from middle part of body, anterior. H, same, gonopods and sternum of seventh segment, anterior. I, same, left first gonopod, anterolateral. J, same, left second gonopod, anterior. K, same, left second gonopod, posterior.

For explanation of lettering see pages 248–249.

attachment to the leg muscles, are extremely small, thin, delicate, triangular, horizontally overlapping lamellae, similar to those of *Julus* as described by Krug (1907), Ziegler (1907), and Ripper (1931). The legs are more slender than those of *Apheloria,* but they have the same segmentation.

The posterior end of the body of the spiroboloid is bluntly rounded (fig. 67 E). There is only one legless segment, which corresponds with the anal segment of *Apheloria* (C, 20) since it bears the lateral and ventral anal lobes. In the Nematophora the anal segment carries two or three pairs of slender, hollow spines that serve as outlets for silk-producing glands. The silk is discharged in the form of threads, and from this fact the Nematophora derive their name.

Eyes of Diplopoda and Chilopoda

The eyes of the diplopods and chilopods are in all cases lateral eyes; they are of particular interest in a comparative study of the optic organs of the arthropods because of their resemblance to the lateral eyes of certain apterygote insects and holometabolous insect larvae. Each eye is formed from a single layer of epidermal cells (fig. 69 A), which grows inward beneath the cuticular lens in the form of a cup (B, C). The outer cells of the cup (B, CgCls) retain the corneagenous function; the inner cells (SCls) become the sensory retinal elements. During the formative stage of the eye, and at each subsequent moult, the corneagenous cells extend in a continuous layer beneath the eye cuticle to form the lens, and after the completion of the lens they retract again to the periphery. The inner ends of the retinal cells directed toward the axis of the eyecup acquire light-sensitive striated borders (B, C) and may become elongated as terminal optic rods of the retinal cells (G, H). In some of the chilopods the retinal cells lose their uniformity of structure characteristic of the simpler types and become differentiated into a basal group of short central cells (D, cSCls) and an outer group of long peripheral cells (pSCl) converging beyond the basal cells. In this type of development the chilopod eye takes on the structure of the eyes of some holometabolous insect larvae.

The eyes of diplopods retain the simple cup-shaped type of retinal structure (fig. 69 F, G), in which the sensory cells are of approximately uniform size and have their receptive ends convergent along the axis of the eye. In some diplopods, as in *Glomeris* (F), the eyes

246

are few in number and well separated, and each eye has a thick, strongly convex lens. In the Julidae, on the other hand, there are numerous closely contiguous eyes in each lateral group (G), having a common corneal covering (*Cor*) with a smooth outer surface. The inner surface of the cornea, however, is produced into thick lens processes (*Ln*) projecting individually into the retinal cups.

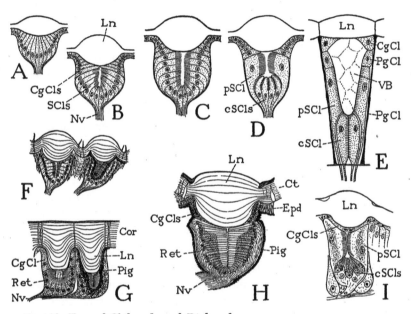

Fig. 69. Eyes of Chilopoda and Diplopoda.

A–D, diagrams of development of the eyecup. E, diagram of eye of *Scutigera* (based on Grenacher, 1880, Adensamer, 1894, and Hemenway, 1900). F, section of two adjacent eyes of *Glomeris* (from Grenacher, 1880). G, section of two eyecups of *Julus* (from Grenacher, 1880). H, section of eye of a scolopendrid, *Heterostoma australicum* (from Grenacher, 1880). I, section of eye of *Lithobius* (from Willem, 1892).

CgCls, cornagenous cells; *Cor,* cornea; *cScls,* central sense cells; *Ct,* cuticle; *Epd,* epidermis; *Ln,* lens; *Nv,* nerve; *PgCl,* pigment cell; *Pig,* pigment; *pScl,* peripheral sense cell; *Ret,* retina; *SCls,* sense cells, retina cells; *VB,* vitreous body.

Among the chilopods the Geophilomorpha are mostly without eyes; in the Scolopendromorpha the Scolopendridae have typically four small eyes on each side of the head, but the Cryptopidae are eyeless; the eyes of Lithobiomorpha have from 1 to 25 or 30 ocular units, those of Scutigeromorpha from 1 to 200. In the eye of a scolopendrid (fig. 69 H) the retinal cells (*Ret*) are of uniform elongate shape, and their distal receptive parts form cylindrical or polygonal optic rods directed horizontally toward the median axis of the eyecup. In *Lithobius* (I) the retinal cells are differentiated into an inner globular group of central cells (*cSCls*), with optic rods convergent distally, and a single layer of large, distal peripheral cells (*pSCl*) disposed radially around the axis of the eye, with their receptive surfaces contingent beyond the central cells. The numerous eyes in a compact group on each side of the head of *Scutigera* (fig. 54 A, B) give the appearance of a pair of large compound eyes. Each individual eye, moreover, has the form of a tubular ommatidium (fig. 69 E) tapering toward the inner end and contains a large, conical vitreous body (*VB*) beneath the lens. The retinal part of the eye resembles that of Lithobius in that it consists of a distal set of long peripheral cells (*pSCl*), 10 to 12 in number according to Hemenway (1900), that embrace the sides of the vitreous body and of a proximal central group (*cSCl*) of 3 or 4 cells. The entire eye is surrounded by a sheath of pigment cells (*PgCl*). The nature of the vitreous body is not known, but the body is composed of a number of polygonal sections, in which, according to Adensamer (1894), nuclei may be seen in immature specimens. There can be little doubt that the compact group of simple eyes of *Scutigera* must function in the manner of an appositional compound eye, but the structural differences between the chilopod eyes and the compound eyes of crustaceans and insects are sufficient to show that there can be no relation between them. On the other hand, the structure of the eye of *Scutigera* is almost a replica of that of the eyes of dytiscid, trichopterous, and lepidopterous larvae among the insects (see Snodgrass, 1935, fig. 281).

Explanation of Lettering on Figures 63–68

aAp, anterior apodeme of a body segment.
An, anus.
Ant, antenna.
Ap, apodeme.
aSp, anterior spiracle of a segment.

Br, brain.

Col, collum.
cvpl, cervical plate.

Cvx, cervix, neck.
Cx, coxa.
cxnd, coxal endite.

d, "mentum" of gnathochilarium.

e, "stipes" of gnathochilarium.
E, eye.
Ephy, epipharynx.
Epst, epistome.

248

f, "lamina lingualis" of gnathochilarium.

flt, a much reduced fultural sclerite.

Flt, fultura, supporting sclerite of hypopharynx.

Fm, femur.

g, "cardo" of gnathochilarium.

Gnch, gnathochilarium.

gnL, gnathal lobe of mandible.

Gpd, gonopod; *1Gpd, 2Gpd*, first and second gonopods.

h, "prebasilare" of gnathochilarium.

H, head.

hAp, head apodeme.

Hphy, hypopharynx.

Hpt, hypoproct.

L, leg.

Lg, intergnathal ligament.

Lm, labrum.

mcR, midcranial ridge.

mdB, base of mandible.

Mth, mouth.

Mtz, metazonite.

opl, opercular lobe of vulva.

OT, organ of Tömösvary.

pAp, posterior apodeme of a body segment.

Papt, paraproct.

Pen, penis.

Ph, phragma.

Pl, pleuron.

Poc, postocciput.

poR, postoccipital ridge.

pos, postoccipital sulcus.

PrC, preoral food cavity.

Prz, prozonite.

pSp, posterior spiracle of a segment.

Ptar, pretarsus.

rp, pore of repugnatorial gland.

S, sternum.

SoeGng, suboesophageal ganglion.

Sp, spiracle.

Stom, stomodaeum.

t, muscle tendon.

T, tergum.

Tar, tarsus.

Tb, tibia.

Tlpd, telopodite.

1Tr, first trochanter.

2Tr, second trochanter, prefemur.

Tra, tracheae.

Vd, vas deferens.

vlv, valvelike lobe of vulva.

* IX *

THE PAUROPODA

THE members of this group are minute, elongate, soft-bodied arthropods of the myriapod type of structure (fig. 70 A, B), but because of their relatively few legs, usually nine pairs in the adult stage, they have been named *pauropods* (Lubbock, 1868). A pauropod of average size is about a millimeter in length, but some species are only half as long, and others reach a length of nearly 2 mm. Probably owing to their small size, the pauropods have no circulatory system and no tracheae or other differentiated organs of respiration. They live in moist places under logs and stones, on the ground among decaying leaves, and in the soil to a depth of several inches. The feeding habits of the pauropods are not well known, but their food has been thought to be humus and decaying plant and animal tissue. Starling (1944) says that mold fungi were observed to be the usual food of *Pauropus carolinensis* and that a "correlation appears to exist between the optimum temperature for mold growth in general and high incidence of pauropod population." He gives reasons for believing that pauropods, where abundant, regardless of their small size, play a significant part in soil formation.

A typical adult pauropod (fig. 70 B) has a relatively small, conical head and an elongate body of 12 segments, counting as segments the first and the last body divisions, which are known respectively as the *collum* (*Col*) and the *pygidium* (*Pyg*). Statements by other writers as to the number of segments may vary, because some do not include the pygidium as a segment and some exclude both the collum and the pygidium, but such differences are merely a matter of definition for a "segment."

250

The number of legs in an adult pauropod, except in one known species, is invariably nine pairs, the first pair being on the second body segment, the last on the tenth (fig. 70 B). Preceding the pygidium is a legless segment, and the pygidium itself never bears appendages. A ten-legged pauropod, *Decapauropus* (F), said to have 13 body segments, has been described by Rémy (1931). In this form the usual eleventh segment appears to be partly subdivided ventrally into two parts, the anterior of which carries the extra pair of legs; but it might be suspected that the apparent division is superficial and that *Decapauropus* differs from other genera only in having legs on the eleventh segment.

The body segmentation of most pauropods appears to be different on the dorsal and ventral surfaces (fig. 70 B, F). In the specimen shown at A all the segments between the collum and the pygidium appear to be fully and equally developed, but generally, as seen in the genus *Pauropus* (B), there are only six tergal plates on the dorsum, though there are ten corresponding segmental divisions of the venter. This condition suggests the diplopod type of segmentation, but in the extended pauropod the narrowed dorsal arcs of segments 4, 6, 8, and 10 are exposed between the tergal plates of the other segments. Hansen (1902) has noted that live specimens of Pauropodidae "are able to elongate their body or to shorten it by contraction in a very high degree." Each of the tergal plates except the first bears a pair of long, slender tactile setae.

The first segment, or collum, of the pauropod body (fig. 70 B) is quite different from the collum of a diplopod; instead of being a large tergal plate, it forms but an inconspicuous fold dorsally behind the head and is much expanded below. It bears ventrally a pair of small papillae (H, g), which have commonly been regarded as vestigial legs, but according to Tiegs (1947) they neither develop in the manner of limbs nor have the structure of legs, and appear rather to be comparable to the exsertile vesicles of Symphyla and apterygote insects.

The pauropod legs are all six-segmented (fig. 70 I), including the small pretarsus (*Ptar*). Except in the legs of the first and the last pairs, the tarsus of the adult (*Tar*) is divided into two subsegments. Writers who do not regard the pretarsus as a segment, state that the leg is either "six-segmented" or "five-segmented" according as the tarsus is divided or not. That the "segmentation" of the tarsus is not a division into true segments is shown by the fact that no muscles are attached on the distal part, and none arises in the proximal part. The entire tarsus is traversed in the usual manner by the tendon of the flexor muscle of the pretarsus (*flptar*), the branches of which arise in the tibia (*Tb*) and the femur (*Fm*). The musculature thus identifies the distal segments of the leg, and there is only one segment (*Tr*) in the trochanteral region between the coxa and the short femur; the leg, therefore, is six-segmented in all cases, regardless of the subsegmentation of the tarsus. The structure of the pretarsus is variable; there is a median claw and usually a ventral lobe, or empodium, but one or two accessory lateral claws may be present (fig. 70 J), and the claws themselves may be padlike on their undersurfaces.

The pauropod head has a unique feature in the presence of a pair of large, smooth, clear, oval areas on the sides (fig. 70 D, E, *e*) that have the appearance of eyes and were formerly supposed to be eyes. The sublying tissue, however, has nothing of the structure characteristic of light receptors, and the organs are now given the meaningless name of "pseudoculi," their function being unknown. It is shown by Tiegs (1947) that the convex, cornealike cuticle of a pseudoculus is separated by an intervening space from a flat layer of large epidermal cells beneath it, which are innervated from the protocephalon. Tiegs suggests that the organs might be responsive to vibrations, but that, since the subcuticular space is apparently filled with liquid, the organs must be supposed to respond, by contact, to vibrations of solid objects rather than to air vibrations.

Characteristic of the pauropods are the relatively large, branched antennae (fig. 70 D). Each antenna consists of a basal stalk, which is four-segmented in the adult stage and of two apical branches, one dorsal, the other ventral. The dorsal branch ends in a single slender multiarticulate flagellum; the ventral branch bears two shorter flagella and, arising between their bases, a club-shaped or globular appendage termed the *globulus* (*f*), supposed to be a special sense organ of some kind. The segments of the antennal stalk and the bases of the branches are individually musculated, but the flagella have no muscles.

The pauropod mouth parts include only a pair of mandibles (fig. 70 H, *Md*) and an underlip structure (*Mx, mxS*), the latter, said to be composed of the appendages and sternum of the first maxillary

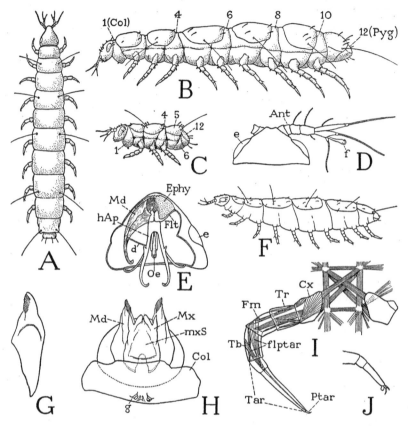

Fig. 70. Pauropoda. (B, C, E from Tiegs, 1947; D, G, H, I from Silvestri, 1902.)

A, a pauropod (unidentified) with regular segmentation on the dorsum. B, *Pauropus silvaticus* Tiegs, fully extended adult. C, same, first instar. D, *Stylopauropus pubescens* Hansen, head and right antenna, dorsal. E, *Pauropus silvaticus* Tiegs, underside of head, with right mandible in place and head apodemes. F, *Decapauropus cuenoti* Rémy (from Rémy, 1931). G, *Allopauropus brevisetus* Silv., mandible. H, same, head and collum segment, ventral. I, same, segmentation and musculature of a leg. J, a pauropod leg with accessory pretarsal claw and empodium.

Ant, antenna; *Col,* collum (first body segment); *Cx,* coxa; *d,* ligament from mandible to head apodeme; *e,* "pseudoculus"; *Ephy,* epipharyngeal surface; *f,* "globulus" of antenna; *flptar,* flexor muscle of pretarsus; *Flt,* futural sclerite; *Fm,* femur; *g,* ventral papillae of collum segment; *hAp,* head apodeme; *Md,* mandible; *Mx,* maxilla; *mxS,* maxillary sternum; *Oe,* oesophagus; *Ptar,* pretarsus; *Pyg,* pygidium; *Tar,* tarsus; *Tb,* tibia; *Tr,* trochanter.

253

segment, appears, therefore, to be comparable with the diplopod gnathochilarium. The pauropod mandibles, however, have no likeness to the jaws of a diplopod; the gnathal lobes are solid extensions from the elongate, tapering bases (G), which, instead of being fully exposed on the sides of the cranium, are deeply sunken into the head (E, *Md*), as are the mandibles of a chilopod. Insofar as the pauropod mandibles are solid, single-piece appendages they resemble in particular the mandibles of a geophilid chilopod. In *Pauropus*, according to Tiegs (1947), the tapering basal part of each mandible is connected by a ligament (E, *d*) to the head apodeme (*hAp*) of the same side and by a lateral ligament (not shown in the figure) to the cranial wall just behind the pseudoculus. This lateral ligament is suggestive of the slender rod connecting the end of the chilopod mandible with the cranial margin (fig. 59 B, *mdr*), which will be met with again in the Protura and Collembola. The ventrally concave distal ends of the pauropod mandibles are provided with combs of delicate curved blades (fig. 70 H), and together with a median elevation of the epipharyngeal surface, Tiegs says, they form the upper wall of a tubular food passage above the grooved floor of the preoral cavity. The mandibular musculature is described by Tiegs as consisting of protractor, retractor, depressor, and adductor muscles, the last attached on the head apodemes.

Associated with the mandibles is a suspensory apparatus of the same nature as the hypopharyngeal fulturae of the chilopods. It consists of a pair of slender, transverse, premandibular sclerites in the ventral head wall attached laterally on the cranial margins (fig. 70 E, *Flt*). From the enlarged mesal ends of the sclerites, as described by Tiegs in *Pauropus*, arises a pair of large head apodemes (*hAp*). The bases of the apodemes unite in a sclerotic ring around the oesophagus (*Oe*); the distal parts are long, two-branched arms, of which the mesal pair extend back into the collum segment, while the lateral branches curve dorsally and are attached on the occipital margin of the head.

The postmandibular component of the pauropod mouth parts (fig. 70 H), representing the gnathochilarium of the diplopods, is said by Silvestri (1902), Tiegs (1947), and other investigators to include the first maxillary appendages (*Mx*) and a triangular median plate (*mxS*), which is the sternum of the first maxillary segment. The parts, however, are much less unified than in the diplopods. In

254

Allopauropus (H), according to Silvestri, the maxillae are free from the median plate only at their distal ends. On the other hand, Hansen (1930) says that in *Stylopauropus* the maxillae are so sharply separated from the median plate that there can be no question of union between them. Tiegs, in his study of *Pauropus silvaticus,* makes no specific statement as to the degree of union of the maxillae with the sternal sclerite, but from his illustration (text fig. 2 B) it would appear that the three parts are quite separate. In any case, there is little doubt that the maxillae and the plate between them in the pauropods are the elements that are united in the diplopod gnathochilarium. The pair of median apical lobes, called "galeae" by Silvestri, are said by Tiegs to arise within the preoral cavity from the mandibular segment, "and seem therefore to be the equivalent of the superlinguae."

The pauropods unquestionably are more primitive in their general organization than are the diplopods; the most important difference between the two groups is in the structure of the mandibles, but otherwise a close relationship is indicated by the segmental composition of the head, the primarily ventral position of the gonads, and the opening of the genital ducts on the third body segment. The development of the pauropod head is said by Tiegs (1947) to show that the adult head contains only one maxillary segment, the appendages and sternum of which form the gnathochilarium. The next segment, representing the second maxillary, or labial, segment of most other arthropods, becomes the first body segment, termed the collum segment. The ovaries of the pauropods remain throughout life between the alimentary canal and the nerve cord, and their ducts discharge through a single median pore into a depression of the venter on the third body segment between the bases of the second legs. A sperm receptacle opens from the dorsal wall of the genital depression. In the male pauropod the testes are formed ventrally during embryonic development, but, as shown by Tiegs (1947), they later acquire secondarily a dorsal position above the alimentary canal. The two ejaculatory ducts open separately through a pair of penes on the third segment mesad of the coxae of the legs. The exit genital ducts of the pauropods, Tiegs says, are not persisting coelomoducts, as in most other arthropods, but are epidermal ingrowths, which fact suggests that the functional ducts may not be the primary ducts, but secondary outlets connecting with the gonads.

The pauropods are anamorphic in their postembryonic development, and, as with the diplopods, the young pauropod when first exposed by rupture of the eggshell is still a quiescent embryo invested
in a cuticular covering in which it completes its embryonic development. In this condition the embryo of *Pauropus amicus* is said by
Harrison (1914) to remain for three days; then the investing cuticle
splits lengthwise on the back, and the young, active, six-legged
pauropod of the first instar becomes free. The young of *Pauropus
silvaticus*, according to Tiegs (1947), has six body segments in front
of the pygidium (fig. 70 C). Tergal plates are present on segments
2, 3, and 5, and the legs are carried by segments 2, 3, and 4. Starling
(1944), however, gives only five body segments, exclusive of the
pygidium, for the six-legged young instar of *Pauropus carolinensis*;
the last segment he says is the definitive eleventh segment. There are
four immature instars, during which the number of body segments
increases and extra legs are added, but different observers are not
entirely in accord as to the order of succession of the newly formed
segments. According to Starling (1944), the prepygidial segment is
a permanent segment from the beginning and becomes the definitive
eleventh segment, while the new segments are generated in front of it.
Silvestri (1902) says that the formation of new segments takes place
between the last segment and the one preceding it, but since elsewhere he does not enumerate the pygidium as a body segment, his
ultimo segmento may refer to the eleventh. The idea of a permanent
preanal segment, however, is said by Tiegs (1947) to be an error,
the new segments being generated between the eleventh segment
and the anal segment, or pygidium.

X

THE SYMPHYLA

THE symphylans are small, slender, soft-bodied, progoneate, centi-pedelike arthropods with long antennae, 12 pairs of legs in the adult stage, and two tapering, unjointed appendages projecting from the rear end of the body (fig. 71 A). They feed mostly on vegetable matter and, in general, are found in moist places under stones and in rotting logs, but they live also in the ground, and one species at least, known as the "garden centipede," becomes in some localities a serious pest of outdoor garden crops and of greenhouse plants.

In a zoological way the Symphyla have become of particular interest to entomologists because they have been much investigated as possible ancestors of the insects. Though in important features the symphylans more closely resemble the diplopods, they do have other characters that are found elsewhere only in the lower insects. In the structure of the mandibles, in the ventral position of the gonads, and particularly in the anterior position of the genital openings on the third body segment, the Symphyla are diplopodan; in the presence of two distinct pairs of maxillary appendages on the head and of eversible vesicles and styli on the venter of the body segments they seem to show some relation to the apterygote insects. Finally, however, as will be shown in the next chapter, the insects themselves have important features that could not have been derived from symphylan ancestors. In short, theoretical arthropod phylogeny becomes so complicated that we had best leave it to the phylogenists. The best presentation of the evidence adduced to support the symphylan claim to insect ancestry has been given by Imms (1936).

There are between 50 and 60 described species of Symphyla

distributed throughout the world, classified by Attems (1926a) in three families, Scolopendrellidae, Scutigerellidae, and Geophilellidae. The most important recent papers on their structure, development, and life history are one by Michelbacher (1938) on the structure, life history, and economic aspect of the "garden centipede," *Scutigerella immaculata* (Newport) and another by Tiegs (1940) on the structure and embryonic development of *Hanseniella agilis* Tiegs. Papers of lesser scope will be mentioned in the text. The following account is based mostly on *Scutigerella immaculata*, with comparative notes from Tiegs on *Hanseniella agilis*. Both species belong to the Scutigerellidae.

General Structure of the Body

Scutigerella immaculata, in common with other Symphyla, has 12 pairs of legs when it becomes an adult (fig. 71 A), but the dorsum of the body presents a series of 15 tergal plates. Three of the back plates, however, are duplicate terga of segments 3, 5, and 7, there being no corresponding ventral segmentation of the body. The nerve ganglia and the embryonic coelomic sacs are said to correspond in number with the legs. In the genus *Geophilella* there are as many as 22 tergal plates, but only the usual 12 pairs of legs. The segment numbered 13 in the figure of *Scutigerella* (fig. 71 A), known as the *preanal segment,* bears a pair of tapering, unjointed appendages (G, *Spn*), which are functionally spinnerets, since silk-forming glands open through them, but developmentally they arise in line with the leg rudiments and are therefore regarded as true segmental appendages. The preanal segment in most species carries also the last pair of legs, and thus might seem to be a double segment. In *Scutigerella immaculata* (G) there is a narrow but distinct circular fold (*12*) of the integument between the preanal segment (*13*) and the preceding segment (*11*), and the legs of the last pair (*12L*) appear to be attached to this fold rather than to the segment of the spinnerets. It is possible, therefore, that the fold represents the true twelfth segment, and the possibility is much increased by the fact that in a species of *Symphylella* (J), described by Tiegs (1940), there is in front of the pregenital segment (*13*) a well-developed extra segment (*12*) carrying the twelfth legs. In species with only 13 body segments, therefore, it is probably true, as said by Tiegs, that during anamorphosis "the twelfth leg-bearing segment, unlike

258

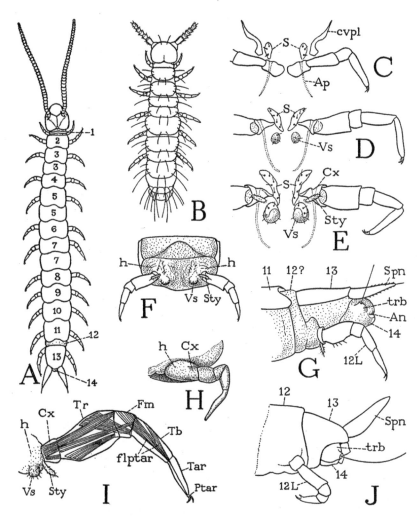

Fig. 71. Symphyla. General structure of the body and the legs.

A, *Scutigerella immaculata* (Newp.), adult. B, same, first instar (from Michelbacher, 1938). C, first legs and associated ventral sclerites. D, same, second legs, with imperfect vesicles. E, same, third legs, with styli and fully developed vesicles. F, same, ventral surface of segment from middle part of body. G, same, posterior end of body. H, *Hanseniella agilis* Tiegs, embryonic leg (from Tiegs, 1947). I, *Scutigerella immaculata* (Newp.), leg of adult, showing principal muscles and associated vesicle and stylus. J, *Symphylella* sp., posterior end of body with fully developed twelfth segment (from Tiegs, 1947).

For explanation of lettering see page 270.

the four that arise before it, fails to separate from the cerus-bearing segment." The last segment of the body is a small *anal segment* (A, G, J, *14*). On each side it bears a conical tubercle supporting a long sensory hair, the structure being known as a *trichobothrium* (*trb*). The anus (G, *An*) is situated apically between slightly protruding dorsal and ventral lobes.

It is a remarkable thing, shown by Tiegs (1940) in his study of the development of *Hanseniella,* that the anal segment in the embryo contains both coelomic sacs and nerve ganglia. The end segment of the symphylan, therefore, *is not the telson,* as it might appear to be; the telson, if present, must then be represented only by the anal lobes. The trichobothria, Tiegs says, arise as blunt outgrowths on the sides of the anal segment and appear to be vestiges of the appendages of this segment. The Symphyla evidently have 14 body segments representing primary somites, with segments 12 and 13 usually united.

The postembryonic development of the symphylans is anamorphic, the first instar having only nine body segments and either six (fig. 71 B) or seven pairs of legs. The preanal segment of the first instar is relatively large, and it is within this segment, as shown both by Michelbacher (1938) and by Tiegs (1940), that the new segments will be formed. From the coelomic sacs of the preanal segment, Tiegs says, arises the mesoderm for the segments that appear, during anamorphosis, between the seventh segment and the definitive preanal segment. The anamorphic development of the Symphyla thus differs from the teloblastic development of segments in other anamorphic arthropods, in which the generative zone lies immediately in front of the telson. It would appear, therefore, that in the symphylans telogenesis must already have produced before hatching the complete material for the segments added during postembryonic growth, and that anamorphosis in these arthropods is merely a delayed differentiation of this material into segments, accompanied by the formation of the new appendages. If the postembryonic development in Symphyla is to be explained in this manner, it evidently represents an intermediate stage between anamorphosis with typical teloblastic growth and epigenesis, or the completion of segmentation before hatching.

The Head and the Mouth Parts

The symphylan head (fig. 72 A, B, E, G) is of the prognathous type of structure insofar as the mouth parts lie horizontally against its undersurface and are directed forward, in which respect it resembles the head of a chilopod, diplopod, or pauropod. The head of *Scutigerella* is somewhat flattened (B); that of *Hanseniella* (E) is thicker and strongly declivous anteriorly. On its dorsal surface the head carries the long, many-segmented antennae, which are set in large membranous areas containing also the postantennal organs of Tömösvary (B, E, *OT*). Each antenna is pivoted on an articular process, or *antennifer* (A, *af*), arising from the cranial wall mesad of its base. Anterior to the antennae is a large epistomal area of the cranium (*Epst*), but there is no specifically demarked clypeus. The bilobed, spiny labrum (C, *Lm*) is scarcely separated from the epistome. The subantennal lateral margin of the cranium is formed by a narrow sclerotic bar (B, E) extending from the angle of the epistome to the posterior part of the cranium. In *Scutigerella* (B) the bar becomes so weak that its posterior cranial connection can be seen only in a well-stained specimen; in *Hanseniella* (E) the bar is somewhat thicker. Below this subantennal bar the cranium is deeply emarginate, as in the diplopods, but in the symphylan a wide membranous space intervenes between the bar and the base of the mandible (B, E) in which is one of the pair of head spiracles (*Sp*), the only tracheal apertures of the Symphyla. The dorsum of the head of *Scutigerella immaculata* (A) is marked by a midcranial sulcus that forks anteriorly to the antennal pivots and forms internally a corresponding Y-shaped ridge. From near the posterior end of the midcranial sulcus two pale lines diverge forward to the antennal membranes behind the organs of Tömösvary, but these lines are not grooves, nor do they form internal ridges. In stained specimens they are seen as faint, clear tracts of the cuticle, apparently of denser texture than the rest, devoid of the small setae that cover the other parts of the head; they are not present in other genera. The lines on the head of *Scutigerella* have nothing to do with moulting. The Symphyla have no eyes.

The antennae of Symphyla have been shown by Imms (1939) to be truly segmented appendages, since each division of the shaft is independently musculated from the one proximal to it. The basal

261

segment contains only a single muscle; each of the others except the apical segment contains four muscles inserted on the segment beyond. According to Michelbacher (1938), the antenna of the first instar of *Scutigerella immaculata* (fig. 71 B) has a basal stalk and four free distal segments, but within the stalk are seven fully formed segments merely enclosed in a sheath. With subsequent moults, and the animal continues to moult throughout life, new segments are added. The average number of antennal segments in the adult is perhaps 30 or 40, but Michelbacher reports an observed maximum number of 60. When part of an antenna, or even the whole appendage, is broken off, the lost part is rapidly regenerated at the moults. From Michelbacher's account of the antennal growth in Scutigerella it would seem that the new or regenerated segments are formed in the basal part of the appendage. Imms (1940), on the other hand, has shown that growth of the segmented antenna of the collembolan *Campodea* is acrogenous, that is, the new segments are cut off from the base of the permanent apical segment. This type of segment multiplication, however, would not allow of regeneration in an antenna lacking the distal part, unless each segment is capable of assuming the role of an apical segment when occasion demands.

The postantennal Tömösvary organs of *Scutigerella* (fig. 72 A, B, E, *OT*) have been well described and illustrated by Pflugfelder. (1933). Each organ has a dome-shaped outer surface with a hole in the center that opens into a large globular chamber lined with a thin cuticle. From the floor of the chamber project numerous fine filaments that contain nerve fibrils given off from the outer ends of a group of sublying sensory nerve cells. In *Hanseniella*, as described by Tiegs (1940), the outer dome of the organ has a wide central opening, and the floor of the cavity beneath it is covered by a thick layer of gelatinous substance. The filaments are grouped in a central bundle that projects through the gelatinous layer; they are said to be cuticular tubules that resist boiling in caustic potash and are shed at each moult. Most writers find that the centripetal nerves of the Tömösvary organs go to the protocerebrum. Many sensory functions have been theoretically attributed to the organs of Tömösvary, but none has been experimentally demonstrated.

The head spiracles give rise to a tracheal system which is most elaborately developed in the head and extends back through only the first three segments of the body. The standard, widely copied

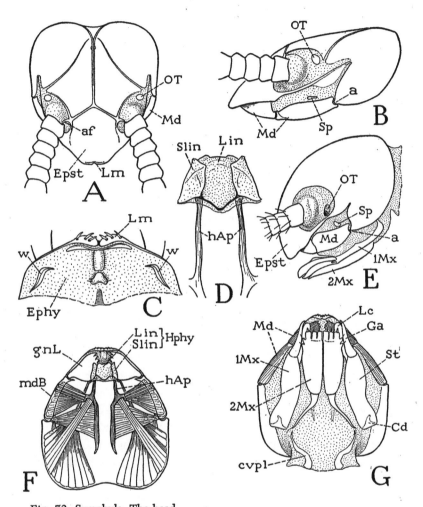

Fig. 72. Symphyla. The head.

A, *Scutigerella immaculata* (Newp.), head, dorsal. B, same, head, lateral.
C, same, epipharyngeal surface and labrum, ventral. D, same, hypopharynx
and head apodemes, dorsal. E, *Hanseniella agilis* Tiegs, head, lateral (outline
from Tiegs, 1940). F, *Scutigerella immaculata* (Newp.), horizontal optical
section of head below level of mandibles, ventral, showing mandibular muscles.
G, same, head, ventral.

For explanation of lettering see page 270.

illustration of the tracheal system of *Scutigerella immaculata* (see Verhoeff, 1933, Michelbacher, 1938) is taken from Haase (1884). According to Tiegs (1940), the tracheal invaginations of *Hanseniella* appear during the ninth day of embryonic development, apparently on the mandibular segment, but the openings become closed and remain so until the first larval moult. Presumably a complete tracheation of the body is not necessary on account of the small size of the animal and the permeability of its soft integument. The intima of the tracheal tubes of *Scutigerella* is said by Richards and Korda (1950) to have reticulate thickenings sometimes taking the form of incomplete taenidia.

The ventrolateral parts of the symphylan head are occupied by the bases of the mandibles (fig. 72 B, E, *Md*), which latter thus have the same position as in the diplopods (fig. 63 D), though they do not occupy the entire lateral emarginations of the cranium. The ventral wall of the head (fig. 72 G) is membranous except as it is occupied by the two pairs of maxillae (*1Mx, 2Mx*), the second pair of which is partly united to form a labium. The mouth lies some distance behind the labrum, so that there is a long intervening epipharyngeal area (C, *Ephy*), as in the diplopods (fig. 63 E) and in *Scutigera* among the chilopods (fig. 54 D). Immediately behind the labrum of *Scutigerella* (fig. 72 C) is a transverse epipharyngeal ridge, followed by a median elevation. Laterally on each side is a strongly sclerotic, sinuous sclerite (*w*), which has a mechanical relation to a small process on the base of the gnathal lobe of the corresponding mandible (fig. 73 A, *c*). The two epipharyngeal sclerites of *Scutigerella*, therefore, are suggestive of the epipharyngeal arms of the fultural sclerites of the chilopods (fig. 54 D, *w*) on which the mandibles have loose articulations.

Projecting from beneath the mouth is a three-lobed hypopharynx (fig. 72 D; F, *Hphy*). Borrowing terms from entomology, the median hypopharyngeal lobe is the *lingua* (*Lin*), the lateral lobes the *superlinguae* (*Slin*), but the superlinguae evidently represent the paragnaths of Crustacea. In *Scutigerella* (D) the broad lingua is overlapped dorsally on its sides by the superlinguae, and the bases of the latter are connected by an angulated bar through the base of the lingua. According to Tiegs (1940), the superlinguae of *Hanseniella* develop on the sternum of the mandibular segment and are innervated from the mandibular ganglion. The lingua ("hypopharynx") when first formed is purely a product of the first maxillary segment, but later "the labial ectoderm seems to become incorporated into its inferior wall." In the Symphyla there are no premandibular sternal sclerites (fulturae) supporting the hypopharynx, but at the sides of the lingua arises a pair of strong head apodemes (D, F, *hAp*) that extend far back in the head cavity and give attachment to adductor muscles of the mouth parts and to ventral muscles from the body. The head apodemes of Symphyla are clearly homologues of the similar apodemes in Chilopoda, Diplopoda, and Pauropoda, though supporting sclerites are absent. The apodemes (hypopharyngeal apophyses) Tiegs says are developed in *Hanseniella* a little median to the bases of the mandibles, and later take a more median position; Tiegs assigns them to the mandibular segment. Hansen (1930, p. 270) mistook the head apodemes of Symphyla for ventral surface sclerites of the head, such as those present in the apterygote hexapods Diplura and Collembola.

The symphylan mouth parts (fig. 73) may be likened to those of insects in that the first maxillae (B) have something of the insect type of structure, and the second maxillae (C) are partly united to form a labiumlike underlip attached to the head. The mandibles (A), however, closely resemble the mandibles of a diplopod, since each consists of a basal plate implanted on the side of the head and of an independently movable gnathal lobe.

The mandibles of Symphyla are relatively shorter than those of the diplopods, and, as noted above, their basal plates do not occupy the entire lateral emarginations of the cranium (fig. 72, B, E). The gnathal lobes (fig. 73 A, *gnL*) are flattened and lie horizontally in the preoral cavity of the head; each is articulated to the basal plate (*mdB*) by the outer angle of its base and has a process (*c*) on its dorsal surface that loosely bears on the corresponding lateral sclerite of the epipharynx (fig. 72 C, *w*). The mandibular musculature includes the same muscles that are attached on the diplopod jaw, but the two ventral adductors (fig. 73 A, *4a, 4b*) both arise on the head apodeme (*hAp*), there being no intergnathal ligament such as that of the diplopods. A dorsal muscle (*2*) from the posterior part of the cranium is attached on the mandibular base; the gnathal lobe has a huge cranial flexor (*1*). Though the movable gnathal lobe of the symphylan mandible, with its cranial flexor muscle, resembles the movable lacinea of a generalized insect maxilla, it is hardly to be supposed that the symphylan or diplopod mandible represents a primitive arthropod jaw. The "segmentation" of the mandible of *Hanseniella*, according to Tiegs, takes place on the ninth day of embryonic growth.

The maxillae of *Scutigerella* are relatively long appendages lying laterally against the underside of the head (fig. 72 G, *1Mx*). Each has an elongate, elaborately musculated basal plate, or *stipes* (fig. 73 B, *St*), attached by its entire length to the head wall, and two free apical lobes which may be termed *galea* (*Ga*) and *lacinia* (*Lc*),

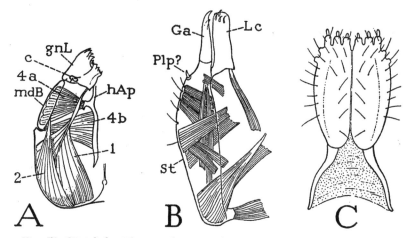

Fig. 73. Symphyla. The mouth parts of *Scutigerella immaculata* (Newp.). A, left mandible and muscles, dorsal. B, left first maxilla, dorsal. C, second maxillae, labium, ventral.

For explanation of lettering see page 270.

since they appear to correspond with the lobes so named in an insect maxilla. The lacinia has a cranial flexor muscle. A small lateral spur (*Plp*) on the stipes near the base of the galea is possibly a remnant of the palpus. The implantation of the basal plates of the mandibles and maxillae on the head wall, Tiegs (1940) observes, "does not appear seriously to impede their movement," for "if an animal is watched under a lens when feeding, the ventro-lateral wall of the head is seen in active movement." The strong musculature of the basal plates of the two appendages would imply that this movement is of considerable importance.

The labium of Symphyla (fig. 73 C) is evidently composed of two head appendages homologous with the second maxillae of the

chilopods or the labial components of insects, though it has no resemblance in form to either. It consists of a broad, oval anterior
plate, divided by a median groove, with six papilla-bearing lobes on
its distal margin, and of two slender, tapering proximal arms diverging posteriorly in the membranous head wall to the anterior ends
of a pair of cervical sclerites (fig. 72 G, *cvpl*). Palpi are entirely absent. In the embryonic development of *Hanseniella*, Tiegs (1940)
says, "the most anterior intersegmental cleft is between the maxillary
and labial segments," the labial segment being at first clearly a part
of the body and only in late embryonic growth becoming a part of
the head. Considering the lack of evidence of a second maxillary segment in the head of the diplopods and pauropods, it seems probable
that the common ancestors of the progoneate myriapods had only
one maxillary segment incorporated into the head.

The Body Appendages

The legs of Symphyla, except those of the first pair, are all alike,
and each is divided into six segments (fig. 71 I). The segments are
generally interpreted as being a small coxa (*Cx*), an unusually large
trochanter (*Tr*), a very short femur (*Fm*), an elongate tibia (*Tb*),
an undivided tarsus (*Tar*), and a clawlike pretarsus (*Ptar*) with an
accessory lateral claw. The four distal segments are fairly well
identified by the distribution of the fibers of the flexor muscle of the
pretarsus (*flptar*), but the identity of the two proximal segments may
be questioned. The large second segment, however, is suggestive
of the long second trochanter of the diplopod leg (fig. 65 H). If the
basal segment of the symphylan leg is the coxa, as it appears to be,
then a first trochanter is absent. The symphylid leg has some resemblance to the leg of a pauropod (fig. 70 I) but none at all to an
insect leg (fig. 83 A), in which the principal segment proximal to
the "knee" is the femur and the trochanter is always small. As before
noted, the relative size of the leg segments among the arthropods
generally is no index of the identity of the segments. The legs of the
first pair in the Symphyla (fig. 71 C) are not only smaller than the
others, but they lack a segment which apparently is the tibia. In some
species these legs are much reduced in size; in others they are almost as long as the other legs.

Each leg of the Symphyla arises from a slightly bulging area of
the lateroventral wall of the supporting body segment (fig. 71 F, I,

h), which bears also on most of the segments a short *stylus* (*Sty*) and an *eversible vesicle* (*Vs*). The leg-bearing mound has been interpreted by Hansen (1930) as the coxa of the leg and by Ewing (1928) as a subcoxa. Tiegs (1940) calls it the "limb-base," and in his figure of an embryonic leg of *Hanseniella* (H) the "limb-base" (*h*) does have the appearance of a basal segment of the limb, but Tiegs asserts that it does not develop as a part of the leg and is therefore not a subcoxal segment of the leg. It represents the area of a "ventral organ," that is, one of the paired segmental thickenings of the embryonic ventral ectoderm from the inner surface of which the nerve tissue is differentiated. The eversible vesicle, according to Tiegs, is the epidermal remnant of the ventral organ protruded at the surface. The stylus arises laterad of the vesicle very close to the base of the coxa, but not on it. Vesicles and styli are absent on the first body segment (C); on the second segment (D) is a pair of imperfect vesicles; on the third (E) and the following segments are both styli and fully developed vesicles. The vesicles are eversible and are said to have retractor muscles, but the styli lack muscles. The function of these organs is not known. The coxae are articulated ventrally on small sternal plates lying before the vesicles (D, E, S), but have no dorsal articulations.

The spinnerets of the preanal segment (fig. 71 G, J, *Spn*) give exit at their tips to a pair of glands, the secretion of which hardens on contact with the air to a silklike thread. Because of their prominence at the end of the body (A) the spinnerets have been thought to be homologues of the insect cerci. The preanal segment of the Symphyla is at most the thirteenth body segment; while it is true that the cerci of many adult insects are attached on the thirteenth postcephalic segment, embryologists are agreed that they are the appendages of the fourteenth segment in the embryo. Hence, though the insect cerci may also contain glands, as in some of the Japygidae, their segmental status is not the same as that of the symphylan spinnerets. Among the diplopods, members of the Nematophora have silk glands that open through spines on the anal segment, but in the arthropods generally silk-spinning is a function of such widely different appendages that it cannot be invoked as evidence of homology in any particular case. If the trichobothria of the anal segment of the symphylans are the appendages of this, the fourteenth, body segment, they, rather than the spinnerets, would be the homologues of the insect cerci. Of just what use the silk spun from the spinnerets is to the symphylans in nature is not known, but Michelbacher (1938) says that, when ejected on a camel-hair brush, the threads form a most convenient means of transferring live specimens without injury from one container to another.

Life History of Scutigerella immaculata

The life history of the garden centipede is probably most fully and correctly described by Michelbacher (1938). The first instar (fig. 71 B) has six pairs of legs; Tiegs (1940) says there are seven pairs of legs in the first instar of *Hanseniella*. According to Michelbacher, *Scutigerella immaculata* goes through seven immature instars, adding a pair of legs at each moult, the adult number of 12 being thus attained by the seventh instar. The definitive number of back plates, however, is present in the sixth instar. During the immature stages the antennae have increased in length by the successive addition of segments until in the young adult there are about 30 antennal segments. The symphylids, however, do not cease moulting on becoming adult. Michelbacher says that evidence indicates the number of moults may be in excess of 50. During the first few adult moults the body may increase in size up to a maximum length of 7 or 8 mm., but the antennae continue to add segments throughout life until each may have as many as 60 segments. Under laboratory conditions, Michelbacher says, the garden centipede lives for several years.

The manner of ecdysis in the Symphyla is the same as that in other myriapods; there is no lengthwise dorsal rupture of the cuticle forking on the head, as is usual with insects. The outer cuticle splits transversely behind the head and the animal crawls out of the body skin. Williams (1907) says this is at least the most common method of ecdysis, and the one observed in a specimen in the act of emergence, but that shed skins are often found to be torn in various ways. Michelbacher (1938) also observes that at ecdysis "the garden centipede frees its head from the capsule and by forward propulsions starts to work its way out of the skin."

Explanation of Lettering on Figures 71–73

a, cranial articulation of mandible.
af, antennifer.
An, anus.
Ap, apodeme.

c, epipharyngeal articulation of mandible.
Cd, cardo.
cvpl, cervical plate.
Cx, coxa.

Ephy, epipharyngeal surface.
Epst, epistome.

flptar, flexor muscle of pretarsus.
Fm, femur.

Ga, galea
gnL, gnathal lobe of mandible.

h, lobe of body supporting the leg.
hAp, head apodeme, hypopharyngeal apophysis.
Hphy, hypopharynx.

L, leg.
Lc, lacinia.
Lin, lingua.

Lm, labrum.

Md, mandible.
mdB, base of mandible.
1Mx, first maxilla.
2Mx, second maxilla.

OT, organ of Tömösvary.

Plp?, small spur of maxilla, possibly vestige of palpus.
Ptar, pretarsus.

S, sternum
Slin, superlingua.
Sp, spiracle.
Spn, spinneret.
St, stipes.
Sty, stylus.

Tar, tarsus.
Tb, tibia.
Tr, trochanter.
trb, trichobothrium.

Vs, eversible vesicle.

w, lateral epipharyngeal sclerite.

THE HEXAPODA

THE Hexapoda are the six-legged arthropods commonly known at present as insects; formerly, however, the name "insect" was given to almost any familiar small arthropod, since the bodies of most of them are more or less "insected," and even today it is difficult to convince some people that spiders, centipedes, and sowbugs are not insects.

The hexapods are divided naturally into three major groups. The first group includes the Protura, Collembola, and Diplura (or Dicellura). These are small wingless forms in which the mandibles and maxillae are enclosed in pockets of the head formed by a union of the labium with the lateral walls of the cranium. From this common character these three orders may be termed the entognathous apterygote hexapods. They have other features, however, that distinguish them from the rest of the hexapods, and some entomologists are reluctant to call them insects. A second hexapod group is the Thysanura, including the well-known families Machilidae and Lepismatidae. The thysanurans also are wingless, but in other respects they are more closely related to the winged insects than to the entognathous apterygotes. The winged insects constitute the third hexapod group, named the Pterygota. The conventional division of the hexapods into Apterygota and Pterygota on the absence or presence of wings is a convenient device for making "keys," but clearly the absence of wings is not in itself an index of relationship. The wingless entognathous orders are less closely related to the wingless ectognathous Thysanura than are the Thysanura to the winged Pterygota. Even the three entognathous groups themselves differ in many respects from one another and are by no means closely related. Inasmuch as the anatomy of the hexapods is described in various readily accessible textbooks on entomology, only a few examples will be treated here, representing the three major groups given above.

DIPLURA

The members of the Diplura, or Dicellura, including the genera *Campodea* (fig. 74 B), *Japyx, Heterojapyx* (A), and others, are the most insectlike of the entognathous apterygote hexapods, and some entomologists class them with the Thysanura. The enclosure of the mandibles and maxillae in head pouches, and particular features of the head structure itself, however, leave no doubt that the diplurans belong with the Protura and Collembola and are no more related to the Thysanura than are these two groups.

As an example of the Diplura we may take the relatively huge *Heterojapyx gallardi* of Australia (fig. 74 A), some individuals of which attain a length of two inches, though the structure will not be essentially different from that of the much smaller, widely distributed species of *Japyx*. The japygid head is angularly ovate and carries a pair of large antennae, but eyes are absent. The elongate body consists of three thoracic segments bearing the three pairs of legs, and of a ten-segmented abdomen. The last three abdominal segments are strongly sclerotized as compared with those preceding, and the large tenth segment is armed with a powerful forceps. There is no trace of a segment beyond the forceps, the anus being situated in a depression between the bases of the pincers.

On the underside of the head (fig. 74 C) the long median labium, with a pair of small palpi (*Plp*), is seen to be united on the sides with the ventrally inflected lateral walls of the cranium. The mandibles and maxillae are thus enclosed in a pair of deep lateral pouches above the labium with only their tips (*Md, Lc*) exposed beyond the latter. The mandibles and maxillae, therefore, are not retracted into the head; they are merely covered below by the labium; their position in the pouches is seen in the cross section at F of the figure (*Md, Mx*).

The mouth of *Heterojapyx* is located anteriorly above the free margin of the labium, and just behind it is a large hypopharynx (fig. 74 D, *Hphy*) composed of two broad superlingual lobes and a small median lingua. A three-lobed hypopharynx is characteristic of the

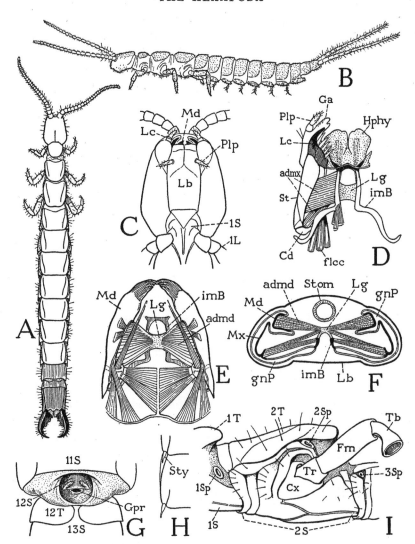

Fig. 74. Hexapoda—Diplura.

A, *Heterojapyx gallardi* Tillyard. B, *Campodea* sp. C, *Heterojapyx gallardi* Tillyard, head and part of prosternum, ventral. D, same, hypopharynx, inter-maxillary brachia, and right maxilla, ventral. E, same, mandibles and their muscles, dorsal. F, same, cross section of head, somewhat diagrammatic. G, same, genital region of male, ventral, exposed by separation of eleventh and twelfth body segments. H, same, right halves of two abdominal sterna, showing styli. I, same, mesothorax and parts of adjoining segments, ventrolateral.

For explanation of lettering see pages 337–339.

entognathous apterygotes and resembles the three-lobed hypopharynx of the isopod *Ligyda* (fig. 49 G). From a skeletal support in the base of the hypopharynx of *Heterojapyx* (fig. 74 D) two long rodlike sclerites (*imB*) extend posteriorly in the mesal walls of the gnathal pouches (F, *imB*) and then turn laterally to end behind the bases of the maxillae (D), which are articulated by the cardines (*Cd*) on them. These posthypopharyngeal sclerites have been regarded as representing the anterior apodemal tentorial arms of Thysanura and Pterygota (see Snodgrass, 1935, p. 118), since they give support to the adductor muscles of the mandibles and maxillae (D, F). However, they are not apodemes, but sclerites of the ventral head wall within the gnathal pouches, present also in Collembola, in which Folsom (1900) has shown that they are formed in the embryo as surface sclerotizations of the sternal wall of the head. Similar rods are present likewise in Protura, but their anterior parts are united in a median sternal bar. In *Heterojapyx* the anterior parts of the sclerites form internal ridges (F, *imB*), but their posterior parts are superficial. These sternal sclerites of the head are peculiar to the entognathous apterygotes among the Hexapoda, but they are exact counterparts of the intermaxillary sternal brachia of the amphipod and isopod crustaceans (fig. 49 H, *imb*), on which are supported the first maxillae.

In the Japygidae and in *Campodea* the parallel parts of the two sternal brachia of the head are connected, inside the head, by an arched membranous bridge (fig. 74 D, E, F, *Lg*), on which are attached adductor muscles of the mandibles and the maxillae. The interbrachial bridge of the Diplura, therefore, evidently represents the intergnathal ligament of the chilopods and diplopods, which in the diplurans has become attached secondarily to the sternal brachia, since the brachia, being intermaxillary in position, cannot belong to the mandibular segment of the head. The entognathous apterygote hexapods have no tentorium corresponding with that of the Thysanura and Pterygota, but in the Collembola the bridge of the sternal brachia is elaborated into a platform for muscle attachments supported on arms of the brachia.

The mandibles of *Heterojapyx* are slender, elongate organs (fig. 74 E) produced distally into simple, toothed gnathal lobes. The base of each jaw is connected with the mesal wall of the gnathal pouch (F, *Md*) by a long oval foramen, and the inner end projects into the pouch as a free point (E). The dipluran mandible has no articulation with the cranium either anteriorly or posteriorly, but in Protura and Collembola the inner end of the mandible is connected by a slender rod in the pouch wall with the lateral cranial margin, just as in the chilopods. The principal muscles of the dipluran mandibles (E) include posterior muscles from the cranial wall, attached dorsally and ventrally on the base of the mandible, and adductors (*admd*) from the interbrachial ligament (*Lg*) inserted into the cavities of the jaws (F).

The maxillae of *Heterojapyx* (fig. 74 D) have the structure typical of insect maxillae, except for the reduction of the palpus (*Plp*). The base of each appendage is divided by a joint into a small proximal cardo (*Cd*) articulated on the sternal brachium and an elongate stipes (*St*) bearing two apical lobes, a galea (*Ga*) and a lacinia (*Lc*). The adductor muscles (*admx*) arise partly on the sternal brachia and partly on the connecting ligament (F). The lacinia has a large cranial flexor (D, *flcc*) characteristic of the insect lacinia. The labium, being united with the lateral walls of the cranium (C, *Lb*), forms practically a part of the ventral wall of the head.

In the thorax of *Heterojapyx* each segment has a distinct tergal plate (fig. 74 A, I), but the pleural and sternal parts are united before the bases of the coxae on each side (I). Precoxal folds of the pleuron somewhat resemble the subcoxal sclerites of the chilopods. The dipluran thorax has an unusual feature in the presence of four spiracles, in most insects there being only two. The first spiracle (I, *1Sp*) has a lateral position between the prothorax and the mesothorax; the second (*2Sp*) lies above the coxa of the mesothorax just below the margin of the tergum; the third (*3Sp*) is on a level with the first in the anterior part of the metathorax; the fourth lies above the metathoracic coxa in the position of the second on the mesothorax. The two more dorsal thoracic spiracles fall in line with the abdominal spiracles. The legs have the six segments characteristic of insect legs (fig. 83 A), but the tarsus is undivided. The pretarsus bears a pair of large lateral claws and a very small median claw; its under part forms an unguitractor plate on which the tendon of the flexor muscle is attached.

The ten segments of the abdomen (fig. 74 A) are alike, except the ninth which is short and the tenth which is longer than the others. On the undersurfaces of the first seven segments are small paired styli arising from the posterior lateral angles of the sternal

275

plates (H). Each stylus is provided with two small muscles. In *Campodea* (B) the styli are much larger and serve to support the abdomen. The tergum of the short ninth abdominal segment of *Heterojapyx*, or body segment 12, underlaps the venter (G, *12T*), and between it and the sternum of the preceding segment (*11S*) is a deeply infolded membranous pocket (exposed by separation of the segments in the figure), which contains the genital opening, or gonopore (*Gpr*), of the male. In front of the gonopore is a small plate bearing a pair of styli, which evidently is the forwardly displaced ninth abdominal sternum (*12S*). The apical forceps is perhaps serially homologous with the paired styli of the preceding segments. In *Campodea* (B) the corresponding appendages are long, multiarticulate filaments resembling the antennae. Each jawlike prong of the forceps of *Heterojapyx* has a large adductor muscle, but no abductor, the pincers remaining open when not muscularly closed. Some interesting observations on the use of the forceps by *Japyx* are given by Kosareff (1935). *Japyx* feeds on other small, soft-bodied arthropods such as collembolans and symphylans. As described by Kosareff, the prey is first seized either by the jaws or by the forceps; in either case the abdomen is turned forward over the back and the prey, grasped in the forceps, may be carried around until a suitable place is found for feeding on it; but again, *Japyx* may devour the prey at once while the latter is held in the forceps.

THYSANURA

The thysanurans, or bristletails, are small wingless insects of two principal types of structure represented by the families Machilidae (fig. 75 A) and Lepismatidae (fig. 76 A). They derive their name from the fact that the caudal end of the body bears three long, multiarticulate filaments; the middle filament is a prolongation of the tergum of the last abdominal segment; the lateral ones are appendages of the same segment corresponding with the cerci of lower winged insects. The most important features by which the Thysanura differ from the entognathous apterygotes are in the structure of the head, the mandibles, and the antennae. The thysanurans are ectognathous, the mandibles and maxillae being fully exposed and the labium a free appendage. The mandibles and maxillae, moreover, are articulated on the lateral margins of the cranium. Sternal sclerites of the head corresponding with the intermaxillary brachia of the Diplura are absent; they are replaced functionally, for muscle attachment, by internal cuticular apodemes that form a true tentorium unquestionably homologous with that of the Pterygota. The antennae consist each of a basal scape and a long, multiarticulate flagellum; their only intrinsic muscles arise in the scape and are inserted on the base of the flagellum (see Imms, 1939). The thysanuran antennae thus have the structure characteristic of the antennae of winged insects and differ radically from the segmented antennae of Diplura, Collembola, and the myriapods. The thysanuran labium has the structure of a generalized pterygote labium. It consists of a large proximal plate, or *postmentum* (fig. 76 H, *Pmt*), broadly attached on the head, and of a small, free *prementum* (*Prmt*) bearing the palpi and apical lobes. The prementum alone is movable; it has a pair of short muscles arising in the postmentum and other muscles from the tentorium.

The thysanuran body consists of 14 distinct segments, of which three belong to the thorax, but there is no constriction or separation between the thoracic and abdominal regions. Some of the abdominal segments bear styli similar to those of the dipluran *Campodea,* and in the machilids eversible vesicles are associated with the styli. The styli evidently do not represent abdominal appendages, since Heymons (1897) says the appendage rudiments present in the embryo of *Lepisma* unite entirely with the sterna, and the styli appear during postembryonic growth. In *Ctenolepisma urbana,* Lindsay (1940) notes that the ninth-segment styli appear in the fourth instar, those of the eighth segment in the ninth instar of the male and the eleventh of the female. The median caudal filament and the cerci arise from the last abdominal segment, but the muscles of the cerci take their origins in the penultimate segment. The male thysanurans have a simple median penis arising ventrally at the base of the ninth abdominal segment (fig. 75 J, *Pen*); the females have a long ovipositor formed of two pairs of slender processes borne on the eighth and ninth segments (K, *1Gon, 2Gon*).

Important differences between the Machilidae and Lepismatidae are in the mechanism of the mandibles, the structure of the hypopharynx, and the relative development of the tentorium. In these features the machilids are clearly more generalized than the lepismatids, in which the mandibles, the hypopharynx, and, to a lesser extent, the tentorium take on the structure of these parts typical of orthopteroid pterygotes.

The Machilidae— The machilids (fig. 75 A) can be distinguished from the lepismatids (fig. 76 A) by the more cylindrical body, the shorter cerci, and the greater number of styli, which support the abdomen like a series of small legs. The head (fig. 75 B) is of the hypognathous type of structure with the mouth parts hanging downward from the cranial margin behind a long, lobelike extension of the head below the antennae. The outer surface of the upper, or epistomal, part of the lobe constitutes a distinct *clypeus* (*Clp*) set off from the facial region, or *frons,* above it by a transverse *epistomal sulcus* (*es*); the lower part is the labrum (*Lm*). On the top of the head is a pair of large compound eyes, and behind them a deep, transverse postocular groove (*gr*). The posterior lateral angles of the cranium are extended posteriorly to give support on each side to the maxilla (*Mx*) and the labium (*Lb*). In the end of the cranial extension, between the bases of the two appendages, is a depression, the *posterior tentorial pit* (*pt*), from which is invaginated a tentorial bar that goes through the back of the head (C, *TB*).

The endoskeleton of the machilid head includes three separate tentorial elements; two may be termed the *anterior tentorial arms* (fig. 75 C, *AT*), the other is the posterior bar, or *tentorial bridge* (*TB*), mentioned above. The anterior arms arise from lateral points of invagination, the *anterior tentorial pits,* on the ventral side of the head behind the base of the clypeus. Between them is the wide base of the hypopharynx. The arms extend first mesally and then turn posteriorly along the sides of the stomodaeum; each arm gives off a slender dorsal branch (C, *DT*) attached on the upper wall of the cranium by short muscle fibers. The anterior tentorial arms of the machilids clearly represent the anterior arms of the pterygote tentorium, but, on the other hand, they are so similar to the head apodemes of the chilopods, the diplopods, the pauropods, and the symphylans as to suggest that the anterior head apodemes are homologous structures in all these groups. The posterior tentorial bridge of the machilids (fig. 75 C, *TB*) is not represented in the myriapods, but in position it corresponds with the posterior arms of the tentorium in the pterygotes.

The machilid hypopharynx is a flattened, three-lobed structure

278

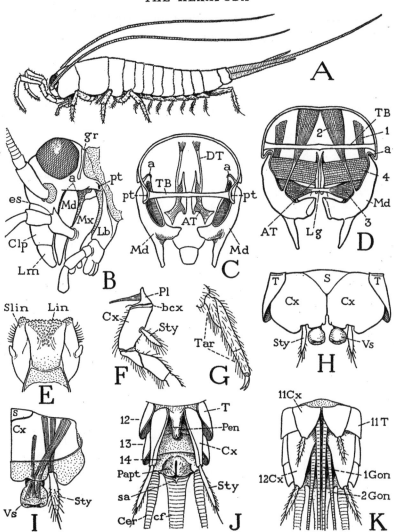

Fig. 75. Hexapoda—Thysanura. Machilidae.

A, *Machilis* sp. B, *Nesomachilis maoricus* Tillyard. C, same, posterior view of head, somewhat diagrammatic, with mandibles in place, showing endoskeleton. D, same, mandibles and their muscles, posterior. E, same, hypopharynx, posterior. F, *Machilis* sp., base of right metathoracic leg, lateral. G, same, tarsus and pretarsal claws. H, *Nesomachilis maoricus* Tillyard, ventral surface of a pregenital abdominal segment. I, same, right half of ventral plates of an abdominal segment, with stylus and vesicle, dorsal. J, same, terminal part of male abdomen, ventral. K, *Machilis* sp., end of female abdomen, with ovipositor, ventral.

For explanation of lettering see pages 337–339.

(fig. 75 E), as is that of the Diplura and Symphyla. The median lobe is known as the *lingua* (*Lin*); the lateral lobes are termed the *superlinguae* (*Slin*) because they arise anterior or dorsal to the lingua.

The mandibles of the Machilidae are elongate jaws (fig. 75 B, *Md*) hanging from single points of articulation (*a*) on the cranial margins behind the clypeolabral lobe. The free lower end of each jaw (C, *Md*) divides into a tapering incisor process and a thick molar process. The mandibular musculature (D) includes four muscles for each jaw, two dorsal in their origin and two ventral. The dorsal muscles are a small anterior rotator (*1*) and a large posterior rotator (*2*), both arising on the dorsal wall of the cranium. Of the ventral muscles, which are adductors, one is a broad, flat bundle of fibers (*4*) arising on the anterior tentorial arm and attached proximally on the mandible; the other (*3*) is a conical muscle with its fibers spreading into the cavity of the mandible from a median ligament (*Lg*) on which are attached the fibers of the corresponding muscle of the opposite jaw. The ligament lies just behind the base of the hypopharynx. The machilid jaw is the most primitive mandible found among the insects. A comparison with the mandible of the generalized crustacean *Anaspides* (fig. 38 E) will show an essential likeness in the structure of the jaws, their articulation, and their musculature, the only difference being the retention of the palpi in *Anaspides*. The same jaw structure is characteristic also of the branchiopod crustaceans. The musculature of the machilid mandible furnishes the basic pattern of the jaw musculature of Lepismatidae and lower Pterygota.

The thoracic region of the machilid body is distinguished by the large size of its three tergal plates (fig. 75 A) as compared with the abdominal terga. The pleural areas of the thorax are concealed beneath the overhanging lateral edges of the terga; in the mesothorax and metathorax each contains a single triangular pleural sclerite (fig. 75 F, *Pl*) bearing a large apodeme, situated immediately over the base of the coxa and forming an articular support for the latter. A faint submarginal groove of the coxa sets off a narrow marginal flange, or *basicoxite* (*bcx*). On the prothorax, the pleural sclerotization, which has been particularly described by Carpentier (1946), is more complex and somewhat resembles the pleural structure in Lepismatidae (fig. 76 I, J).

The legs are six-segmented, the long tarsus (fig. 75 G, *Tar*) is divided into three subsegments, and the pretarsus bears only a pair of lateral claws. A special feature of the machilid legs is the presence in most species of a styluslike appendage on the outer side of the coxa (F, *Sty*) of the second and third legs. These thoracic styli have no muscles.

The machilid abdomen is broadly joined to the thorax (fig. 75 A), and tapers somewhat posteriorly. It consists of 11 segments, but the last segment, which bears the caudal filament and the cerci, is mostly concealed from above by the tenth. On the undersurface of the abdomen are eight pairs of styli pertaining to the second to the ninth segments, inclusive. The ventral surface of each of the first seven segments in the female and the first eight in the male is covered by a pair of large, contiguous lateral plates (H, *Cx*) and a small median basal plate (*S*). The lateral plates, except those of the first segment, bear each a stylus (*Sty*) and are regarded as representing the coxae of primitive abdominal appendages, the median plate (*S*) being interpreted as the true sternum. In *Nesomachilis* each of the first seven coxal plates bears also an eversible vesicle (*Vs*) just mesad of the base of the stylus on the stylus-bearing segments. In some of the machilids several of the abdominal segments may have two pairs of vesicles. The abdominal styli and the vesicles are provided with muscles arising on the supporting coxal plates (I), the styli of the abdomen thus differing from those of the thorax, which have no muscles. On the ninth segment both the coxal plates and the styli are much longer than those of the preceding segments, a sternal plate is absent, and the long, narrow coxal plates are entirely separate (J, *Cx*). On this segment in the male (J) the median tubular penis (*Pen*) arises between the bases of the coxal plates. In the female (K) the coxal plates of both the eighth and ninth abdominal segments (body segments 11 and 12) are separated at their bases, and each bears a long, slender genital endite, or *gonapophysis* (*1Gon*, *2Gon*). The four rodlike gonapophyses are normally closely connected to form a tubular *ovipositor*. The opening of the oviduct lies between the bases of the first pair of gonapophyses. In the males of some species of *Machilis* similar but much shorter gonapophyses are present, as in the female, on both the eighth and the ninth abdominal segments, but they are entirely dissociated.

The thirteenth body segment (tenth abdominal) in each sex is a simple annulus (fig. 75 J, *13*) without either coxal plates or styli.

Following it is the short fourteenth segment (*14*), which carries the caudal filament (*cf*) and the cerci (*Cer*). On its undersurface is a pair of soft valvelike lobes (*Papt*) enclosing the anus, behind which is a small median lobe (*sa*), apparently on the base of the caudal filament. The paired lobes probably are the paraprocts of other insects; the postanal lobe (*sa*) may be the epiproct. The three anal lobes are regarded as representing the telson of more generalized arthropods.

The Lepismatidae— The lepismatids in their general appearance (fig. 76 A) are similar to the machilids, but in several respects they differ from the Machilidae, on the one hand, and, on the other, approach more closely to the Pterygota. The mouth parts are attached on the lateral margins of the cranium (B), but the mandibles (*Md*) have a long, approximately horizontal connection with the head and are doubly articulated. The primary articulation (*a*) of each jaw is posterior in a notch of the cranial margin; the secondary anterior articulation is by means of a small condyle on the ventrally inflected angle of the gena below the base of the antenna (C, E, *c*), a short distance behind the clypeus (C, *Clp*) and just outside the anterior tentorial pit (*at*). The mandibles thus swing transversely on horizontal anteroposterior axes (G, *a-c*), as do the jaws of the amphipods and isopods among the crustacea, and the jaws of the biting-and-chewing type in most of the pterygote insects. It is to be noted, however, that the doubly articulated mandible of the crustaceans and the pterygotes has its anterior articulation on the epistome, or clypeus.

The musculature of the lepismatid mandible includes the same muscles that operate the pendent mandible of the machilids, but the different articulation brings about a change in the action of the dorsal muscles on the jaw. In *Ctenolepisma urbana* (fig. 76 G) the anterior dorsal muscle of the machilid mandible (fig. 75 D, *1*) is represented by two muscles (fig. 76 G, *1a, 1b*) which are functionally dorsal abductors. The posterior dorsal muscle (*2*) becomes a dorsal adductor. Of the two ventral adductors, the proximal one (*4*) has its origin as in Machilidae on the anterior arms of the tentorium (C, *4*), but the distal muscle (G, *3*) of each mandible is attached separately on the base of the hypopharynx (D, *3*), as it is in all pterygote insects in which this muscle is retained.

The tentorium of the Lepismatidae resembles that of *Machilis*

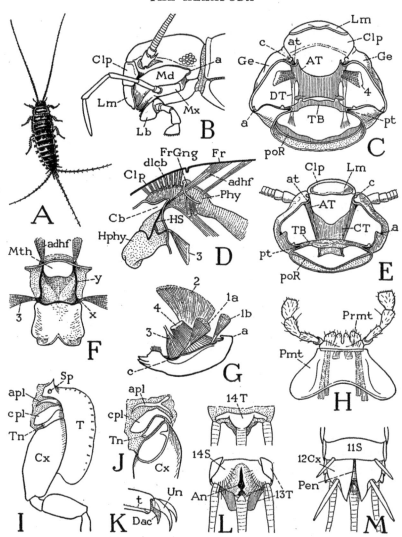

Fig. 76. Hexapoda—Thysanura.

A, *Thermobia* sp. B, *Ctenolepisma urbana* Slabaugh. C, same, section of head below level of tentorium. D, same, hypopharynx, cibarium, and pharynx, lateral. E, *Lepisma* sp., section of head below level of tentorium. F, *Thermobia* sp., hypopharynx, anterior. G, *Ctenolepisma urbana* Slabaugh, left mandible and muscles, dorsolateral. H, same, labium, ventral. I, same, base of left meso-thoracic leg and adjoining parts, ventral. J, same, base of right mesothoracic coxa and pleuron, mesal. K, same, end of tarsus and pretarsus. L, same, eleventh abdominal segment (body segment 14), dorsal and ventral. M, same, end of male abdomen, ventral.

For explanation of lettering see pages 337–339.

insofar as the anterior and posterior parts are not united, but it approaches the orthopteroid type of tentorial structure in that the anterior arms are confluent in a large central plate (fig. 76 C, E). In *Ctenolepisma* (C) the plate is relatively short and only touches on the posterior bridge, but in *Lepisma* (E) the plate widely overlaps the supporting bridge beneath it. In *Thermobia* the median part of the bridge is a delicate band so closely adherent to the undersurface of the plate that the writer formerly (1951) described the two parts as united, though Chaudonneret (1950) had shown that they are separate. The invagination points of the anterior tentorial arms (*at*), as already noted, are in the ventrally inflected, anterior lower angles of the genae (*C, Ge*) just mesad of the articular condyles (*c*) of the mandibles; those of the posterior bridge (*pt*) are in the posterior part of the cranium between the attachments of the maxillae and the labium.

The hypopharynx of the lepismatids is of particular interest because it has, fully developed, the typical structure of the hypopharynx in the lower pterygotes. The organ is a simple lobe projecting below the mouth (fig. 76 D, *Hphy*), supported by a U-shaped suspensorium (*HS*) on the proximal half of its anterior, or dorsal, surface, with oral arms (F, *y*) going through the mouth angles to give attachment to muscles from the frons (D, F, *adhf*), and a pair of lateral arms (F, *x*) on which the distal adductor muscles (F, D, 3) of the mandibles are attached. The preoral food cavity of the head runs back into a cibarial pocket (D, *Cb*) over the base of the hypopharynx, on the dorsal wall of which are attached a double row of dilator muscles (*dlcb*) arising on the clypeus (*Clp*) anterior to the frontal ganglion (*FrGng*) and its brain connectives. This entire structure is duplicated in the cockroach (fig. 79 F, G), but there is no suggestion of it in the Machilidae.

On the thorax each segmental pleural area of the lepismatids contains three fairly distinct superposed sclerites, or sclerotized folds, over the coxa (fig. 76 I, J). The uppermost sclerite is termed the *anapleurite* (*apl*), the middle one the *catapleurite,* or *coxopleurite* (*cpl*), and the one adjoining the coxa the *trochantin* (*Tn*). The nature and homologies of the pleural sclerites of Collembola and Thysanura have been the subject of considerable discussion (see Carpentier, 1946, 1947, and Barlet, 1950), but whether or not the pleural, or "subcoxal," sclerites as developed in the apterygote in-

284

sects and the chilopods are anything more than local sclerotization patterns is something the morphologists have not decided; it may be said, however, that there is no demonstrated reason for regarding them as remnants of a primitive subcoxal segment of the leg. The pretarsus of the lepismatid leg has a well-developed median dactyl (K, *Dac*), on which is attached the tendon (*t*) of the flexor muscle, and a pair of relatively long lateral claws (*Un*).

The abdomen of Lepismatidae has the same segmentation as that of Machilidae, but styli are present only on the eighth and ninth segments, or in some species on the seventh, eighth, and ninth. Eversible vesicles are absent. The ventral surfaces of the first eight segments in the male, and the first seven in the female, are covered by simple plates with no division into coxal and primary sternal plates as in the machilids. According to Heymons (1897), however, appendage rudiments in the form of lateral lobes of the sterna are present in the embryo of *Lepisma* on all the segments, but in the course of development they flatten out and finally unite completely with the sterna, except those of the eleventh segment, which elongate and become the cerci of the adult. On the ninth abdominal segment (twelfth body segment) of the adult male (fig. 76 M) is a pair of large, independent stylus-bearing plates (*12Cx*), evidently homologous with the coxal plates of the machilids (fig. 75 J, *Cx*), and between their bases is a simple, median penis (fig. 76 M, *Pen*). In the female similar coxal plates are present on both the eighth and the ninth segments and bear gonapophyses that form an ovipositor as in the machilids. The tenth abdominal segment (thirteenth body segment) is well developed in the lepismatids; its tergum (L, *13T*) is produced into a broad lobe over the base of the caudal filament. Beneath it is concealed the small tergal plate of the last segment (*14T*), which bears the caudal filament. Behind the ventral region of this segment (*14S*) are the lateral anal lobes and a small median postanal lobe on the base of the caudal filament. The abdominal structure is thus seen to be essentially the same in both the Lepismatidae and the Machilidae.

PTERYGOTA: THE COCKROACH, *PERIPLANETA*

The cockroach is here presented as a representative of the great group of winged insects known as the Pterygota, though in various anatomical respects it is not a typical pterygote. Most cockroaches that have fully developed wings are known to fly, some better than others, but they do not have the usual flight mechanism, and it is a problem to understand how they fly at all. On the other hand, the locomotor function is highly developed in connection with the legs, and the pleural and sternal parts of the thorax have acquired an atypical structure apparently to accommodate freedom of leg action. The head and the feeding apparatus, however, are excellent subjects for a study of the generalized structure of the insect head and the fundamental mechanisms of the pterygote mouth parts. But again, since the female cockroach encloses her eggs in a capsule, the ovipositor does not have the typical form and mechanism of this organ in most egg-laying insects, and the external genitalia of

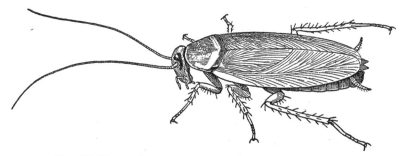

Fig. 77. Hexapoda—Pterygota. *Periplaneta americana* (L).

the male cockroach represent a special type of development of the organs of copulation and insemination. Yet, the cockroach is regarded as a generalized insect, and modern species appear to differ but little from their ancestors of Carboniferous times, which are among the oldest of known insects. In studying the cockroach, therefore, we are dealing with an ancient type of insect, and the problems it presents in comparative insect anatomy make it all the more interesting and instructive. Moreover, the cockroach has become a favorite subject for the study of insect physiology and for experimental work in determining the effects and mode of action of insecticides. *Periplaneta americana,* described in the following pages, in particular thrives and multiplies well in the confinement of breeding cages and is now almost everywhere available for classroom study.

The cockroaches are often called simply "roaches," but this abbreviation is not etymologically justified, because "cockroach" is a phonetic derivation from the Spanish *cucaracha*, and a roach is properly a kind of fish. In classical Latin the name *blatta* was applied to various insects, including cockroaches, but Linnaeus made it specifically a cockroach name, and from it the whole cockroach tribe is now called the Blattoidea by modern entomologists. Near relatives of the cockroaches are the mantids and the termites.

Periplaneta americana (fig. 77) is the largest of our common domestic cockroaches in the United States, but it is not so frequently found in houses as are some other species; it seems to prefer more roomy accommodations such as are offered by bakeries, mills, and restaurants, where there is also plenty of food and warmth. Our commonest home cockroach is the much smaller *Blattella germanica*, known as the waterbug, or Crotonbug, though the large black cockroach, *Blatta orientalis,* is a not infrequent visitor. In spite of their geographical names, the original homeland of these three cockroach species was probably Africa (see Rehn, 1945). In recent years another small cockroach with domestic habits, *Supella supellectilium*, the brown-banded cockroach, has been introduced and is already rather widely spread in many of the states. The giant cockroach, *Blaberus craniifer*, of Central America, occurs now in southern Florida. Besides the imported, so-called "domestic" cockroaches, there are also various native "wild" species that live out of doors, mostly in wooded places.

The Head

A cockroach at rest tucks its head back beneath the projecting margin of the shieldlike pronotum, so that from above little of the head is exposed. Museum specimens keep the head in this position, and consequently in pictures the cockroach nearly always has its head pulled back against the body with the face directed downward. Observation of a live cockroach in activity, inspecting a food source or in the act of feeding, however, will show that the head may be turned horizontally forward and that it is highly mobile in all directions. The head, in fact, is attached to the body by a large flexible neck and is pivoted on the ends of sclerites in the lateral neck walls.

In describing the cockroach head we may arbitrarily orient it

287

in the hypognathous position, in which the facial aspect will be forward, and this direction we can then call "anterior" and the opposite "posterior."

The head of *Periplaneta* is ovate in facial view (fig. 78 A) but is flattened anteroposteriorly (G). The antennae (A, *Ant*) arise from large, circular, dark-rimmed, widely separated membranous areas, or "sockets," on the upper part of the face, which are surrounded laterally and dorsally by the compound eyes (*E*). The top of the head between the eyes is called the *vertex* (*Vx*), the region between and below the antennae is the *frons* (*Fr*), and a ventral extension from the frons between the bases of the mandibles represents the *clypeus* (*Clp*). In most insects the clypeus is set off from the frons by a transverse epistomal sulcus; in the cockroach, what may be taken to be the upper limit of the clypeus is marked by a pair of short lateral grooves (*at*) above the bases of the mandibles, which are the roots of the anterior arms of the tentorium. The distal part of the clypeal region is membranous and supports the *labrum* (*Lm*). At the sides of the clypeus are the mandibles (*Md*), which close behind the labrum, and behind the mandibles are seen the maxillae (*Mx*). In a nymphal cockroach the vertex of the head is divided by a median line that forks downward to the antennal sockets. This inverted Y-shaped line, present in most young insects, has commonly been called the "epicranial suture" and has been regarded as an important structural character of the head. Actually, however, it is a line of weakness in the nymphal cuticle where the latter will split at moulting to permit the ecdysis of the succeeding instar. At the last moult the cleavage line is not renewed in *Periplaneta,* though in some insects a trace of it may remain on the adult head.

The filamentous antennae of the cockroach consist each of three parts characteristic of the insect antenna in general. The large basal segment is the *scape,* beyond which is a much smaller segment known as the *pedicel,* followed by the long, multiarticulate *flagellum.* The scape is supported on a pivotal process, or *antennifer* (fig. 78 G, *af*), from the ventral rim of the antennal socket and is provided with three thick muscles arising on the tentorium, inserted on the base of the scape at three sides of the pivot, so that the antenna can be moved freely in all directions. The only muscles within the antenna are a dorsal and a ventral muscle arising in the scape, which are inserted on the pedicel and serve as a levator and depressor of the

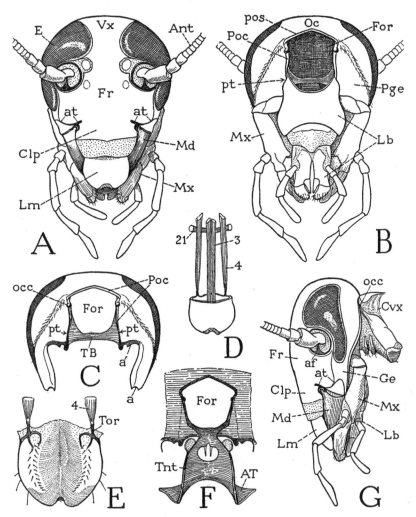

Fig. 78. Hexapoda—Pterygota. *Periplaneta americana* (L.). The head.

A, head, anterior. B, head, posterior. C, posterior wall of cranium, appendages removed, showing articulation of mandible at *a*, and of maxilla at *a'*. D, labrum and its muscles, anterior; *21*, transverse muscle of frons. E, labrum, inner surface. F, posterior wall of cranium surrounding the neck foramen, and tentorium, anterior. G, head and neck, lateral.

For explanation of lettering see pages 337–339.

flagellum. The pedicel contains a chordotonal sense organ that probably registers the flagellar movements.

The large compound eyes of the cockroach have a lateral position on the head, but they are widened anteriorly above the bases of the antennae. On the upper part of the face, in the angles between the eyes and the rims of the antennal sockets, are two small, pale, oval areas in the position of the usual lateral ocelli of other adult insects, which are often called the "ocellar spots" of the cockroach. Beneath each cuticular disc is a small cellular body connected by a nerve with the protocerebral part of the brain. The structure of these organs in *Blatta orientalis* has been described by von Reitzenstein (1905) and by Haller (1907), but the accounts by the two writers are not entirely in agreement. In *Periplaneta* the cornealike cuticle is slightly convex outwardly but is of uniform thickness; in *Blatta* the central part of the disc is slightly thickened and thus has a lenslike form, as shown by von Reitzenstein, but in a dissection no such plug-shaped inner projection, such as that shown by Haller, is to be seen. Von Reitzenstein says that each organ is formed by invagination and has the structure of the "inverted" eyes of spiders; according to Haller, the structure of the adult organ does not suggest an origin by invagination. Both writers agree, however, that the body beneath the cuticular disc includes a central mass of cells produced individually into nerve fibers that come together to form the nerve trunk going to the brain, and that there is an entire absence of the usual elements of an ocular organ, such as rhabdoms and pigment. Apparently no experiments have been made on the function of the organs, but their position and brain connections leave little doubt that they represent lateral ocelli that have not developed in the usual manner. The mantis has three ocelli of ordinary structure.

Below the ocellar spots may be seen two other superficially similar but smaller spots. These spots, however, are merely the attachment points of a short transverse muscle on the inner surface of the frons (fig. 78 D, *21*).

On the side of the head (78 G) a narrow area behind and below the eye, separated from the frons by a subocular groove, is the *gena* (*Ge*). The maxillae (*Mx*) and the labium (*Lb*) are seen to hang from the back of the head. Above their bases the head is attached to the body by the large neck (*Cvx*) and is supported on the ends of a pair of lateral neck plates.

290

On the back of the head (fig. 78 B) the most prominent feature is the large, rectangular foramen (*For*), generally called the *occipital foramen*, corresponding to the *foramen magnum* of the vertebrate skull, which connects the cavity of the head with that of the body. The rear part of the vertex over the foramen is known as the *occiput* (*Oc*), but in the cockroach the occipital region is not anatomically demarked. The areas of the cranium at the sides of the foramen are the *postgenae* (*Pge*). Very closely surrounding the foramen dorsally and laterally is a groove, the *postoccipital sulcus* (*pos*), which sets off a narrow marginal rim of the foramen termed the *postocciput* (*Poc*). Near the upper ends of the lateral parts of the postocciput are two small knobs, the *occipital condyles* (C, *occ*), by which the head is articulated on the lateral neck plates (G). Ventrally the neck foramen is closed only by the large basal plate of the labium (B, *Lb*), which is suspended by its lateral angles from the postocciput. On each side at the base of the labium is a slitlike depression (B, *pt*) in the lower end of the postoccipital sulcus. The two depressions are the *posterior tentorial pits*, marking the sites of the invaginations that formed the posterior part of the tentorium (C, *TB*).

When the labium and the maxillae are removed from the head (fig. 78 C), the posterior part of the tentorium is seen extending like a bridge (*TB*) through the back of the head between the two posterior tentorial pits. From the lower ends of the postocciput the cranial margins curve laterally and downward to the bases of the mandibles, bearing on each side near the foramen a small knob (*a'*) on which the maxilla is articulated, and ventrally a socket (*a*) for the posterior articular condyle of the mandible. It should be noted now that the maxillae hang from the cranial margins *anterior* to the postoccipital sulcus, and that the labium is suspended by its basal angles from the postocciput *behind* the sulcus. Between the two on each side is a posterior tentorial pit. The dorsal arc of the maxillary segment, therefore, would appear to be represented by the postgenae and the occiput, while the postocciput should be a remnant of the labial segment. If so, the posterior bridge of the tentorium is an intersegmental invagination.

The tentorium of the cockroach is formed as in most other insects by the union of a pair of anterior arms with the posterior bridge, and the bridge itself is formed by the union of two posterior arms.

291

The invagination points of the anterior arms vary in different insects, but the posterior arms are always ingrowths from the postoccipital sulcus. In the cockroach, as in Lepismatidae (fig. 76 E), the anterior tentorial arms are confluent in a large central plate (fig. 78 F, *Tnt*), but the union is not complete since there is left an oval aperture for the passage of the nerve connectives from the brain to the ventral ganglion of the head. Since the insect head has no sternal sclerotization, the tentorium serves as a substitute for bracing the lower edges of the cranial walls, and for the attachment of the ventral muscles of the mouth parts. The tentorial plate of the cockroach slopes steeply upward and backward between its anterior and posterior supports. The pharynx lies above it and turns down to the mouth between the anterior arms.

The neck of the cockroach is a complex structure enabling the insect to move the head freely on the body. On each side are two large neck sclerites flexibly joined to each other, which may be termed the dorsal and ventral *lateral cervical plates* (fig. 81 D, *lcvpls*), though the two of the lower pair almost come together on the ventral side of the neck. On the dorsal surface are two small, weakly sclerotized *dorsal plates*, one behind the other, the first of which is attached to the postoccipital margin of the head. On the ventral surface are two narrow *gular sclerites* (*gu*), which are mere transverse ridges. The upper ends of the dorsal lateral plates are articulated on the occipital condyles of the head (fig. 78 G, *occ*) and are the fulcral supports on which the head moves. The neck musculature, as described by Carbonell (1947), includes muscles that retract the head against the thorax, muscles that protract the head by extending the neck, and muscles that tilt the head up or down on the neck fulcra, or turn it from side to side. In the retracted position of the head, the lateral neck plates take an oblique position against the prothorax, and the two of each side are strongly elbowed on each other. Extension of the neck is produced evidently by muscles from the head and from the pronotum attached on the lateral neck plates, the pull of which flattens the angle between the plates of each pair and thus protrudes the head. Few animals except insects can actually stretch their necks. Muscles from the thorax attached dorsally on the postoccipital ridge of the cranium, and others attached ventrally on the tentorium, serve antagonistically to tilt the head up and down on the cervical fulcra and probably produce also lateral movements. Carbonell lists ten pairs of neck muscles in *Periplaneta* concerned with movement of the head.

The nature of the insect neck is somewhat of a morphological problem. The anatomical relations of the posterior parts of the cranium seem to indicate that the postoccipital ridge and the posterior arms of the tentorium are inflections between the maxillary and the labial segments. If so, the narrow postocciput is a sclerotic remnant of the labial segment, and the primitive intersegmental line between the labial and prothoracic segments should be somewhere in the neck. The head muscles, however, go from the prothorax to the postoccipital ridge and the tentorium, and thus would seem to be continuous through two primary segments. There is no evidence whatever that the neck itself represents a segment, and the segmental status of the cervical sclerites is not clear.

The Feeding Apparatus

The parts of the insect concerned with the acquisition and ingestion of food include the three pairs of segmental appendages of the head associated with the mouth, and also the labrum, the inner, or so-called "epipharyngeal," surface of the clypeus, and the hypopharynx. The appendicular organs are the mandibles, the maxillae, and the labium.

The Labrum— The labrum (fig. 78 A, *Lm*) is a flat, hollow lobe of the head suspended from the clypeus. It is movable by muscles inserted on its base, and in the cockroach it is freely retractile because of the wide membranous connection with the clypeus. The labral muscles (D) arise on the upper part of the frons, one pair (3) being median and inserted anteriorly on the base of the labrum, the other pair (4) inserted laterally and posteriorly. The posterior muscles are attached on special sclerites, known as the *tormae* (E, *Tor*), of the inner wall of the labrum. The varied movements of the labrum by a live cockroach suggest that the muscles may act as antagonists in different combinations. In most insects the labrum is compressible by a pair of interior muscles between its outer and inner walls; in the cockroach these muscles take a transverse position.

The Mandibles— The mandibles of *Periplaneta* are strongly toothed jaws (fig. 79 A) suspended from the lateral margins of the cranium (B) so that they close upon each other within the preoral food cavity of the head between the labrum in front and the hypopharynx behind. On the mesal surface at the base of each mandible is a small, flat molar area, and proximal to it is a thin membranous flap. When the jaws are closed with the molar areas in contact, the teeth of the left mandible overlap anteriorly those of the right, and the two basal flaps project over the food trough, or sitophore (F, Sit), on the base of the hypopharynx.

Each mandible is attached to the head by an articular membrane all around its base, but it is specifically hinged to the cranial margin by an anterior and a posterior point of articulation on the outer side of its base (fig. 79 B, c, a). The mandibular articulations are on the free surface of the mandible and are of the ball-and-socket type of structure, but the parts are reversed in the two, the condyle of the anterior articulation (c) being on the cranium, that of the posterior articulation (a) on the mandible. The posterior articulation represents the single primary articulation of a mandible such as that of the Machilidae (fig. 75 B, a).

The musculature of the cockroach mandibles includes *four* distinct muscles for each jaw (fig. 79 C), two arising dorsally on the cranium, one on the tentorium, and one on the hypopharynx. Of the dorsal muscles, one is a relatively small abductor (27) arising on the side of the head and inserted by a slender tendon in the membrane at the base of the outer side of the mandible; the other (28) is a huge adductor of several bundles of fibers converging upon a broad tendon (t) attached at the inner angle of the base of the mandible. The power of the adductor muscle is increased by its long leverage mesad of the mandibular hinge on the cranium. The hypopharyngeal muscle of the mandible (29) has a narrow origin on a small branch (x) of the hypopharyngeal suspensorium, and its fibers spread to the inner surface of the lateral wall of the jaw. The fourth muscle is a small bundle of fibers (30) from the underside of the anterior arm of the tentorium (AT) to the posterior wall of the mandible. The function of the hypopharyngeal and tentorial muscles of the insect mandibles is not clear; these muscles, however, represent the important ventral adductors of a primitive mandible, which, though much reduced, are still retained in most of the lower insects, but the functions of adduction and abduction have been largely taken over by the dorsal muscles, which are the rotators of the primitive, singly articulated jaw. In the higher insects with biting and chewing mandibles, the ventral muscles have been eliminated,

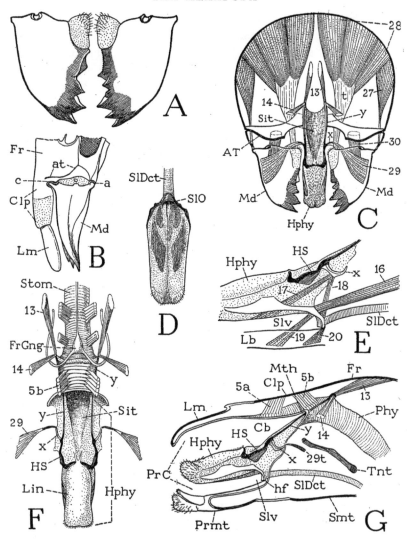

Fig. 79. Hexapoda—Pterygota. *Periplaneta americana* (L). The mandibles and the hypopharynx.

A, mandibles, anterior. B, left mandible and lower part of head, lateral. C, diagrammatic cross section of head, anterior, showing tentorium, mandibles and their muscles, and hypopharynx. D, lingual lobe of hypopharynx, posterior, showing opening of salivary duct on its base. E, basal part of hypopharynx supported on labium, showing hypopharyngeal muscles and salivary duct. F, hypopharynx and pharyngeal part of stomodaeum, with frontal ganglion on lower end of pharynx. G, diagrammatic longitudinal section of lower part of head, showing position of hypopharynx in preoral cavity.

For explanation of lettering see pages 337–339.

and the jaws are operated entirely by the dorsal muscles. The same evolution of the mandibular mechanism has taken place in the amphipods and isopods among the Crustacea. The mandibles of the cockroach can only bite and chew, they cannot reach out and grasp the food; it is the function of the maxillae to bring food back to the mandibles.

The Maxillae— The insect maxilla retains in its structure something of its leg origin. The large base of the appendage (fig. 80 A), representing the coxa, supports a jointed palpus, which in the cockroach has five segments comparable to those of an arthropod leg with two trochanters, a knee bend between the femur and the tibia, and a tarsus, but lacking the apical segment, or pretarsus. The basal part of the maxilla bears mesad of the palpus two large endites, an inner *lacinia* (*Lc*) and a lateral *galea* (*Ga*), and is itself divided by an elbowlike joint into a proximal *cardo* (*Cd*) and a distal *stipes* (*St*). The cardo and stipes clearly do not represent segments of the maxilla; there are no muscles between them, and the two parts have a common, wide, mesal opening from the head. The single articular support of the maxilla on the head is by means of a small process on the base of the cardo (D, *a'*) that articulates with the cranial margin on the back of the head (fig. 78 C, *a'*); otherwise both cardo and stipes have only a wide membranous connection with the head (fig. 80 A), which allows the maxilla a free movement on the cardinal articulation. The elbow joint between the cardo and the stipes is merely a mechanical device that permits the maxilla as a whole to be protracted and retracted.

The lacinia (fig. 80 A, *Lc*) is a rigid, flattened lobe tapering distally and ending with two sharp, incurved teeth, proximal to which is a weak, subapical process. The inner margin bears a fringe of long hairs, from which the lacinia gets its name. The galea (*Ga*) is a relatively soft, thick lobe with a hoodlike apical pad that partially encloses the end of the lacinia. The galea is so named from its fancied resemblance to a helmet. The galea and the lacinia are each individually movable. The galea has a large muscle (B, 42) from the base of the stipes; a muscle (41) attached on the base of the lacinia has its origin on the side of the stipes. The hinge of the lacinia on the stipes (A, *h*), however, allows the lacinia only an anteroposterior movement on the stipes, while keeping it firmly in position against the galea. The maxillary palpus is movable as a whole by two basal

296

muscles (C, 35, 36) arising in the stipes; the segments are musculated as shown in the figure.

The muscles of the maxillary base correspond with the muscles of the mandible insofar as they include dorsal muscles arising on the cranial wall and ventral muscles arising on the tentorium. In *Periplaneta* (fig. 80 B) there is a small dorsal muscle (31) inserted on the inner end of the cardo just beyond the cranial articulation (*a'*) and a much larger and longer dorsal muscle (32) inserted at the base of the lacinia. The ventral muscles of the maxilla, arising on the undersurface of the tentorium, include several compact bundles of fibers (33) going to the outer end of the cardo, and a large group of muscles (34a, b, c) attached on the posterior margin of the stipes. The ventral muscles of the maxilla represent the adductors of a generalized appendage, but the distal parts of the insect maxillae lie close against the sides of the hypopharynx (F, *Hphy*). Consequently the mesal pull of the tentorial muscles (E, 33, 34) flattens the angles between the cardines and stipites, with the result that the maxillae, instead of being adducted, are protracted beyond the end of the hypopharynx. Retraction then is effected by the contraction of the long cranial muscles (B, 32) of the stipites.

The maxillary movements are easily observed in a live cockroach while feeding. They are exactly comparable to the movements of a pair of human arms flexed outward at the elbows, with the hands held together, and successively protruded and retracted. The movements are always the same regardless of the stimulus; whatever is placed between the maxillary lobes elicits the same response. During feeding, particles of food are grasped by the lacinial teeth, enclosed by the galeal lobes, and brought back to the mandibles for mastication. In addition to their function in feeding, the maxillae are used for cleaning the antennae, the palpi, and the front legs.

The Labium— The insect labium, or at least a part of it, represents the second maxillae of other arthropods. In its generalized form, as we have seen it in the Thysanura (fig. 76 H), the labium consists fundamentally of two parts, an immovable basal part implanted on the back of the head, and a distal part, which bears the palpi and is movably suspended on the base. In the cockroach the labial structure (fig. 80 G) is the same as in the thysanurans, except that the free distal part of the organ has a wide membranous connection with the base. The labium as a whole is thus divided into a proximal

297

part conventionally termed the *postmentum* (*Pmt*) and a distal part called the *prementum* (*Prmt*). In the cockroach and various other insects, the postmental region contains two plates distinguished as the *mentum* (*Mt*) and the *submentum* (*Smt*). The submentum of the cockroach is a large plate of the posterior head wall just below the neck foramen (fig. 78 B); the mentum is a small, weakly sclerotized plate (fig. 80 G, *Mt*) in the membranous distal part of the postmental region. In some insects the mentum occupies the entire space between the prementum and the submentum, while in others there is only a single postmental plate. The prementum of the cockroach (*Prmt*) much resembles a pair of maxillae united at their bases, except for the lack of cardines. Each half of the body of the prementum (*St*) evidently represents the stipes of a maxilla. The prementum bears a pair of three-segmented palpi (*Plp*) and four apical lobes, the two median lobes being the *glossae* (*Gl*), the outer lobes the *paraglossae* (*Pgl*). The four lobes together are sometimes termed the *ligula.*

The labial musculature (fig. 80 H) includes muscles that move the prementum as a whole and individual muscles of the palpi, the glossae, and the paraglossae, which take their origins in the stipital lobes of the prementum. The premental muscles include two pairs of long slender lateral muscles (*43, 44*) from the posterior part of the tentorium, inserted distally and proximately on the prementum, and a pair of median muscles (*45*) from the submentum to the base of the prementum. The median muscles serve as retractors of the prementum; they always cross over the mentum, showing that this plate belongs to the postmental region of the labium. The mentum and the submentum have no muscles of their own, and they seem to have no counterparts in the maxillae. Some writers have regarded the submentum as the sternal plate of the labial segment of the head, but embryologists say it is derived from the embryonic labial appendages. Again, it has been supposed that the postmental sclerotization represents the cardines of the maxillae, but the cardines are movable parts of the maxillae and have their own muscles from both the cranium and the tentorium.

The functional underlip of the insect is the so-called prementum. It serves passively to close the preoral food cavity of the head behind the maxillae and the hypopharynx but takes no particularly active part in feeding, except that the glossae and paraglossae prevent

298

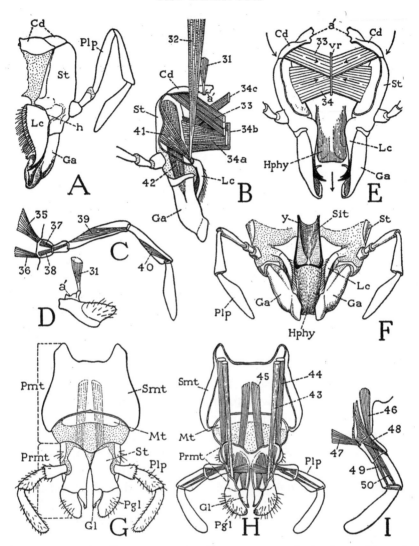

Fig. 80. Hexapoda—Pterygota. *Periplaneta americana* (L.). The maxillae and the labium.

A, right maxilla, posterior. B, right maxilla and muscles, anterior. C, maxillary palpus and muscles. D, maxillary cardo and muscle. E, diagram showing relation of maxillae to hypopharynx and protractor action of adductor muscles of maxillae, anterior. F, maxillae in position of retraction against sides of hypopharynx, anterior. G, labium, posterior. H, labium and its muscles, anterior. I, labial palpus and its muscles.

For explanation of lettering see pages 337–339.

the loss of food particles from the mandibles. The strongly musculated palpi (fig. 80 I), however, are extremely active, though their function appears to be mainly sensory.

To anyone who holds that the application of scientific terms should be consistent with their meanings, it is obvious that our current nomenclature for the parts of the insect labium is highly inconsistent. Since *mentum* means "chin," and *labium* means "lip," the term *labium* should be restricted to the true underlip of the insect, which is the part called prementum (literally, "prechin"). The postmentum then might not inappropriately be termed the *mentum*, or *mentum* and *submentum* if the insect has a "double chin," but anatomically it is quite incongruous that the "chin" should be a part of the "lip."

The Preoral Food Cavity and the Hypopharynx— The functional mouth cavity of the insect, into which the food is first received and where it is masticated by the jaws, is merely the space enclosed between the clypeus and labrum in front, the prementum of the labium behind, and the mandibles and maxillae on the sides. In an anatomical sense this space (fig. 79 G, *PrC*) is not a "mouth cavity," or "buccal cavity," as it is often called; it is properly a *preoral food cavity*, since the true mouth (*Mth*) is at its inner end and opens directly into the stomodaeal pharynx (*Phy*). The inner wall of the food cavity in the cockroach slopes downward and posteriorly from the mouth to the base of the prementum, and from it projects the large, tongue-like hypopharynx (*Hphy*). The mandibles close upon each other in the food cavity between the labrum and the hypopharynx. The part of the cavity proximal to the mandibles, therefore, serves as a food receptacle, or *cibarium* (*Cb*), for the masticated food passed back from the jaws. Between the base of the hypopharynx and the base of the prementum is a pocket, the *salivarium* (*Slv*), into which opens the salivary duct (*SlDct*).

The hypopharynx of the cockroach (fig. 79 F, *Hphy*) has a long sloping base on the oblique inner wall of the preoral cavity (G), and only its distal part is a free lingual lobe (F, *Lin*). In the lateral walls of the lobe is a pair of elongate sclerites, the tapering inner ends of which curve posteriorly and unite with each other *behind* the orifice of the salivary duct (D, *SlO*) in an arc that rests on the base of the prementum (G, *hf*) and serves as a fulcrum for the movements of the hypopharynx. Though it is commonly said that the opening of the salivary glands lies between the base of the hypopharynx and the base of the prementum of the labium, in the cockroach and some other lower insects the salivary duct opens actually on the base of the hypopharynx. This fact was noted in the cockroach by Miall and Denny (1886, p. 45), who says: "The common duct of the salivary glands enters the lingua, and opens on its hinder surface." As is well known, in such insects as Hemiptera, Diptera, and Siphonaptera the duct traverses the hypopharynx to open at its tip.

The anterior wall of the basal part of the hypopharynx, between the free lingual lobe and the mouth (fig. 79 F), is margined by a pair of rodlike sclerites (HS) that curve mesally at their lower ends in the base of the lingua, while their upper parts are continued as slender *oral arms* (y) that run through the angles of the mouth (F, G) to give insertion each to a pair of muscles, one muscle (13) arising dorsally on the frons, the other (14) ventrolaterally. The hypopharynx is thus hung from the cranium by a long U-shaped *suspensorium*. Small lateral branches (x) of the suspensorial arms give attachment to the hypopharyngeal muscles of the mandibles (C, F, 29). Between the suspensory arms the surface of the hypopharynx is depressed, forming a troughlike food channel, or *sitophore* (F, Sit), widening upward into the mouth.

The principal movements of the hypopharynx are production and reduction of the lingual lobe on the labial fulcrum (fig. 79 G, hf). The productor muscles include the dorsal frontal muscles of the suspensorial arms (F, G, 13) and a pair of long ventral muscles (E, 16) from the tentorium attached to the proximal parts of the lateral lingual sclerites, and perhaps also the pair of muscles (17) from the lateral branches (x) of the suspensorial arms to the posterior wall of the lingua. Antagonistic to the productors are the ventral frontal muscles of the suspensorium (F, G, 14) and a muscle from the labium (E, 19) attached on the lingual sclerites opposite the tentorial muscles (16). If the base of the hypopharynx is brought against the inner wall of the clypeus, the cibarium (G, Cb) will become a closed pocket, which can be expanded by the compressor muscles of the clypeus (5a, 5b) and contracted by the transverse muscles on the inner clypeal wall (F, G). There is little doubt that the cibarium thus acts as a sucking pump when the cockroach drinks liquids. It would seem that by some similar activity of the cibarium food stored on the sitophore of the hypopharyngeal base must be forced back

301

into the mouth, but the exact mechanism of food ingestion is not clear and has not been observed. The hypopharynx is not retractile toward the mouth, except to the extent that the base of the lingua moves upward with its forward movement on the labial fulcrum. The membranous lobes on the bases of the mandibles (A) very probably serve to press the masticated food back into the sitophore of the hypopharynx, but the food is still a long way from the mouth. In most of the insects that feed entirely on liquid food, the preoral cibarium has been developed into an efficient sucking pump operated by the clypeal muscles.

The Thorax

When the ancestors of the flying insects developed wings, the thorax was already differentiated as the locomotor center of the body, but it now had new responsibilities thrust upon it, and to meet them it had to undergo a considerable amount of reconstruction. Hence we find that the thoracic segments of flying insects, particularly the wing-bearing segments, have many structural features that make them quite different from the ordinary leg-bearing segments of other arthropods and from the simplified legless segments of the insect abdomen. Insect wings are merely flat outgrowths of the integument from the lateral parts of the dorsum of the mesothorax and the metathorax, and as such they may be compared to the tergal folds of the carapace that cover the gills in decapod crustaceans (fig. 41 D, *tf*); if the crustacean branchiostegites were extended outward from the body they would have the position of insect wings. In either case, the part of the body wall above the fold becomes specifically the *tergum,* and the part between the fold and the base of the leg is called the *pleuron.* In the decapod thorax the simple pleuron is the inner wall of the gill chamber, on which the legs are articulated; in the insect the pleura of the mesothorax and the metathorax, in addition to supporting the legs, furnish also the supports for the wings and give attachment to important wing muscles. The general similarity of the propleuron to the pleura of the winged segments might suggest that the fundamental structure of the thoracic pleura was established before wings were developed. On the other hand, it may be that in the glider stage of the wing evolution there were wing folds (paranotal lobes) also on the prothorax, since some ancient fossil insects do have such folds on the prothorax in addition to fully

302

developed wings. However, there is no end to the possibilities of speculation as to how insects acquired their wings, but there is no definite information on the subject, inasmuch as the very oldest of known winged insects already possessed two pairs of perfect wings. The tergum of a winged segment has undergone more extensive adaptational modifications than have the pleura, because the wings are movably supported on its lateral margins and the tergum itself becomes an important part of the mechanism of wing movement.

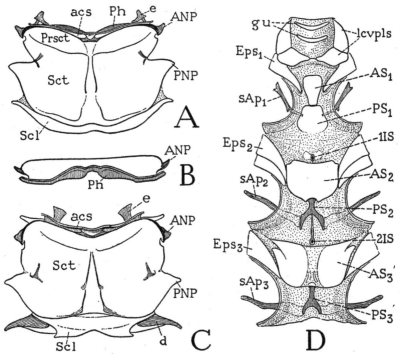

Fig. 81. Hexapoda—Pterygota. *Periplaneta americana* (L.). Tergal and sternal plates of the thorax.

A, mesonotum, dorsal. B, metanotum, anterior. C, metanotum, dorsal. D, ventral surface of neck and thorax.

For explanation of lettering see pages 337–339.

The thorax of the cockroach is in some respects more generalized than that of most other flying insects because the wing-bearing terga play a smaller part in the mechanism of flight; but, on the other hand, the pleura of the three thoracic segments have specialized features adaptive to a free movement of the legs. The thoracic musculature

303

of the cockroach is strongly developed in relation to the legs, but some of the usual flight muscles of other insects are much reduced or absent. A fully illustrated account of the thoracic musculature of *Periplaneta* is given by Carbonell (1947).

The Thoracic Terga— Though the word *tergum*, of Latin origin, is the general term given to the back plate of an arthropod body segment, the thoracic terga of insects are commonly called *nota*, from the Greek *noton*, because *notum* more properly combines with the Greek prefixes *pro*, *meso*, and *meta* that designate the three thoracic segments.

The pronotum of *Periplaneta* is a large, slightly convex, triangular plate with rounded corners set like a shield on the back of the prothorax. Its wide free margins overlap the retracted head anteriorly, the bases of the wings posteriorly, and the prothoracic pleura on the sides. The central disc of the shield gives attachment to muscles of the head, the neck, and the prothoracic pleura, coxae, and trochanters.

The mesonotum and the metanotum of *Periplaneta* are similar to each other in size and shape. Each plate is a wide, almost flat, rectangular sclerotization of the dorsum of its segment, with irregular lateral margins (fig. 81 A, C), and is divided into a large antèrior plate and a smaller posterior plate. The major part of the anterior plate is termed the *scutum* (*Sct*), but a short anterior part set off by a pair of weak transverse grooves is designated the *prescutum* (*Prsct*). The scutum is strengthened internally by a narrow, median V-shaped ridge indicated by two anteriorly convergent grooves on the outer surface. The narrow posterior plate behind the scutum of each tergum is the *scutellum* (*Scl*). The front margin of the prescutum is deflected into a deep, transverse groove (*acs*), termed the *antecostal sulcus* because in general it forms an internal submarginal ridge, or *antecosta*, of the tergum, on which the longitudinal intersegmental dorsal muscles are attached. In the wing-bearing segments of insects with large dorsal muscles the antecostae are produced into deep, usually bilobed plates, termed *phragmata*. The dorsal muscles of the cockroach, however, are very small, and each phragma is a relatively low, bilobed infolding from the antecostal sulcus (A, B, *Ph*). The antecostal sulci mark the true intersegmental lines of the thorax, so that the narrow anterior lip of each groove really belongs to the preceding segment, as does also the functional "intersegmental membrane" before it. In most strong-flying insects the precostal lip of both the metanotum and the first abdominal tergum is enlarged to form a wide plate, the *postscutellum*, or *postnotum*, which takes the place of the conjunctival membrane and firmly connects the adjoining terga. In the cockroach each intertergal membrane contains a pair of small connective sclerites (A, C, *e*) between the consecutive terga.

The structure of the lateral margins of the mesonotum and the metanotum of *Periplaneta* is characteristic of that of winged insects in general. Anteriorly on each side is a small but strongly developed *anterior notal wing process* (fig. 81 A, C, ANP), behind which is a small incision of the tergal margin, followed by a deeper indentation, and then a large triangular projection (*PNP*), which is the *posterior notal wing process*. The relation of these parts to the wing will be shown in the section on the wings. The scutellum of each segment (*Scl*) is produced laterally into tapering arms with which are connected the posterior margins of the wings. In the metathorax the scutellar arms (C, *d*) are detached sclerites supporting the basal membranes of the wings (fig. 84 D).

The Thoracic Pleura— The pleural areas of the thoracic segments of the cockroach do not closely conform in structure with the typical pleuron of a wing-bearing segment, shown diagrammatically at A of figure 82, but there is little difficulty in identifying their parts by comparison with the diagram. The sclerotic pleural wall is always marked by a deep, vertical or oblique groove, the *pleural sulcus* (A, PlS), that forms a strong ridge on the inner surface between the coxal articular process (*a*) below and the wing-supporting process (WP) above. The part of the pleuron in front of the groove and its ridge is the *episternum* (*Eps*), the part behind it, the *epimeron* (*Epm*). The episternum may be extended ventrally before the coxa to the sternum (S). Between the coxa and the precoxal arm of the episternum is a separate sclerite, the *trochantin* (*Tn*), closely connected with the episternum dorsally and articulated with the anterior margin of the coxa ventrally (*c*). Resting on the upper end of the episternum before the wing process is a small sclerite (*Ba*) termed the *basalare*, and in the pleuroalar membrane behind the wing process is a sclerite distinguished as the *subalare* (*Sa*). The

pleural structure in the prothorax is simpler than that of the mesothorax and metathorax because of the absence of the wing process
and the epipleural sclerites.

In the prothorax of *Periplaneta* (fig. 82 B) the pleural sulcus (*PlS*)
runs obliquely upward and forward from the coxal articulation (*a*)

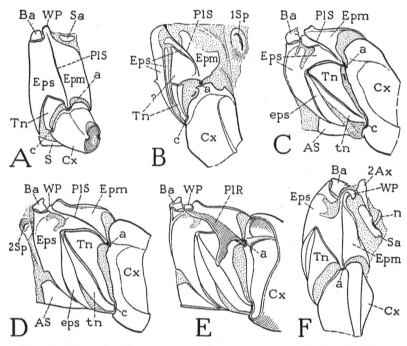

Fig. 82. Hexapoda—Pterygota. *Periplaneta americana* (L.). The thoracic
pleura.

A, diagram of principal parts of a generalized pleuron of a winged insect.
B, propleuron of *Periplaneta* with base of coxa and first spiracle, left side. C,
mesopleuron and base of coxa, left side. D, metapleuron with base of coxa and
second spiracle, left side. E, right metapleuron, inner surface. F, left metapleuron and base of coxa, dorsal, with subalar sclerite and second axillary of
wing base.

For explanation of lettering see pages 337–339.

to the tergum. Behind the sulcus is an irregular epimeral sclerotization (*Epm*); in front of it is the episternum and the trochantin. The
episternal sclerotization (*Eps*), however, is broken up into several
parts, and the trochantin (*Tn*) appears to be divided into a large,
triangular dorsal plate partially united with the episternum and a
narrow ventral sclerite articulating below with the coxa. At least, the

prothoracic trochantin is thus interpreted in the cockroach by Crampton (1927) and in the mantis by Levereault (1936). On the other hand, Fuller (1924) attributes the much smaller dorsal sclerite of the winged termite to the episternum. The cockroach itself is noncommittal, but the trochantinal muscles of *Periplaneta* are shown by Carbonell (1947) to be all inserted on the lower sclerite.

In the mesothorax (fig. 82 C) the pattern of the sclerotization is somewhat different from that in the prothorax, but the episternum (*Eps*) and the epimeron (*Epm*) are to be identified as the parts lying respectively before and behind the oblique pleural sulcus (*PlS*). The episternal surface is more continuously sclerotized than is that of the prothorax and is prolonged ventrally in a broader extension to the sternum. The trochantin (*Tn*) is a long, triangular sclerite; its anterior part is set off as a narrow marginal band (*tn*) that may be termed the *trochantinal apotome*. A similar *episternal apotome* (*eps*) is cut off from the posterior margin of the ventral extension of the episternum and partly overlaps the trochantinal apotome. The wing process of the mesopleuron is not fully visible in the lateral view (C), but in front of it is a distinct basalar sclerite (*Ba*) flexibly attached on the upper end of the episternum. A subalar sclerite lies in the pleuroalar membrane above the epimeron but is not shown in the figure.

The pleuron of the metathorax (fig. 82 D) closely resembles that of the mesothorax and will need no special description. It may be noted, however, that the pleural sulcus becomes increasingly oblique in the successive thoracic segments. A dorsal view of the metapleuron (F) gives a better view of the wing process (*WP*) than the side view and shows something of the relation of the pleuron to the wing base. Closely connected with the wing process is the second axillary sclerite of the wing (*2Ax*), with its posterior arm attached to the elongate subalar sclerite (*Sa*). Also attached on the subalare is a ventral stalk (*n*) of the third axillary, but these features will be more fully described in connection with the wing mechanism. On the inner surface of the pleuron (E) is seen the strong pleural ridge (*PlR*) along the line of the pleural sulcus on the outer surface (D, *PlS*), forming at its lower end a condyle for the coxal articulation (*a*) and ending above in the wing process (*WP*). Near the coxa the ridge is produced into a large apodemal arm. A similar pleural ridge and arm are present also in the prothorax and the mesothorax.

The Thoracic Spiracles— The cockroach, in common with most other insects, has ten pairs of spiracles, two pairs of which are on the thorax and eight on the abdomen. The two sets of spiracles, however, are quite different in structure, particularly in the nature of the closing mechanism. The spiracles of the first thoracic pair (fig. 82 B, *1Sp*) lie between the pleural areas of the prothorax and the mesothorax just in front of the bases of the forewings; those of the second pair (D, *2Sp*) are similarly situated, though at a somewhat lower level, between the mesothorax and the metathorax. The spiracles of each pair on the thorax are presumed to belong to the segment behind them, since a truly intersegmental position is an unlikely place for a respiratory orifice.

The first spiracles of *Periplaneta* are much larger than any of the others. Each of these spiracles appears externally (fig. 85 G) as an elongate oval mound with an oblique median slit; the wider anterior lip is entirely soft; the narrower posterior lip has a beveled edge with a strongly sclerotized margin. The outer slit opens into an atrial chamber from which the tracheae are given off. Below the middle of the spiracle the thickened edge of the posterior lip is produced in an obliquely transverse arm through the floor of the atrium to the base of the inner wall of the anterior lip, where it gives insertion to a fan-shaped muscle (*mcl*) arising anteriorly at the base of the spiracle. This muscle serves to close the spiracle by drawing the beveled edge of the posterior lip against the soft edge of the anterior lip. Four large tracheae are given off directly from the atrium above the muscle arm, and below it arises a single trachea that immediately divides into two branches.

The second spiracle (fig. 85 H) is smaller than the first and appears externally as a somewhat circular mound with an obliquely longitudinal crescentic slit. The larger and relatively rigid dorsal lip is hood-shaped with a strong, concave margin; the posterior lip is a soft, thick flap with a rounded margin. At the lower, anterior angle of the spiracular cleft is attached the narrow end of a fan-shaped muscle (*mcl*) arising ventrally at the base of the spiracle. The pull of this muscle closes the spiracle by bringing the ventral lip against the dorsal lip, as may be demonstrated by pressing down with a needle on the point of the muscle insertion. The spiracular orifice leads into a cup-shaped atrium, from which arise several tracheal trunks.

308

The Thoracic Sterna— The ventral surface of the thorax in the cockroach is largely membranous (fig. 81 D), but in each segment there are to be distinguished two sternal plates, one anterior, the other posterior. The plates have been called respectively the sternum, or eusternum, and the sternellum, or, in the more commonly used terms of Crampton (1927), the anterior plate is the basisternum, the posterior plate the furcasternum. The name "furcasternum" is based on the fact that in most of the higher insects the second plate carries a two-pronged apodemal fork. In the cockroach and in various other of the more generalized insects, however, the sternal apodemes are widely separated arms (D, *sAp*) that in no sense constitute a "furca." To avoid ambiguity in describing the cockroach, therefore, we may call the anterior sternal plate of each segment the *antesternite* (*AS*) and the posterior plate the *poststernite* (*PS*). In the prothorax the antesternite (*AS₁*) is a relatively small, pear-shaped plate, the somewhat larger poststernite (*PS₁*) carries the sternal apodemes (*sAp₁*) on its lateral margins. In the mesothorax and the metathorax the antesternites (*AS₂, AS₃*) are large, shield-shaped plates, and the poststernites (*PS₂, PS₃*) are small Y-shaped sclerites, each attached by the stem to the preceding antesternite and bearing the long sternal apodemes on the ends of its divergent arms. Between the prothorax and the mesothorax, and between the mesothorax and the metathorax, are two very small intersegmental sternites (*1IS, 2IS*), commonly called the *spinisternites* because each supports a small spinelike median apodeme.

The Legs

The legs of the cockroach are typical of the legs of insects in general. Each leg (fig. 83 A) has six segments. There is only one segment in the trochanteral region; a sharp knee bend takes place at the femorotibial joint; the tarsus is subdivided into five tarsomeres; the pretarsus bears a pair of lateral claws and a median adhesive lobe. Only in the dragonflies among the insects are there two trochanteral segments; there are never more than five tarsal subsegments, but the number may be fewer; the pretarsus undergoes various modifications, but a median claw is present in adult insects only in the apterygotes.

The legs of *Periplaneta* are essentially all alike in structure. When the insect is at rest (fig. 77), the coxae lie back against the sides of the body, with the first legs directed forward, the hind legs stretched

309

out posteriorly, and the middle legs taking whatever position is convenient. During activity the slenderer and more mobile fore coxa are turned downward and the first legs are directed forward. These legs appear to determine the course of the insect when walking or running, while the others serve as the chief organs of locomotion. The front legs hold up the fore part of the body and the head during feeding, and when an antenna needs cleaning, one of them reaches up and pulls the base of the antenna down to the lobes of the maxillae. The coxae of the middle and the hind legs show little activity during ordinary walking, the principal movements of these legs being at the coxotrochanteral and the femorotibial joints. Yet the coxae are strongly musculated and they can turn forward and backward on their pleurotrochantinal hinges (fig. 82 C, D, a, c). Furthermore, as has been well explained by Carbonell (1947), each coxa can be adducted on its pleural articulation (a) by the trochantin, which swings mesally and forward on its dorsal connection with the episternum, while the trochantinal apotome folds beneath the episternal apotome. The trochantin thus appears to be an important part of the mechanism for the coxal movement. Carbonell has shown that there are six muscles from the tergum and the episternum attached on the trochantinal apotome. The greater mobility of the fore coxa is due to the freely flexible connection of the coxa with the pleuron by the small lower trochantinal sclerite (fig. 82 B), on which all the muscles of the trochantin are attached.

For a study of the structural details and the mechanism of the cockroach legs we may take one of the hind legs because of its greater size. In the usual position of the hind leg, as it is stretched out posteriorly in a horizontal plane (fig. 77), the true anterior surface is ventral and the posterior surface dorsal, but for descriptive purposes the leg will be assumed to be extended laterally from the body, so that "anterior" refers to the surface ordinarily turned downward and "posterior" to the opposite surface.

The large flat coxa of the hind leg of *Periplaneta* (fig. 83 A, *Cx*), is hinged almost transversely on the body between its pleural and trochantinal articulation (a, c). On each coxa are attached ten muscles from the body, most of which are either promotors or remotors, but the coxal muscles include a large muscle from the upper angle of the coxa (the *meron*) to the subalar sclerite of the wing base, which evidently is an important wing muscle. The movements of the coxa regulate the position of the leg on the body, but the principal movement of the leg as a whole is at the coxotrochanteral joint.

The trochanter, though a very small segment closely attached to the base of the femur (fig. 83 A, *Tr*), is the most strongly musculated segment of the leg. It rocks on an anteroposterior hinge on the end

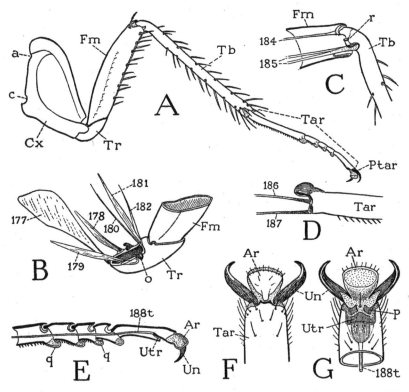

Fig. 83. Hexapoda—Pterygota. *Periplaneta americana* (L.). Structure of a leg.

A, left hind leg, anterior. B, trochanter and base of femur, with tendons of trochanteral muscles. C, the knee joint between femur and tibia. D, base of first tarsomere and its muscle tendons. E, tarsus and pretarsus, with tendon of pretarsal muscle. F, end of tarsus, and pretarsus, dorsal. G, same, ventral.

For explanation of lettering see pages 337–339.

of the coxa between strong anterior and posterior points of articulation (*B, o*), and its movements cause a flexion or extension of the leg in the plane of the coxa. As shown by Carbonell (1947), there are three trochanteral depressor muscles and three levators. The principal depressor consists of five branches, three of which are body muscles arising on the tergum, the basalar sclerite of the pleuron, and the apodeme of the sternum, all of which are inserted on a large, flat, spatulate apodemal tendon (B, *177*) attached on the proximal end of the trochanter. The two other depressor muscles arise anteriorly and posteriorly in the coxa and are inserted on separate anterior and posterior tendons (*178, 179*). The three antagonistic levator muscles have their origins in the coxa and are inserted on three slender tendons (*180, 181, 182*) attached to the trochanter distal to the axis of rotation. The depression of the leg at the coxotrochanteral joint is the force that lifts the body when the tarsus is pressed against a support, or enables the leg to give a forward thrust against the body by flattening the knee joint.

It should be noted that the tendons of the trochanteral muscles are not attached literally on the wall of the trochanter but on the articular membrane close to the trochanteral margin; the same is true of most other insect tendons, otherwise they would be too rigid. The so-called "tendons" of arthropod muscles are really apodemes on which the muscle fibers are inserted; they are ingrowths of the body wall with a cuticular core. Muscles without tendons are inserted directly on the integument.

The femur is set closely on the oblique distal end of the trochanter (fig. 83 B, *Fm*) and has little movement other than a slight posterior flexion on the trochanter. A single flat muscle, the *reductor femoris,* arises in the trochanter and is inserted on the posterior margin of the femoral base. The obliquity of the trochanterofemoral joint in the dorsoventral plane of the leg allows the muscles of the trochanter to be effective in lifting or depressing the leg as a whole beyond the coxa. The femur is a long, thick segment; its size is due to the fact that it contains the large and important muscles of the tibia that move the part of the leg beyond the knee.

At the femorotibial joint the tibia has a dicondylic articulation on the end of the femur (fig. 83 C) with an anteroposterior axis, so that the tibial movement at the knee is one of levation and depression in the plane of the femur. The fibers of the tibial muscles completely fill the cavity of the femur, and their tendons of attachment on the tibia are easily seen. The tendon of the single levator muscle (*184*) pulls on the head of the tibia well above the articular axis (*r*); the tendons (*185*) 'of two depressor muscles arise close together from wide bases in the ventral articular membrane of the joint. An ample infolding of the ventral membrane allows the tibia to be closely flexed ventrally against the femur.

The tibiotarsal joint differs from the other joints of the leg in that the tarsus articulates on the tibia by a large dorsal knob on its base (fig. 83 D) that fits into an overhanging concavity on the end of the tibia. The tarsus thus has a free movement on the tibia, but it is flexible principally in a forward or ventral direction. It is provided with only two muscles, a depressor with its tendon (*187*) attached ventrally on the tarsus and a promotor (*186*) attached anteriorly.

The subsegments of the tarsus (fig. 83 E) are freely movable on each other, but they are not connected by muscles, which fact may be taken to mean that the tarsomeres are not true leg segments. The entire tarsus is traversed by the tendon of the pretarsal muscle (*188t*), the fibers of which arise in the tibia and the femur; the pull on the tendon very evidently must cause a ventral flexion of the tarsus when the latter is not held taut by the pretarsal claws.

The pretarsus projects from the end of the last tarsomere. The body of this apical segment of the leg is a soft, hollow lobe, termed the *arolium* (fig. 83, E, F, G, *Ar*); from the sides of its base arise the paired lateral claws, or *ungues* (*Un*). The dorsal wall of the arolium (F) contains a spatulate plate; the broad distal end (G) presents an oval disc with a smooth membranous surface devoid of setae. The arolium is an adhesive organ for holding to smooth surfaces that the claws cannot grasp. The large decurved claws are hollow outgrowths from the base of the arolium; dorsally they are articulated on the end of the tarsus (F). The claws of the insect foot are commonly called by entomologists "tarsal claws," but they clearly belong to the pretarsus; their articulation on the tarsus means nothing more than does the articulation of the tarsus on the tibia. That the pretarsus is a true end segment of the insect leg corresponding with the clawlike apical segment, or dactylopodite, of a myriapod or crustacean leg is evident from the fact that it has its own muscle.

In the articular membrane at the base of the undersurface of the pretarsus (fig. 83 G) is a large plate, the *unguitractor* (*Utr*), on which is attached the tendon (*188t*) of the pretarsal muscle. Intervening between the unguitractor plate and the bases of the claws are two small accessory sclerites (*p*), termed *auxilliae*, and the proximal end of the unguitractor is deeply inserted into a pocket of the tarsus (E, G) to allow of its retraction. The pull of the pretarsal

313

tendon on the unguitractor is directly transmitted to the bases of the claws, so that the latter are turned downward until their points hold on the irregularities of any ordinary surface. When the insect finds itself on a smooth, hard surface, however, the claws can offer no resistance, they then turn forward and the pull of the tendon is now transferred to the arolium, which is turned downward until its end disc is pressed flat against the surface. The arolium maintains its hold by simple adhesion; it leaves no smear when lifted or moved from one position to another, showing that there is no exuded substance to give it its adhesive properties. The same may be observed in flies and other insects walking on the undersurface of a glass slide.

The Wings

The insects are the only flying animals, except the winged creatures of fiction, that have not paid for their wings by the functional loss of a pair of legs. The wings of insects are new organs added to the primitive locomotor equipment; they are flat integumental folds of the back, strengthened by radiating cuticular thickenings that form hollow, branching ribs known as the *veins*. Blood circulates freely through each wing along definite courses, in general going outward from the body in the channels of the anterior veins and returning through those of the posterior veins, the transfer from one system to the other being by way of cross veins between the main branches (see Yeager and Hendrickson, 1934, and Clare and Tauber, 1939).

The mechanism of the wing movement is fairly simple, and the motor power in most insects is derived from muscles that probably were present in the wingless ancestors of the hexapods. The base of each wing is pivoted from below on the pleuron and is hinged to the edge of the tergum somewhat mesad of the pleural support. All that is needed to produce an up-and-down movement of the wings, therefore, is an alternate depression and elevation of the wing-bearing tergum. The tergal movements in most insects are brought about by dorsoventral tergosternal muscles and longitudinal intertergal muscles. The dorsoventral muscles depress the tergum and cause the elevation of the wings; the longitudinal muscles arch the tergum upward and produce the downstroke of the wings. It must be noted, however, that the arching effect of the intertergal muscles on the back plates is dependent on a close union between the two wing-bearing terga and between the second wing-bearing tergum and the first abdominal tergum. Otherwise the force of the muscle contraction would be expended in pulling the consecutive terga together. In most flying insects these two sets of *indirect wing muscles* are greatly enlarged in each of the wing-bearing segments. Since this mechanism of wing movement is as fully developed in the Ephemeroptera (Dürken, 1907) and Plecoptera (Grandi, 1948, 1949) as in the higher insects, it would seem probable that it gave the first up-and-down movement to the primitive wing folds when the latter became flexible on the body. In order that the wings may act as propellers, however, it is necessary that each wing should have a torsion on its long axis. Anterior and posterior movements of deflection are produced by muscles inserted on the basalar and subalar sclerites of the pleuron closely associated with the wing base respectively before and behind the pleural fulcrum.

The wings of most insects have, in addition to the flight mechanism, a mechanism of flexion and extension in a horizontal plane, by which the wings when not in use are folded posteriorly over the back and brought forward preliminary to flight. A piece of stiff paper attached by one edge to a support cannot be turned horizontally without crumpling; conversely, the wing has a crumpling device in its base which automatically turns the wing posteriorly in a horizontal plane. If the wing is broad, the same mechanism may cause it to fold lengthwise on itself. The flexing apparatus is operated by a special flexor muscle; extension probably results from the pull of the basalar muscle on the anterior angle of the wing base, but the subalar muscle must contribute to the extension movement by restoring the crumpled area of the wing base to a flat condition.

The dragonflies appear to have a wing mechanism different from that of other flying insects, inasmuch as their wings are moved by muscles attached directly on the wing bases laterad and mesad of the pleural fulcra. In a study of the dragonfly muscles it has been shown by Clark (1940), however, that nearly all the wing muscles can be identified with muscles of ordinary flying insects, suggesting that the flight mechanism of the dragonflies is merely a specialization of the primitive musculature and not something radically different from that of other flying insects. Dragonflies do not flex the wings, and consequently all the muscles associated with the wings can be used for flight. The dragonfly's mechanism of flight appears

315

to be the simplest and most efficient way of moving the wings, but it could not be derived so directly from the structure of a nonflying ancestor as that of other insects. The thorax of a dragonfly is highly specialized to accommodate its wing mechanism (see Sargent, 1937, 1951). Though the dragonflies are among the oldest of known insects, being fully developed in Carboniferous times, they are not primitive insects and are not ancestral to any of the other insect orders.

By contrast with the dragonflies, the cockroaches, the mantids, and the termites have a well-developed wing-flexor mechanism, but a very poorly developed mechanism of flight. Since the mayflies have the usual flight muscles of other insects, but no wing-flexing mechanism, it would seem that flexion of the wings was a secondary innovation. It follows, then, that the cockroaches, mantids, and termites must be descended from insects that had a fully developed flight mechanism; their diminished power of flight is due to a great reduction or absence of the dorsoventral and longitudinal dorsal muscles of the wing-bearing segments. Still these insects are able to fly, and in the following study of the wing mechanism of *Periplaneta* we shall attempt to understand how they do it.

The two pairs of wings of *Periplaneta* are very different in size and form. The large hind wings undoubtedly are the principal organs of flight, the forewings, or *tegmina*, serving as coverings over the more delicate hind wings when the latter are flexed and folded over the back. The tegmina show by their venation that they are wings of the same type of structure as the hind wings, but the wing structure will be most easily understood from a study of one of the metathoracic wings.

A hind wing of the cockroach when spread out flat (fig. 84 A) has the appearance of two wings stuck together along a line of folding (*vf*). The part before the fold is the wing area principally effective in flight and may be termed the *remigium* (*Rm*); the large, fanlike expansion behind the fold is sometimes termed the *neala,* implying that it is a secondary enlargement of the wing, but its shape suggests the name *vannus* (*Vn*), which is Latin for "fan." The narrow area mesad of the vannus, separated from the latter by a second fold (*jf*), is the *jugal region*. The venation is entirely different in the two major parts of the wing, but the principal veins of the remigial region are those characteristic of the wings of insects in general.

316

The first two veins are the *costa* (fig. 84 A, *C*) and the *subcosta* (*Sc*), which in the cockroach are unbranched and together form a strong marginal thickening along the basal half of the wing. The third vein is the *radius* (*R*), which has two main stems and numerous terminal branches. The fourth vein is the *media* (*M*) with only a few terminal branches; in the adult wing the media is united basally with the radius, but the corresponding trachea in a nymphal wing has an independent origin (see fig. 117 in Comstock, 1918). The next vein is the *cubitus* (*Cu*), and is the most profusely branched vein of the wing. Behind the cubitus are three long, slender, unbranched veins termed the "plical veins." Only the third one, however, is a true plical vein, since it lies in the vannal fold; it is sometimes called the *vena dividens*. The other two may be designated simply *postcubitals* (D, *Pcu*), though in most insects there is only one postcubital in front of the fold. The veins of the fan are the *vannal veins*, and those of the jugal region the *jugal veins*. The vannal veins all arise from an arcuate vein in the base of the fan and thus form a very distinct group of veins separated from those of the remigium by the vannal fold. The postcubitals and the vannal veins together are the *anal veins* of the Comstock-Needham system of vein nomenclature, but in the nymphal wing of the cockroach a postcubital trachea is independent of the vannal group, and in the mechanism of the adult wing the postcubitals belong to the remigium. Details of the branching of the veins will be found to differ somewhat in different specimens, and between the main branches are small secondary veins and numerous cross veins not shown in the figures.

The wing veins have definite relations to small sclerites, the *axillaries*, in the base of the wing (fig. 84 D), which determine the wing movements of extension, flexion, and folding. There are three principal axillaries, generally termed the *first axillary* (*1Ax*), the *second axillary* (*2Ax*), and the *third axillary* (*3Ax*), but a *median plate* (*m*) associated with an arm of the third axillary is also an important element of the wing-base mechanism.

The first axillary (fig. 84 B, D, *1Ax*) is a triangular sclerite produced anteriorly in a long neck that rests on the anterior wing process of the notum and abuts against the head of the subcostal vein (*Sc*). By its mesal margin the body of the first axillary is closely hinged to the edge of the notum and is thus the principal hinge plate of the wing. The second axillary (2Ax) is an elongate sclerite lying just laterad of the first (D); anteriorly it is connected with the head of the radial vein (R) and posteriorly with the third axillary, but it has a long ventral process (B, g) closely attached to the large subalar sclerite (Sa) in the pleuroalar membrane below the wing. The body of the second axillary rests on the pleural wing process, and this axillary is therefore the pivotal plate of the wing base. The third axillary (B, D, 3Ax) has a complex structure; mesally it presents a large, concave disc (B, h) on which is attached a muscle from the pleural apodeme; laterally it is extended in a posterior arm (k) associated with the bases of the cubital and postcubital veins of the wing (D), anteriorly it gives off a larger arm (l) closely hinged to the edge of the median plate (m) of the wing base. The crescent-shaped muscle disc (B, h) articulates by its anterior horn (i) with the posterior end of the dorsal part of the second axillary (f), as seen at D, and by its posterior horn (B, j) with the posterior wing process of the notum (D). Ventrally, the third axillary is strongly connected by a stalklike arm (B, n) with the posterior part of the subalar plate (Sa).

The third axillary, being the only plate of the wing base on which a muscle is attached, is the active agent for the flexing and folding of the wing. The downward pull of the pleural muscle on the muscle disc of this axillary revolves the axillary on its articular points (B, i, j) and turns the lateral arms (k, l) upward. The posterior arm (k) pulls directly on the bases of the cubital and postcubital veins (D), the anterior arm (l) forms a sharp upward fold with the median plate (m); the whole remigium in consequence swings posteriorly on the hinge of the subcosta with the first axillary, which latter turns vertically on the edge of the notum. As the remigium is thus turned posteriorly and mesally by the revolution of the third axillary, the wing is folded along the line of the vannal plica and the hinge between l and m, so that the remigium goes over the vannus, and the latter is turned upside down beneath it (C), until finally the folded wing takes a longitudinal position over the back of the abdomen.

Extension of the wing is perhaps initiated by the tension of the basalar muscle transmitted through the basalar sclerite to the humeral angle of the wing, but an extensor action of the basalare is not easy to demonstrate on a dead cockroach. It seems probable, how-

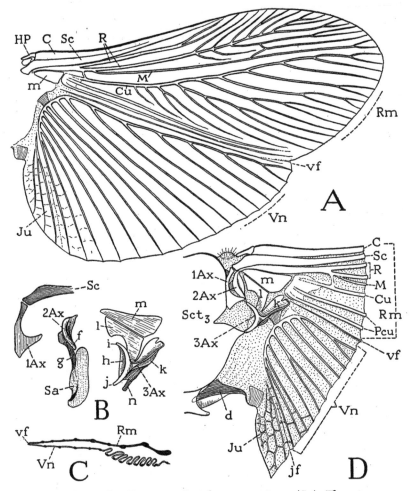

Fig. 84. Hexapoda—Pterygota. *Periplaneta americana* (L.). The wing.

A, right hind wing fully spread out. B, axillary sclerites, subalar sclerite, and median plate of wing base, dorsal. C, section of folded right wing. D, base of right wing and attachment on metanotum.

d, alar sclerite of metanotum; *f,* dorsal process of second axillary; *g,* ventral process of second axillary connecting with subalar sclerite; *h,* muscle disc of third axillary; *i,* articular point of third axillary with *f* of second axillary; *j,* articular point of third axillary with posterior notal wing process; *k,* posterior lateral arm of third axillary; *l,* anterior lateral arm of third axillary; *m,* median plate of wing base; *n,* ventral subalar process of third axillary. For explanation of other lettering see pages 337–339.

ever, that, since the third axillary is strongly connected with the subalar plate, the pull of the subalar muscle would restore this axillary to its position in the expanded wing and thus automatically bring about an extension of the wing.

The flight mechanism of the cockroach is not well understood. Most winged cockroaches fly; some are fairly good flyers, and even *Periplaneta* is known to fly on occasions, but the cockroach ordinarily puts its trust in the locomotor efficiency of its legs rather than in its wings. The dorsoventral muscles that in most insects produce the upstroke of the wings by depressing the notum are entirely absent in the cockroach. Three pairs of oblique dorsal muscles from the middle of the metanotum to the first abdominal tergum can hardly be supposed to have a depressor action on the metanotum, but a pair of large flat tergopleural muscles attached on the lateral edges of the notum should pull down on the notal margins and thus serve to elevate the wings on the pleural fulcra. Also, there are attached on the notum numerous muscles of the legs that could effect an elevation of the wings.

The mechanism of the important downstroke of the wings, on the other hand, is difficult to understand, considering the small size of the dorsal intertergal muscles and the fact that the wing-bearing nota of the cockroach are connected by intervening membranes, so that it would seem whatever force the dorsal muscles may exert would be expended in merely pulling the back plates together. As we have already noted, however, both the second axillary and the third axillary of the wing base are strongly connected with the subalar sclerite, and the latter gives attachment to a huge muscle from the meron of the coxa. This muscle, therefore, must have some important action on the wing. Probably in most insects the subalar muscle contributes to the downstroke of the wings. In the cockroach the subalar muscle has a very oblique, almost horizontal, position, and yet it appears to be the only muscle capable of a depressor action on the wing. The thoracic musculature of strong-flying cockroaches, such as *Parcoblatta*, appears to be the same as that of *Periplaneta*.

The thoracic musculature of the mantis, as described by Levereault (1938) for *Stagmomantis carolina*, is very similar to that of the cockroach, except that in the mantis there are no dorsal longitudinal muscles in either of the winged segments, and there is present in the mesothorax a pair of slender tergosternal muscles. Yet the mantis is a relatively strong flyer, though its progress on the wing is slow and is not sustained for long distances. The downstroke of the wings of the mantis Levereault suggests may be produced by certain muscles that pull backward on the basalar sclerites and thus presumably arch "the tergum over the wing fulcra," but Levereault does not mention the subalar muscle as a possible wing depressor, though this muscle is strongly developed in the mantis and is less oblique than in the cockroach. In the winged termites the thoracic musculature, as described by Fuller (1924), is again much the same as that of the cockroach, though longitudinal dorsal muscles appear to be entirely absent, and in the mesothorax there is a small anterior tergosternal muscle. Fuller does not discuss the wing mechanism of the termite, but he notes that the third axillary is closely associated with the subalar sclerite, and that on the latter is attached a huge muscle from the meron of the coxa.

The mesothoracic wings, or tegmina, of *Periplaneta* are a little longer and not so broad as the outspread hind wings, but they have the same general plan of structure and pattern of venation. When the tegmen is flexed, however, there is no folding between the remigium and the vannus; but the triangular jugal region folds completely beneath the vannus, allowing the rest of the tegmen to take a flat position over the folded hind wing (fig. 77). The parts of the axillary mechanism that produce the folding of the hind wing, such as the lateral arms of the third axillary and the median plate of the wing base, are consequently much reduced in the tegmen, but the connection of the third axillary with the subalare is even stronger than in the metathorax. When the wings are folded, the left tegmen partly overlaps the right; the same is true in the mantids and the termites.

The Abdomen

The abdomen of the cockroach consists of ten segments with distinct dorsal plates and includes a group of terminal lobes around the anus that probably represent an eleventh segment. Inasmuch as entomologists commonly designate the segments of the insect abdomen numerically as abdominal segments distinct from the thoracic segments, the abdominal segments will be numbered in the following descriptions as segments 1–11, and not as body segments 4–14.

General Structure of the Abdomen— The abdomen of *Periplaneta* (fig. 85 A) is broad and somewhat flattened anteriorly, narrowed and more rounded posteriorly (D). On the back there are ten fully formed tergal plates, but the ninth tergum in the male (B, *9T*) and both the eighth and the ninth terga in the female (C) are narrow and may be largely overlapped by the tergum in front. The tenth tergum is a terminal shield-shaped plate in each sex (B, C, *10T*) with a deep posterior emargination; at the sides of its base arise the cerci. On the venter there are nine sternal plates in the male (A) and only seven in the female (C). The first sternum is reduced to a small median plate (A, *1S*) in an otherwise membranous ventral part of its segment. The ninth sternum of the male carries a pair of slender styli (A, *Sty*), the only styli present in the cockroach. The seventh sternum of the female (C, *7S*) supports a pair of large, oval apical lobes (*otL*) turned upward against the anal region. The rounded shape which these lobes give to the end of the abdomen at once distinguishes the female of *Periplaneta* from the male. In the male (A) a group of genital structures (*Phl*) projects from the end of the abdomen.

The lateral margins of the tergal and sternal plates are closely adjacent, but they are separated on the sides by infoldings of the integument between them (fig. 85 D) that allow a dorsoventral expansion and contraction of the abdomen. When the tergum and the sternum of a segment are pulled apart (E), it is seen that the walls of the infold on each side contain small sclerites, two adjoining the tergum, which may be termed *laterotergites* (*ltg*), and one closely connected with the sternum, which may be designated a *laterosternite* (*lst*). The laterotergites of each side include a small anterior plate and a long, narrow posterior plate, with a spiracle (*Sp*) between them. Some writers regard these lateral sclerites of the abdomen as "pleurites," perhaps because the spiracles have a pleural position on the thorax, but in most adult insects the abdominal spiracles are in the lateral parts of the terga.

Though the infoldings along the margins of the abdominal segments allow of a dorsoventral expansion of the abdomen, the cockroach makes no perceptible abdominal movements of respiration, such as those of most other orthopteroid insects. At times, however, there may be observed a lengthwise protraction and retraction of the segments on each other.

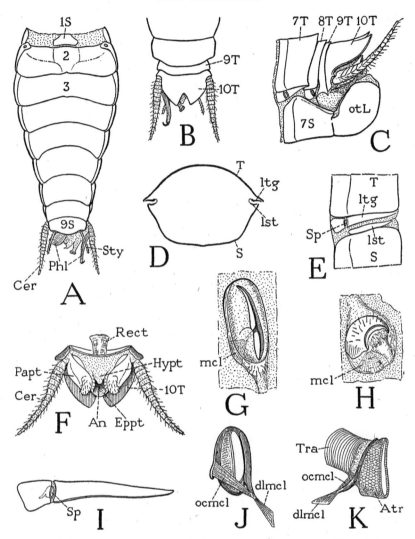

Fig. 85. Hexapoda—Pterygota. *Periplaneta americana* (L.). The abdomen and the spiracles.

A, abdomen of a male, ventral. B, end of male abdomen, dorsal. C, end of female abdomen, left side. D, cross section of posterior part of abdomen. E, adjacent parts of tergum and sternum of fifth abdominal segment separated to show intervening laterotergites and laterosternite, left side. F, terminal part of abdomen, ventral. G, first thoracic spiracle, left. H, second thoracic spiracle, left. I, laterotergites and spiracle of an abdominal segment, left. J, inner view of an abdominal spiracle, with trachea removed. K, lateral view of a left abdominal spiracle with stump of trachea attached.

For explanation of lettering see pages 337–339.

The infolded membrane between the fifth and sixth abdominal terga of the male of *Periplaneta* forms a deep pocket, and at the bottom of it may be seen two small, transverse, slitlike pouches with thick, soft walls. These apparently glandular organs probably correspond with the more highly developed dorsal glands of some other cockroaches (see Oettinger, 1906), which are said to produce a secretion sought after by the female at the time of mating.

Beneath the tenth tergum is the short, conical eleventh segment of the cockroach abdomen (fig. 85 F), known as the *proctiger* because it contains the anus (*An*), which latter is surrounded by four small apical lobes. In the lateral walls of the segment are a pair of thick sclerites, the *paraprocts* (*Papt*), the setigerous ends of which form the lateral anal lobes. Above the anus is a short, rounded lobe, the *epiproct* (*Eppt*), and below it the ventral wall of the segment is produced into a small *hypoproct* (*Hypt*). At the bases of the paraprocts arise the cerci (*Cer*), which in the cockroach evidently belong to the eleventh abdominal segment, as they do in the Thysanura (figs. 75 J; 76 L), though in many insects in which the eleventh segment is reduced or obliterated the cerci are taken over by the tenth, and in the cockroach the cerci have basal articulations on the tenth tergum (C).

The Cerci— The long, tapering cerci are movable by muscles attached on their bases, and each appendage is divided into about 15 short rings, so that it is freely flexible and not easily broken. The somewhat flattened upper surface is smooth except for the presence of minute hairs; the convex undersurface, however, is covered with longer, slender hairs. Each cercus is traversed by a large nerve from the last abdominal ganglion. It has been shown by Pumphrey and Rawdon-Smith (1936) that the cerci of *Periplaneta* and *Gryllus* bear sense organs receptive to sound. On applying to a cercal nerve fine electrodes connected with an amplifier and a recording apparatus, these investigators demonstrated that exposure of the cerci to sound from a loud speaker produces an electric response in the nerve that varies with the pitch of the sound. Covering the ventral hairs of the cerci of the cockroach with vaseline abolished the nerve response, indicating that these hairs are the sound receptors. A live cockroach usually carries its cerci fully exposed in an erect position, and Pumphrey and Rawdon-Smith add that it seems probable that the function of the cercus in the cockroach "as a wind-gauge may be equal in importance to its function as an acoustic organ."

The Abdominal Spiracles— The spiracles of the abdomen are much smaller than those of the thorax, and each presents externally merely a vertical slit in a slightly protruding, soft, oval marginal rim. The spiracles of the first pair lie dorsally in small membranous areas at the sides of the first abdominal tergum. The other seven are situated between the two laterotergal plates (fig. 85 E, I, *Sp*) of their respective segments. The spiracles of the second segment are exposed ventrally (A) but the others are concealed by the overlapping posterior angles of the preceding sterna.

The simple opening of an abdominal spiracle is directed posteriorly and leads into a cup-shaped atrium (fig. 85 K, *Atr*); the atrium opens through a short membranous ring into a single, large tracheal trunk (*Tra*). The walls of the atrium are hexagonally reticulated, but the inner margins are thickened. From the marginal thickening on the outer (anterior) side of the atrial wall there projects forward and downward a tapering, fingerlike process, and from the inner (posterior) wall a shorter and wider process. Between these two processes is stretched an occlusor muscle (J, *ocmcl*), and on the end of the outer process is attached an opening, or dilator, muscle (*dlmcl*) that arises on the small laterotergal plate before the spiracle. The abdominal spiracles are thus closed and opened by a valvelike mechanism that controls the atrial entrance into the trachea, and in this way they differ mechanically from the thoracic spiracles that are closed by bringing the outer lips together.

The Male Genitalia— At the end of the abdomen of a male *Periplaneta* (fig. 85 A) are seen a number of prongs, hooks, and lobes projecting from beneath the tenth tergum (B) and above the ninth sternum (A). These structures are the external genital organs of the male. By removing the tenth tergum (fig. 86 A) it will be seen that the various parts pertain to three separate organs, two of which are dorsal, one right (*rPhm*), the other left (*lPhm*), and the third ventral (*vPhm*). The complex genitalia of the male cockroach serve principally for clasping and holding the female during copulation; they do not form an intromittent organ for the injection of sperm. A spermatophore is introduced into the genital chamber of the female and attached to the mouth of the spermatheca. The term "penis

325

valves" often given to the genital organs of the male cockroach is, therefore, inappropriate; if the group of organs is termed the *phallus,* the three components are *phallomeres* (i.e., phallic parts).

The entire phallic complex of *Periplaneta* can be traced back in its development, as shown by Qadri (1940), to a single pair of small *primary phallic lobes* that are present on the venter of the tenth abdominal segment in first and second instar nymphs (fig. 86 F, *phL*). Between the lobes an ectodermal ingrowth forms the beginning of the ejaculatory exit duct, which later unites with the terminal ampullae (*Amp*) of the vasa deferentia (*Vd*) and thus becomes the outlet of the mesodermal reproductive organs. In the third instar the primary phallic lobes have increased in size and each has divided into two parts (G). Those of the right lobe, according to Qadri, become the right phallomere and the ventral phallomere of the adult; the parts of the left lobe (*lPhm*), however, are not completely separated and together form the duplex left phallomere of the adult. The gonopore (*Gpr*), or opening of the ejaculatory duct, lies between the two dorsal phallomeres and above the base of the ventral phallomere. The early developmental history of the genitalia in all the orthopteroid insects, as shown by Qadri, is similar to that in *Periplaneta,* but the final development into the adult organ may be quite different in the several orders.

From the work of a number of investigators it is now known that in most of the insects above the Orthoptera the external genital organs of the male take their origin, as in the Orthoptera, from a pair of phallic lobes at the sides of an ingrowth that will become the ejaculatory duct. Each lobe, moreover, divides into two branches, but one branch, which may be termed the *mesomere,* is mesal, and the other, known as the *paramere,* is lateral. From here on the development is quite different from that in the Orthoptera. The mesomeres unite to form a tube, the lumen of which is continuous with that of the ejaculatory duct, and the structure so formed is the intromittent organ known as the *aedeagus* (or the *mesosome* in Diptera). The parameres and the aedeagus may retain a common base, or the parameres may become disconnected from the aedeagus and form independent clasping organs. On the other hand, genital claspers may be the styli of the ninth segment, as they certainly are in the Ephemeroptera and perhaps in some other insects, or again they may be the cerci of the eleventh segment. Finally, the copulatory apparatus

326

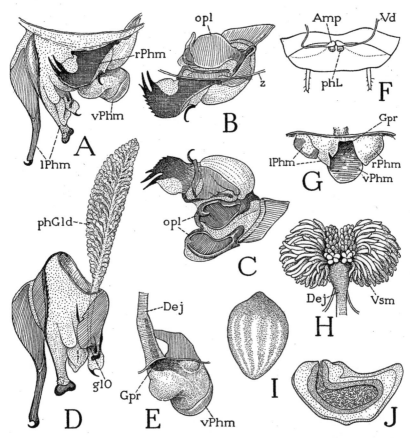

Fig. 86. Hexapoda—Pterygota. *Periplaneta americana* (L.). The male genitalia and the spermatophore.

A, genital organs as seen from above, projecting from anterior wall of genital chamber. B, right phallomere, dorsal, ventral parts exposed by removal of wall of genital chamber, cut off at *z*. C, right phallomere, dorsal, turned forward to expose ventral plates (*opl*), which are artificially opened. D, left phallomere and phallic gland, dorsal. E, ventral phallomere and ejaculatory duct, dorsal. F, primary genital lobes of second instar (from Qadri, 1940). G, genitalia at later stage, showing division of primary lobes into four parts. H, accessory glands and seminal vesicles on anterior end of ejaculatory duct. I, spermatophore, diameter 1.8 mm. (from Gupta, 1947a). J, section of spermatophore (from Gupta, 1947a).

For explanation of lettering see pages 337–339.

in many insects becomes complicated by the development of various secondary accessory structures. Since the aedeagus and the parameres have a common origin from the primary genital lobes, they are both phallic organs; other structures of a different origin may be grouped as periphallic organs. Because of their endless diversity of structure, the male genitalia of insects furnish excellent characters for the separation of species, but, in spite of the large amount of work that has been done on them, the homologies of their parts and the primary nature of the organs are still far from certain. Some writers have regarded the phallic lobes as representing a pair of segmental limb appendages, but the supposed homology with primitive abdominal legs has not been demonstrated. There is likewise no specific evidence that the lobes represent paired penes, such as those of the Ephemeroptera, since at no time in their known development are they traversed by exit ducts.

The right phallomere of *Periplaneta,* as seen from above in the undisturbed condition beneath the base of the tenth tergum (fig. 86 A, *rPhm*), presents a strongly sclerotized plate extending to the left and ending in a pair of sharp prongs, and on the right is a small accessory hook. When the membranous body wall at its base is removed, however, other parts are exposed (B), which are best examined by turning the pronged plate upward and forward (C). There will now be exposed a deep cavity in the ventral part of the phallomere, open posteriorly, between two strong horizontal plates (*opl*) resembling the valves of a clam shell. In the figure the plates are artificially opened; they are hinged to each other on the left by incurved, fingerlike processes, and on the right they are articulated on the ends of a supporting sclerite. The two plates closed upon each other, as seen when exposed from above, are shown at B.

The left phallomere (fig. 86 D) is a complex of several parts with a common base. On the extreme left is a long, slender arm with a strongly curved terminal hook, to the right of it is a second shorter and thicker arm having a transverse, hammer-head enlargement at the end, and from the mesal part of the phallomere there project several soft lobes, one bearing a small hook. Through the left phallomere opens a long, flat, feather-shaped *phallic gland* (*phGld*), the duct of which traverses the body of the phallomere to open distally near the right side (*glO*) between the edge of a mesal plate and the base of the hook-bearing lobe.

The ventral phallomere is a rather large, simple lobe with a rounded free end projecting to the right from beneath the right phallomere (fig. 86 A, *vPhm*). Its ventral surface is a smooth plate (E), but its dorsal integument is soft and is produced to the left in a large rounded lobe. The ejaculatory duct (*Dej*) opens from the left (*Gpr*) into a depression at the base of the lobe.

The Spermatophore— The spermatophore of *Periplaneta americana* is shown by Gupta (1947a) to be a pear-shaped capsule 1.3 mm. in diameter (fig. 86 I). The wall of the completed spermatophore is composed of three layers of noncellular secretion material (J). The innermost wall, as described by Gupta, is formed in the enlarged upper end of the ejaculatory duct (H, *Dej*) by secretion from the long lateral accessory glands opening into the duct. The capsule is at first open at its upper end and receives the spermatozoa from the seminal vesicles (*Vsm*) together with a liquid from the small median tubules of the accessory glands. The inseminated capsule is then closed except for an opening at its larger upper end, and as it passes down the ejaculatory duct it receives the second layer of its wall from the epithelial cells of the duct. Finally, it is attached during mating to the spermathecal papilla of the female, and a secretion from the phallic gland is now poured over it, which hardens to form the outer coating.

The Female Genitalia— The external genital structures of the female cockroach are entirely concealed within a large cavity in the end of the abdomen closed posteriorly by the upturned apical lobes of the seventh abdominal sternum. The cavity contains the ovipositor and the opening of the oviduct; if the female is carrying an egg packet, or ootheca, it is held in the posterior space between the sternal lobes. In a longitudinal section of the abdomen, as shown diagrammatically at A of figure 87, it will be seen that the short ovipositor (*Ovp*) arises from the dorsal wall of the cavity and that the oviduct (*Ovd*) opens on the floor of a flattened anterior pocket (*GC*) above a transverse fold of the ventral wall reflected back from the end of the seventh sternum (7S). Since this pocket lies proximal to the base of the ovipositor, it may be regarded as the true *genital chamber* of the female cockroach corresponding with that of other insects having the ovipositor fully exposed. The posterior part of the main cavity is the *oothecal chamber* (*otC*), termed also the *vestibulum*, a secondary enclosure within the lobes of the seventh sternum (*otL*), in which the ootheca is formed. In the dorsal wall of the genital chamber is the aperture of the spermatheca (*Spt*), and just behind the base of the ovipositor is the opening of the female accessory glands (*AcGlds*). The eggs issuing from the oviduct are therefore inseminated in the genital chamber and guided back by the ovipositor into the oothecal chamber, where they are covered by the secretion from the accessory glands, which hardens to form the shell-like ootheca.

The ovipositor and associated structures are best seen in a ventral view (fig. 87 C) as exposed by removal of the seventh sternum and its apical lobes. In the figure, the floor of the genital chamber containing the gonopore (*Gpr*) has been turned forward in the plane of the dorsal wall, as indicated by the arrows. The gonopore is a long median slit opening from a wide but shallow pouch that represents the unpaired oviduct. The lips of the aperture are strengthened by a pair of plates (*a, a*), at the sides of which are two elongate plates (*b, b*) that, in the normal position, diverge posteriorly at the sides of the ovipositor. In the roof of the genital chamber are two large lateral plates (*c, c*) and a small median plate (*d*). The median plate has a short anterior neck that ends in a small membranous lobe known as the *spermathecal papilla*, shown more enlarged at F, which contains the aperture of the spermatheca (C, *sptO*). The spermatheca (G) consists of two tubes, one longer than the other, with a short common outlet duct. The distal part of the longer tube becomes gradually thickened toward the end and is thrown into various loops; the shorter tube has an apical enlargement at the end of a long slender duct. In the wide part of each tube the axial duct is conspicuous by its dark color; numerous fine striations radiating from it are intracellular ductules of the thick glandular epithelium. Surrounding the epithelium, within an outer tunic, according to Gupta (1948), is a layer of muscle fibers. Each spermathecal tube of the inseminated female of *Periplaneta* is said by Gupta to contain spermatozoa.

The general anatomy and the musculature of the genital region of the adult female cockroach, as shown by Ford (1923), suggest that the plates on the floor of the genital chamber at the sides of the gonopore (fig. 87 C, *a, b*) represent the eighth abdominal sternum deeply inflected above the seventh sternum. However, it is said by Gupta (1948) and by earlier students of the development of the

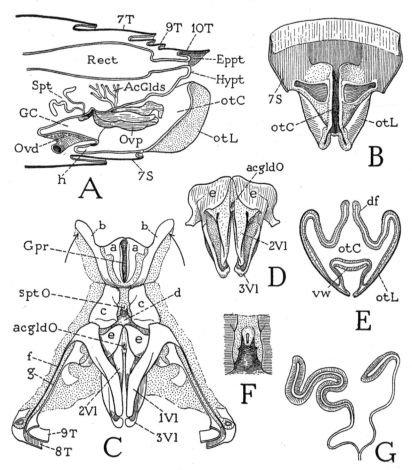

Fig. 87. Hexapoda—Pterygota. *Periplaneta americana* (L.). The female genitalia.

A, longitudinal section of end of abdomen, showing the oothecal chamber (*otC*) containing the ovipositor, and the genital chamber (*GC*) into which the oviduct opens ventrally and the spermatheca dorsally. B, seventh abdominal sternum and its apical oothecal lobes (*otL*) enclosing the oothecal chamber, dorsal. C, the ovipositor, and plates of dorsal wall of genital chamber surrounding orifice of spermatheca (*sptO*), ventral, with floor of genital chamber turned forward as indicated by arrows. D, second and third valvulae of ovipositor, ventral. E, cross section of oothecal chamber and enclosing lobes, posterior. F, spermathecal papilla, ventral. G, spermatheca.

a, b, plates of ventral wall of genital chamber; *c, d,* plates of dorsal wall of genital chamber; *e,* basal plates of second and third valvulae; *f, g,* connections of ovipositor with eighth and ninth terga. For explanation of other lettering see pages 337–339.

genital parts of the cockroach, that the oviduct is formed as an ingrowth in the intersegmental membrane between the seventh and
the eighth sterna, and that the female gonopore of the cockroach
thus retains a primitive position behind the seventh sternum, while
in most of the higher insects it has secondarily come to lie behind
the eighth. According to Gupta, the plates in the dorsal wall of the
genital chamber are derived from the eighth sternum, but, if so, it
is a very unusual thing for the spermatheca to open on the eighth
sternum.

The short ovipositor of the cockroach consists of the same three
pairs of elongate processes that form the ovipositor of most insects.
The processes are known as *valvulae* and are distinguished as the
first, second, and *third,* or as the *ventral, intermediate,* and *dorsal*
valvulae. In *Periplaneta* the first valvulae (fig. 87 C, *1Vl*) have an
independent origin anterior to the other valvulae behind the lateral
plates (*c, c*) of the dorsal wall of the genital chamber. They are long
arms converging posteriorly beneath the second and third valvulae;
the widened base of each one is connected by a very slender bar
(*f*) in the lateral wall of the oothecal chamber with the outer end
of the eighth tergum (*8T*), showing that the first valvulae belong to
the eighth abdominal segment. The second and third valvulae (C,
D, *2Vl, 3Vl*) have a common base formed of two large plates
(*e, e*). The second, or intermediate, valvulae are slender, tapering
arms (D, *2Vl*) arising from the basal plates. The third valvulae
(*3Vl*) are direct continuations from the basal plates; they are longer
and thicker than the second valvulae, with concave undersurfaces
and the ends turned downward over the tips of the second valvulae.
The second and third valvulae belong to the ninth abdominal segment, and each basal plate is connected by a lateral extension with
the expanded anterior end of a slender bar (C, *g*) from the ninth
tergum (*9T*). The anterior end of this bar, however, is attached also
to the base of the first valvula.

The ovipositor of most insects has a well-developed mechanism
by which the prongs slide back and forth on each other, but no such
mechanism is evident from the structure of the cockroach ovipositor,
and it is therefore difficult to identify its basal parts with those of
a more typical egg-laying organ. The ovipositor of the cockroach
serves merely to conduct the eggs back to the oothecal chamber,

332

and, because of its concealment, whatever activity it may have has not been observed.

The female genital accessory glands of *Periplaneta*, two in number with a common opening, are long, dichotomously branched tubes, disposed in a great tangled mass in the abdominal haemocoele beneath the rectum. The aperture of the glands lies between the bases of the second valvulae (fig. 86 C, D, *acgldO*). The secretion furnishes the material for the construction of the ootheca.

The oothecal chamber is enclosed beyond the ovipositor between the apical lobes of the seventh abdominal sternum (fig. 87 B, *otC*). The outer surfaces and the incurved dorsal parts of the lobes (B, E, *otL*) are strongly sclerotized plates, but the inner walls are soft and covered by a delicate cuticle. The upper margin of each lobe is produced into a thick dorsal fold (E, *df*). The lower edges of the two lobes are connected by a bridge of soft integument (*vw*) forming the floor of the oothecal chamber (*otC*), which is ordinarily deeply inflected but allows of a great lateral expansion of the chamber when an egg capsule is being formed within the latter. When the oothecal chamber is empty, the dorsal folds of the lobes may be variously disposed, but when an ootheca is being formed they stand up and embrace the upper part of the capsule, which is molded between them. A more detailed account of the structure of the posterior part of the female abdomen of *Periplaneta* is given in a recent paper by Brunet (1951).

The relation of the male and female genital structures of *Periplaneta* during mating has been described by Gupta (1947b), who shows that the various parts of the male phallic apparatus have specific copulatory functions. The hooked arm of the left phallomere (fig. 86 D) pulls down on the apical lobes of the seventh sternum of the female abdomen and opens the genital cavity. The end of the hammer-headed arm is then thrust into the opening of the oviduct, turned transversely, and thus anchored beneath the gonopore plates (fig. 87 C, *a, a*). The "clam-shell" plates of the right phallomere, called the *opposing lobes* by Gupta (fig. 86 C, *opl*), grip the distal ends of the anterior valvulae of the ovipositor, the prongs of the dorsal plate grasp the base of the right first valvula, and the curved spine on the right holds the base of the left valvula. The spermatophore is then discharged upon the dorsal surface of the ventral phallomere of the male and is attached to the spermathecal papilla of the female.

The Ootheca— The ootheca of *Periplaneta* is a hard-walled, dark reddish-brown, purse-shaped capsule (fig. 88 C), about 9 mm. in length, having a straight-edged crest along its upper side with a finely serrate margin. It is somewhat compressed from side to side and is pear-shaped in cross section (D). Within the capsule is a double row of eggs, seven or eight on each side, set vertically with

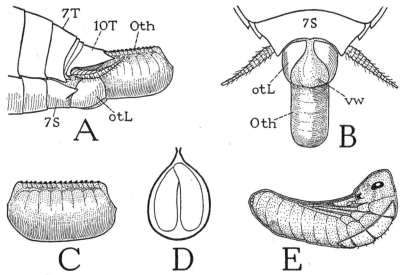

Fig. 88. Hexapoda—Pterygota. The ootheca and newly hatched young of a cockroach.

A, *Periplaneta americana* (L.), end of abdomen of female carrying a fully formed ootheca, lateral. B, same, ventral. C, the ootheca. D, cross section of ootheca showing position of eggs in two rows. E, *Blattella germanica* (L.), newly hatched young beginning to emerge from embryonic cuticle (from Pettit, 1940).

the ventral surfaces of the contained embryos toward each other, and the head ends upward. Lines on the outer surface of the ootheca mark the position of the eggs, which are separated by slight internal ridges. The capsule is closed along the margin of the crest, but a gentle manipulation reveals a median cleft where the two edges come together (D), and, when the young cockroaches are ready to emerge, the ootheca normally opens along this line. In the formative stage of the ootheca, however, the serrated upper edges of the lateral walls are firmly attached to each other, and there is no natural opening here into the interior of the capsule. The structure of the ootheca of different species of cockroaches is described by Lawson (1951).

As already noted, the egg case is formed within the oothecal chamber of the female from the secretion of the accessory glands, which hardens as it is poured over the issuing eggs. Since the eggs in the ovary have the head ends forward, the ovipositor somehow sets them on end in the two opposing rows. At first the lobes of the seventh sternum stand erect and close the distal end of the oothecal chamber, but as the number of eggs increases the ootheca protrudes from the body of the female, while the lobes are depressed and spread apart. The fully formed capsule is held only by its anterior end in the grasp of the tenth tergum above, the oothecal lobes on the sides, and the everted soft ventral wall of the oothecal chamber below (fig. 88 A, B), until finally it is dropped and left for the embryos to mature and the young cockroaches to break out.

The texture of the ootheca might suggest that the cockroach egg capsule is a chitinous structure, but the secretion of the accessory glands is nonchitinous and is of a different composition in the two glands. It has been shown by Pryor (1940) in *Blatta orientalis* that the left gland secretes a water-soluble protein, while the other produces a dihydroxyphenol. The phenol is oxidized in air to a quinone that combines with the protein, which then becomes a scleroprotein resistant to most chemical reagents and enzymes.

The chorion of the eggs within the ootheca is a delicate membrane. On hatching, the cleft along the upper edge of the capsule opens, probably by the activity of the young cockroaches; the escape of the young of *Blattella germanica* has been interestingly described and illustrated by Pettit (1940). When the ootheca opens, the young cockroaches thrust out their heads and begin to swallow minute bubbles of air, which accumulate in the alimentary canal until the insects are almost doubled in size and are forced partly out of the capsule; they finally liberate themselves by their own activities. Each individual, however, is still enclosed in a thin, baglike embryonic cuticle, but the envelope soon splits along the back (fig. 88 E) and the free nymph runs off. At first the young cockroach is almost transparent because of the contained air, but on release of the air the normal size is restored, and a dark pigmentation soon follows.

335

Female cockroaches often carry the ootheca projecting from the abdomen a varying length of time before it is dropped. From this habit of carrying an egg-case full of embryos, it is only an evolutionary step to retaining the case in the body until the eggs hatch, and, in fact, there are various genera of so-called "viviparous" cockroaches. The structural modifications of the female organs that provide for the retention of the ootheca have been described in *Diploptera dytiscoides* by Hagan (1941, 1951). In the female of *Periplaneta* it is to be noted that there is a small anterior pocket of the oothecal chamber below the overhanging posterior edge of the floor of the genital chamber (fig. 87 A, *h*). In *Diploptera* and other viviparous species this pocket is enlarged into a huge sack projecting forward in the abdomen as far as the thorax, usually on the left side. Within this pouch the ootheca is lodged until the hatching of the eggs. The ootheca of *Diploptera*, according to Hagan, is a thin membrane, which only partly envelops the egg mass, and it never turns brown. The left accessory gland, the protein-secreting gland of *Blatta*, Hagan says, continues its secretory activity after the formation of the ootheca and "may be a nutrient organ for the embryos."

Since the ootheca is built up from behind forward in the oothecal chamber, it remains to be explained how it gets into the brood pouch. In the viviparous *Gomphadorhina laevigata*, Chopard (1950) has observed that the forming ootheca is first slowly extruded in the usual manner from the body of the female, but that, when it is held only by its extremity, a reverse movement takes place by which the capsule is slowly drawn back into the abdomen and stored in the brood pouch. After the retraction of the ootheca, Chopard says, the ovipositor is completely reversed in position, the prongs being turned forward instead of posteriorly. It would appear, therefore, that in *Gomphadorhina* the ovipositor withdraws the fully formed ootheca and inserts it into the brood chamber. On the other hand, in the viviparous *Diploptera dytiscoides*, Hagan (1951) says the eggs "are directed by the ovipositor from the genital chamber ventrally into the open end of the uterus." In his chapter on *Diploptera*, Hagan gives a full account of the female reproductive organs and the embryonic development of this viviparous cockroach, with a special description of the pleuropodia and adenopodia and a discussion of the possible functions of these abdominal outgrowths.

Explanation of Lettering on Figures 74–88

a, posterior articulation of mandible, or pleural articulation of coxa.

a', cranial articulation of maxilla.

acgldO, orifice of accessory genital glands.

AcGlds, accessory genital glands.

acs, antecostal sulcus.

adhf, adductor muscle of hypopharynx.

admd, adductor muscle of mandible.

admx, adductor muscle of maxilla.

af, antennifer.

Amp, ampulla of vas deferens.

An, anus.

ANP, anterior notal wing process.

Ant, antenna.

apl, anapleurite.

Ar, arolium.

AS, antesternite.

at, anterior tentorial pit.

AT, anterior arm of tentorium.

Atr, atrium of spiracle.

Ax, axillary sclerite of wing base (*1Ax, 2Ax, 3Ax*, first, second, and third axillary).

Ba, basalare, anterior epipleurite.

bcx, basicoxite.

c, anterior articulation of mandible, or trochantinal articulation of coxa.

C, costa, first vein of wing.

Cb, cibarium.

Cd, cardo.

Cer, cercus.

cf, caudal filament.

Clp, clypeus.

cpl, catapleurite, coxapleurite.

CT, central plate of tentorium, corpotentorium.

Cu, cubitus, fifth vein of wing.

Cvx, cervix, neck.

Cx, coxa.

d, alar arm of metascutellum.

Dac, dactyl, median claw of pretarsus.

Dej, ductus ejaculatorius.

df, dorsal fold.

dlcb, dilator muscles of cibarium.

dlmcl, dilator muscle of spiracle.

DT, dorsal arm of tentorium.

e, intertergal sclerites of thorax.

E, compound eye.

Epm, epimeron.

Eppt, epiproct.

eps, episternal apotome.

Eps, episternum.

es, epistomal sulcus.

flcc, cranial flexor muscle of lacinia.

Fm, femur.

For, neck foramen of head, occipital foramen.

Fr, frons.

FrGng, frontal ganglion.

Ga, galea.

Ge, gena.

GC, genital chamber.

Gl, glossa.

glO, orifice of phallic gland.

gnP, gnathal pouch.

Gon, gonapophysis (*1Gon, 2Gon*, first and second gonapophysis).

Gpr, gonopore.

gr, postocular groove.

h, hinge of lacinia on stipes.

hf, labial fulcrum of hypopharynx.

HP, humeral plate of wing base.

Hphy, hypopharynx.

337

HS, hypopharyngeal suspensorium.
Hypt, hypoproct.

imB, intermaxillary brachium.
IS, intersternite, spinisternite.

jf, jugal fold of wing.
Ju, jugal region of wing.

L, leg.
Lb, labium.
Lc, lacinia.
lcvpls, lateral cervical plates.
Lg, intermandibular, or interbrachial, ligament.
Lin, lingua.
Lm, labrum.
lPhm, left phallomere.

m, median plate of wing base.
M, media, fourth vein of wing.
mcl, muscle.
Md, mandible.
Mt, mentum.
Mth, mouth.
Mx, maxilla.

o, anterior coxotrochanteral articulation.
Oc, occiput.
occ, occipital condyle.
ocmcl, occlusor muscle of spiracle.
Odl, lateral oviduct.
opl, opposing ventral plates of right phallomere.
otC, oothecal chamber.
Oth, ootheca.
otL, oothecal lobe.
Ovd, oviduct.
Ovp, ovipositor.

p, auxiliary sclerites of unguitractor plate.
Papt, paraproct.
Pcu, postcubital veins of wing.
Pen, penis.

Pge, postgena.
Pgl, paraglossa.
Ph, phragma.
phGld, phallic gland.
phL, primary phallic lobe.
Phm, phallomere.
Phy, pharynx.
Plp, palpus.
PlR, pleural ridge.
PlS, pleural sulcus.
Pmt, postmentum.
PNP, posterior notal wing process.
Poc, postocciput.
PoR, postoccipital ridge.
pos, postoccipital sulcus.
PrC, preoral food cavity.
Prmt, prementum.
Prsct, prescutum.
PS, poststernite.
pt, posterior tentorial pit.
Ptar, pretarsus.

q, euplantulae of tarsus.

r, anterior femorotibial articulation.
R, radius, third vein of wing.
Rect, rectum.
Rm, remigium.
rPhm, right phallomere.

S, sternum.
Sa, subalare, posterior epipleurite.
sAp, sternal apodeme.
Sc, subcosta, second vein of wing.
Scl, scutellum.
Sct, scutum.
Sit, sitophore.
SlDct, salivary duct.
Slin, superlingua.
SlO, orifice of salivary duct.
Slv, salivarium.
Smt, submentum.
Sp, spiracle.
Spt, spermatheca.
sptO, orifice of spermatheca.
St, stipes.

Stom, stomodaeum.

Sty, stylus.

t, tendon.

T, tergum.

Tar, tarsus.

Tb, tibia.

TB, tentorial bridge.

tn, trochantinal apotome.

Tn, trochantin.

Tnt, tentorium.

Tor, torma.

Tr, trochanter.

Tra, trachea.

Un, unguis, lateral claw of pretarsus.

Utr, unguitractor plate.

Vd, vas deferens.

vf, vannal fold of wing.

Vl, valvula of ovipositor (*1Vl, 2Vl, 3Vl,* first, second, and third valvula).

Vn, vannus, fan of wing.

vPhm, ventral phallomere.

vr, ventral ridge of tentorium.

Vs, eversible vesicle of abdomen.

Vsm, vesiculae seminales.

vw, ventral wall of oothecal chamber.

Vx, vertex.

WP, pleural wing process.

x, loral arm of hypopharyngeal suspensorium.

y, oral arm of hypopharyngeal suspensorium.

REFERENCES

Adensamer, T. 1894. Zur Kenntnis der Anatomie und Histologie von Scutigera coleoptrata. Verh. zool.-bot. Gesell., Wien, 43: 573–578, 1 pl.

Andrews, E. A. 1906. The annulus ventralis. Proc. Boston Soc. Nat. Hist., 32: 427–479, 6 pls.

Andrews, E. A. 1908. The sperm receptacle of the cray-fishes Cambarus cubensis and C. paradoxus. Proc. Washington Acad. Sci., 10: 167–185, 12 figs.

Andrews, E. A. 1911. Male organs for sperm-transfer in the cray-fish, Cambarus affinis: their structure and use. Journ. Morph., 22: 239–291, 31 text figs., 4 pls.

Annandale, N. 1909. The habits of the Indian king crabs. Rec. Indian Mus., 3: 294–295.

Arthur, D. R. 1946. The feeding mechanism of Ixodes ricinus L. Parasitology, 37: 154–167, 15 figs.

Arthur, D. R. 1951a. The capitulum and feeding mechanism of Ixodes hexagonus Leach. Parasitology, 41: 66–81, 16 figs.

Arthur, D. R. 1951b. The bionomics of Ixodes hexagonus Leach in Britain. Parasitology, 41: 82–90, 6 figs.

Attems, C. G. 1926a. Symphyla. In Kükenthal and Krumbach, Handbuch der Zoologie, 4, first half: 11–19, figs. 6–19. Berlin and Leipzig.

Attems, C. G. 1926b. Diplopoda. In Kükenthal and Krumbach, Handbuch der Zoologie, 4, first half: 29–238, figs. 29–273. Berlin and Leipzig.

Attems, C. G. 1926c. Chilopoda. In Kükenthal and Krumbach, Handbuch der Zoologie, 4, first half: 239–402, figs. 274–477. Berlin and Leipzig.

Baerg, W. J. 1928. The life cycle and mating habits of the male tarantula. Quart. Rev. Biol., 3: 109–116, 3 figs.

REFERENCES

Barlet, J. 1950. La question des pièces pleurales du thorax des Machilides (Thysanoures). Bull. & Ann. Soc. Ent. Belgique, 86: 179–190, 4 figs.

Barrows, W. M. 1925. Modification and development of the arachnid palpal claw, with especial reference to spiders. Ann. Ent. Soc. America, 18: 483–516, 9 pls.

Bartels, M. 1930. Über den Fressmechanismus und chemischen Sinn einiger Netzspinnen. Rev. Suisse Zool., 37: 1–42, 11 figs.

Beck, E. J. 1885. Description of the muscular and endoskeletal systems of Scorpio. Trans. Zool. Soc. London, 11: 339–360, 3 pls.

Benham, W. B. S. 1885. Description of the muscular and endoskeletal systems of Limulus. Trans. Zool. Soc. London, 11: 314–338, 5 pls. (See Lankester, Benham, and Beck, 1885.)

Bertkau, P. 1885. Über den Verdauungsapparat der Spinnen. Archiv. mikr. Anat., 24: 398–451, 2 pls.

Bock, F. 1925. Die Respirationsorgane von Potamobius astacus Leach. (Astacus fluviatilis Fabr.) Zeitschr. wiss. Zool., 124: 51–117, 27 figs.

Börner, C. 1901. Zur aüsseren Morphologie von Koenenia mirabilis Grassi. Zool. Anz., 24: 537–556, 10 figs.

Börner, C. 1903. Die Beingliederung der Arthropoden, Mitt. 3. Sitz.-Ber. Gesells. naturf. Freunde Berlin, Jahrg. 1903, No. 7: 292–341, 7 pls.

Bristowe, W. S., and Millot, J. 1932. The liphistiid spiders. With an appendix on their internal anatomy by J. Millot. Proc. Zool. Soc. London for 1932: 1015–1057, 11 text figs., 6 pls.

Brown, R. B. 1939. The musculature of Agelena naevia. Journ. Morph., 64: 115–166, 12 pls.

Brunet, P. C. J. 1951. The formation of the ootheca by Periplaneta americana. I. The microanatomy and histology of the posterior part of the abdomen. Quart. Journ. Micr. Sci., 92: 113–127, 5 figs.

Buxton, B. H. 1913. Coxal glands of the arachnids. Zool. Jahrb., Suppl. 14: 231–282, 43 pls.

Buxton, B. H. 1917. Notes on the anatomy of arachnids. Journ. Morph., 29: 1–31, 8 text figs., 3 pls.

Calman, W. T. 1896. On the genus Anaspides and its affinities with certain fossil Crustacea. Trans. R. Soc. Edinburgh, 38: 787–802, 2 pls.

Calman, W. T. 1909. Crustacea. In Lancaster, A treatise on zoology, pt. 7. Appendiculata, 3d Fas., Crustacea, 346 pp., 194 figs. London.

Calman, W. T. 1917. Notes on the morphology of Bathynella and some allied Crustacea. Quart. Journ. Micr. Sci., 62: 489–514, 14 figs.

Carbonell, C. S. 1947. The thoracic muscles of the cockroach Periplaneta americana (L). Smithsonian Misc. Coll., 107, No. 2, 23 pp., 8 pls.

Carpentier, F. 1946. Sur la valeur morphologique des pleurites du thorax des Machilides (Thysanoures). Bull. & Ann. Soc. Ent. Belgique, 82: 165–181, 6 figs.

Carpentier, F. 1947. Quelques remarques concernant la morphologie thoracique des Collemboles (Apterygotes). Bull. & Ann. Soc. Ent. Belgique, 83: 297–303.

Chappuis, P. A. 1915. Bathynella natans und ihre Stellung im System. Zool. Jahrb., Syst., 40: 147–176, 17 text figs., 1 pl.

Chaudonneret, J. 1950. La morphologie céphalique de Thermobia domestica (Packard). Ann. Sci. Nat., Zool., Ser. 11, 12: 145–302, 79 figs.

Chopard, L. 1950. Sur l'anatomie et le développement d'une blatte vivipare. 8th Internat. Congr. Ent., 218–222, 6 figs.

Clare, S., and Tauber, O. E. 1939. Circulation of haemolymph in the wings of the cockroach Blatella germanica L. I. In normal wings. Iowa State College Journ. Sci., 14: 107–127, 3 figs.

Clark, H. W. 1940. The adult musculature of the anisopterous dragonfly thorax (Odonata, Anisoptera). Journ. Morph., 67: 523–565, 7 figs.

Clarke, J. M., and Ruedemann, R. 1912. The Eurypterida of New York. New York State Mus. 65th Ann. Rep., Mem. 14, vol. 3, text, 439 pp.; vol. 4, plates.

Cole, L. J. 1901. Notes on the habits of pycnogonids. Biol. Bull., 2: 195–207, 5 figs.

Cole, L. J. 1906. Feeding habits of the pycnogonid Anoplodactylus lentus Wils. Zool. Anz., 29: 740–741.

Comstock, J. H. 1910. The palpi of male spiders. Ann. Ent. Soc. America, 3: 161–185, 25 figs.

Comstock, J. H. 1918. The wings of insects, 430 pp., 427 figs. Ithaca, N.Y.

Comstock, J. H., and Gertsch, W. J. 1948. The Spider Book, revised by W. J. Gertsch, 729 pp., 770 figs. Ithaca, N.Y.

Crampton, G. C. 1925. The external anatomy of the head and abdomen of the roach, Periplaneta americana. Psyche, 32: 197–225, 3 pls.

Crampton, G. C. 1927. The thoracic sclerites and wing bases of the roach Periplaneta americana and the basal structures of the wings of insects. Psyche, 34: 59–72, 3 pls.

Demoll, R. 1914. Die Augen von Limulus. Zool. Jahrb., Anat., 38: 443–464, 15 figs.

Dillon, L. S. 1952. The myology of the araneid leg. Journ. Morph., 90: 467–480, 14 figs.

Dohrn, A. 1881. Die Pantopoden des Golfes von Neapel und der angrenzenden Meeresabschnitte. Fauna und Flora Golfes Neapel, 3, 252 pp., 17 pls.

Douglas, J. R. 1943. The internal anatomy of Dermacentor andersoni

Stiles. Univ. California Pubs. Ent., 7, No. 10: 207–272, 7 text figs., 19 pls.

Dürken, B. 1907. Die Tracheenkiemenmuskulatur der Ephemeriden unter Berücksichtigung der Morphologie des Insektenflügels. Zeitschr. wiss. Zool., 87: 435–550, 30 text figs., 3 pls.

Effenberger, W. 1907. Die Tracheen bei Polydesmus. Zool. Anz., 31: 782–786, 4 figs.

Ellis, C. H. 1944. The mechanism of extension in the legs of spiders. Biol. Bull., 86: 41–50, 3 figs.

Engelhardt, V. von. 1910. Beiträge zur Kenntnis der weiblichen Copulationsorgane einiger Spinnen. Zeitschr. wiss. Zool., 96: 32–117, 49 text figs., 1 pl.

Ewing, H. E. 1918. The life and behavior of the house spider. Proc. Iowa Acad. Sci., 25: 177–204, 12 figs.

Ewing, H. E. 1928. The legs and leg-bearing segments of some primitive arthropod groups, with notes on leg-segmentation in the Arachnida. Smithsonian Mis. Coll., 80, No. 11, 41 pp., 12 pls.

Fahlander, K. 1938. Beiträge zur Anatomie und systematischen Einteilung der Chilopoden. Zool. Bidrag Uppsala, 17: 1–148, 36 text figs., 18 pls.

Falke, H. 1931. Beiträge zur Lebensgeschichte und zur postembryonalen Entwicklung von Ixodes ricinus L. Zeitschr. Morph. Ökl. Tiere, 21: 567–607, 25 figs.

Folsom, J. W. 1900. The development of the mouth-parts of Anurida maritima Guer. Bull. Mus. Comp. Zool., 36, No. 5: 87–157, 8 pls.

Ford, Norma. 1923. A comparative study of the abdominal musculature of orthopteroid insects. Trans. R. Canadian Inst., 14: 207–319, 17 pls.

Fraenkel, G. 1929. Atmungsmechanismus des Skorpions. Zeitschr. vergl. Physiol., 11: 656–661, 2 figs.

Fuller, C. 1924. The thorax and abdomen of winged termites. With special reference to the sclerites and muscles of the thorax. Union S. Africa, Dep. Agric., Entomology Memoirs No. 2: 47–78, 43 figs.

Gerhardt, U., and Kästner, A. 1937, 1938. Araneae. In Kükenthal and Krumbach, Handbuch der Zoologie, 3, second half: 394–656, figs. 484–854.

Gertsch, W. J. 1949. American Spiders, 285 pp., 64 pls. Toronto, New York, London.

Grandi, Marta. 1948, 1949. Contributi allo studio dei Plecotteri. I. Reperti di morphologia e di miologia thoracica di Perla marginata Panz. Boll. Ist. Ent. Univ. Bologna, 17: 130–157, 11 figs.

Grenacher, H. 1880. Über die Augen einiger Myriapoden. Archiv micr. Anat., 18: 415–467, 2 pls.

Gupta, P. D. 1947a. On the structure and formation of spermatophore in the cockroach, Periplaneta americana (Linn.). Indian Journ. Ent., 8, pt. 1: 79–84, 3 figs.

Gupta, P. D. 1947b. On copulation and insemination in the cockroach Periplaneta americana (Linn.). Proc. Nat. Inst. Sci. India, 13, No. 2: 65–71, 2 figs.

Gupta, P. D. 1948. On the structure, development and homology of the female reproductive organs of orthopteroid insects. Indian Journ. Ent., 10, pt. 1: 75–123, 25 figs.

Haase, E. 1884. Das Respirationssystem der Symphylen und Chilopoden. Zool. Beiträge (Schneider), 1, Heft 2: 65–96, 3 pls.

Haase, E. 1890. Beiträge zur Kenntnis der fossilen Arachniden. Zeitschr. Deutsch. Geol. Gesell., 42: 629–657, 2 pls.

Hagan, H. R. 1941. The general morphology of the female reproductive system of a viviparous roach, Diploptera dytiscoides (Serville). Psyche, 48: 1–9, 2 text figs., 1 pl.

Hagan, H. R. 1951. Embryology of the viviparous insects, 472 pp., 160 figs. New York.

Haller, B. 1907. Über die Ocellen von Periplaneta orientalis. Zool. Anz., 31: 255–262, 4 figs.

Hansen, H. J. 1902. On the genera and species of the order Pauropoda. Vidensk. Medd. naturh. Foren. Copenhagen, 53: 323–424, 6 pls.

Hansen, H. J. 1925. Studies on Arthropoda. II. On the comparative morphology of the appendages in the Arthropoda. A. Crustacea, 176 pp., 8 pls. Copenhagen.

Hansen, H. J. 1930. Studies on Arthropoda. III. On the comparative morphology of the appendages in the Arthropoda. B. Crustacea (supplement), Insecta, Myriapoda, and Arachnida, 376 pp., 16 pls. Copenhagen.

Hansen, H. J., and Sörensen, W. 1897. The order Palpigradi Thorell (Koenenia mirabilis Grassi) and its relationship to other Arachnida. Ent. Tidskr., 18: 223–240, 1 pl.

Hanström, B. 1928. Vergleichende Anatomie dis Nervensystems der wirbellosen Tiere, 628 pp., 650 figs. Berlin.

Harrison, L. 1914. On some Pauropoda from New South Wales. Proc. Linn. Soc. New South Wales, 39: 615–634, 2 pls.

Hedgpeth, J. W. 1947. On the evolutionary significance of the Pycnogonida. Smithsonian Misc. Coll., 106, No. 18, 53 pp., 16 text figs., 1 pl.

Helfer, H., and Schlottke, E. 1935. Pantopoda. In Bronns Klassen und Ordnungen des Tierreichs, 5, Abt. 4, Buch 2, Lief. 1, 2, 314 pp., 223 figs.

Hemenway, Josephine. 1900. The structure of the eye of Scutigera (Cermatia) forceps. Biol. Bull., 1: 205–213, 2 figs.

REFERENCES

Hennings, C. 1904, 1906. Das Tömösvarysche Organ der Myriopoden. Zeitschr. wiss. Zool., 76: 26–52, 1 pl.; 80: 576–641, 2 pls.

Henriksen, K. L. 1926. The segmentation of the trilobite head. Saert. Meddel. Dansk Geol. For., 7: 1–32, 27 figs.

Heymons, R. 1897. Entwicklungsgeschichtliche Untersuchungen an Lepisma saccharina L. Zeitschr. wiss. Zool., 62: 583–631, 2 pls.

Heymons, R. 1901. Die Entwicklungsgeschichte der Scolopender. Zoologica. Orig.-Abhandl. Gesammtg. Zool., 33, 244 pp., 42 text figs., 8 pls.

Hobbs, H. H., Jr. 1940. On the first pleopod of the male Cambari (Decapoda, Astacidae). Proc. Florida Acad. Sci., 5: 55–61, 2 pls.

Hobbs, H. H., Jr. 1942. The crayfishes of Florida. Biological Science Series, Univ. of Florida, 3, 179 pp., 23 pls.

Hobbs, H. H., Jr. 1945. Notes on the first pleopod of the male Cambarinae (Decapoda, Astacidae). Quart. Journ. Florida Acad. Sci., 8: 67–70, 1 pl.

Howell, F. B., Frederickson, E. A., Lochman, C., Raasch, G. O., and Rasetti, F. 1947. Terminology for describing Cambrian trilobites. Journ. Paleontol., 21: 72–76, 1 fig.

Huxley, T. H. 1880. The crayfish: an introduction to the study of zoology, 371 pp., 81 figs. New York.

Imms, A. D. 1936. The ancestry of insects. Trans. Soc. British Entomology, 3: 1–32, 11 figs.

Imms, A. D. 1939. On the antennal musculature in insects and other arthropods. Quart. Journ. Micr. Sci., 81: 273–320, 25 figs.

Imms, A. D. 1940. On growth processes in the antennae of insects. Quart. Journ. Micr. Sci., 81: 585–593, 1 fig.

Iwanoff, P. P. 1933. Die embryonale Entwicklung von Limulus moluccanus. Zool. Jahrb. Anat., 56: 163–348, 78 text figs., 3 pls.

Kästner, A. 1929. Bau und Funktion der Fächertracheen einiger Spinnen. Zeitschr. Morph. Ökol. Tiere, 13: 463–558, 56 figs.

Kästner, A. 1940. Scorpions. In Kükenthal and Krumbach, Handbuch der Zoologie, 3, second half, Lief. 14: 116–240, figs. 88–222.

Keim, W. 1915. Das Nervensystem von Astacus fluviatilis (Potamobius astacus L.). Zeitschr. wiss. Zool., 113: 485–545, 28 figs.

Kosareff, G. 1935. Beobachtungen über die Ernährung der Japygiden. Mitt. Konig. naturv. Inst., Sofia, 8: 181–185, 3 figs.

Krug, H. 1907. Beiträge zur Anatomie der Gattung Iulus. Jenaische Zeitschr. Naturwiss., 42: 485–522, 8 text figs., 3 pls.

Lalicker, C. G. 1935. Larval stages of trilobites from the Middle Cambrian of Alabama. Journ. Paleontol., 9: 394–399, 1 pl.

Lamy, E. 1902. Recherches anatomiques sur les trachées des Araignées. Ann. Sci., Nat., Zool., Ser. 8, 15: 149–280, 71 text figs., 4 pls.

REFERENCES

Lankester, E. R. 1884. On the skeleto-trophic tissues and coxal glands of Limulus, Scorpio, and Mygale. Quart. Journ. Micr. Sci., 24: 129–162, 7 pls.

Lankester, E. R. 1885. Comparison of the muscular and endoskeletal systems of Limulus and Scorpio, and considerations of the morphological significance of the facts recorded. Trans. Zool. Soc. London, 11: 361–384, pls. 80–83.

Lankester, E. R., Benham, W. B. S., and Beck, E. J. 1885. On the muscular and endoskeletal systems of Limulus and Scorpio, with some notes on the anatomy and generic characters of scorpions. Trans. Zool. Soc. London, 11: 311–384, 12 pls.

Lankester, E. R., and Bourne, A. G. 1883. The minute structure of the lateral and the central eyes of Scorpio and Limulus. Quart. Journ. Micr. Sci., 23: 177–212, 3 pls.

Lawson, F. A. 1951. Structural features of the oothecae of certain species of cockroaches. Ann. Ent. Soc. America, 44: 269–285, 2 pls.

Lees, A. D. 1948. The sensory physiology of the sheep tick, Ixodes ricinus L. Journ. Exp. Biol., 25: 145–207, 31 figs.

Lees, A. D., and Beament, J. W. L. 1948. An egg-waxing organ in ticks. Quart. Journ. Micr. Sci., 89: 291–332, 13 text figs., 1 pl.

Levereault, P. 1936, 1938. Morphology of the Carolina mantis. Bull. Univ. Kansas, 37, No. 14: 205–259, 8 pls.; 39, No. 11: 577–633, 12 pls.

Lindsay, Eder. 1940. The biology of the silverfish, Ctenolepisma longicaudata Esch., with particular reference to its feeding habits. Proc. Roy. Soc. Victoria, 52: 35–83, 12 text figs., 2 pls.

Locy, W. A. 1886. Observations on the development of Agelena naevia. Bull. Mus. Comp. Zool., 12: 63–103, 12 pls.

Lubbock, J. 1868. On Pauropus, a new type of centipede. Trans. Linn. Soc. London, 26: 181–190, 1 pl.

McClendon, J. F. 1904. On the anatomy and embryology of the nervous system of the scorpion. Biol. Bull., 8: 38–55, 13 figs.

Manton, S. M. 1930. Notes on the habits and feeding mechanisms of Anaspides and Paranaspides (Crustacea, Syncarida). Proc. Zool. Soc. London for 1930: 791–800, 4 plates in color.

Mellanby, K. 1935. The structure and function of the spiracles of the tick, Ornithodoros moubata Murray. Parasitology, 27: 288–290, 2 figs.

Miall, L. C., and Denny, A. 1886. The structure and life-history of the cockroach (Periplaneta orientalis), 224 pp., 125 figs. London and Leeds.

Michelbacher, A. E. 1938. The biology of the garden centipede, Scutigerella immaculata. Hilgardia, 11: 55–148, 29 figs.

346

Miley, H. H. 1927. Development of the male gonopods and life history studies of a polydesmid millipede. Ohio Journ. Sci., 27: 25–43, 2 pls.

Millot, J. 1931. Les glandes venimeuses des aranéides. Ann. Sci. Nat., Ser. 10, Zool., 14: 113–147, 23 figs.

Millot, J. 1949. Ordre des Araneides (Araneae). *In* Grassé, Traité de Zoologie, 6: 589–743, figs. 349–528.

Montgomery, T. H. 1903. Studies on the habits of spiders, particularly those of the mating period. Proc. Acad. Nat. Sci. Philadelphia, 55: 59–149, 2 pls.

Montgomery, T. H. 1909a. On the spinnerets, cribellum, colulus, tracheae and lung books of araneads. Proc. Acad. Sci. Philadelphia, 61: 299–320, 4 pls.

Montgomery, T. H. 1909b. Further studies on the activities of araneads, II. Proc. Acad. Sci. Philadelphia, 61: 548–569.

Nelson, J. A. 1909. Evolution and adaptation in the palpus of male spiders. Ann. Ent. Soc. America, 2: 60–64, 1 pl.

Nordenskiöld, E. 1909. Zur Anatomie und Histologie von Ixodes reduvuis, II. Zool. Jahrb. Anat., 27: 449–464, 1 pl.

Nuttall, G. H. F., Cooper, F. W., and Robinson, L. E. 1908. On the structure of the spiracles of a tick—Haemaphysalis punctata Canestrine and Fanzago. Parasitology, 1: 347–351, 1 pl.

Oettinger, R. 1906. Über die Drüsentaschen am Abdomen von Periplaneta orientalis und Phyllodromia germanica. Zool. Anz., 30: 338–349, 8 figs.

Osterloh, A. 1922. Beiträge zur Kenntnis des Kopulationsapparates einiger Spinner. Zeitschr. wiss. Zool., 119: 326–421, 42 text figs., 1 pl.

Pavlovsky, E. N., and Zarin, E. J. 1926. On the structure and ferments of the digestive organs of scorpions. Quart. Journ. Micr. Sci., 70: 221–261, 7 text figs., 2 pls.

Petrunkevitch, A. 1925. External reproductive organs of the common grass spider, Agelena naevia Walckenaer. Journ. Morph., 40: 559–572, 1 pl.

Petrunkevitch, A. 1933. An inquiry into the natural classification of spiders, based on a study of their internal anatomy. Trans. Connecticut Acad. Arts Sci., 31: 299–389, 13 pls.

Petrunkevitch, A. 1947. Scorpion. Encyclopaedia Britannica, 3 pp., 2 figs.

Petrunkevitch, A. 1949. A study of Palaeozoic Arachnida. Trans. Connecticut Acad. Arts. Sci., 37: 75–315, I–XI, 83 pls.

Pettit, L. 1940. A roach is born. New England Naturalist, No. 7: 15–18, 8 figs.

Pflugfelder, O. 1932. Über den Mechanismus der Segmentbildung bei Embryonalentwicklung und Anamorphosis von Platyrrhacus amauros Attems. Zeitschr. wiss. Zool., 140: 650–723, 40 figs.

Pflugfelder, O. 1933. Über den feineren Bau der Schläfenorgane der Myriapoden. Zeitschr. wiss. Zool., 143: 127–155, 10 figs.

Pocock, R. I. 1901. The Scottish Silurian scorpion. Quart. Journ. Micr. Sci., 44: 291–311, pl. 19.

Pocock, R. I. 1902. Studies on the arachnid endosternite. Quart. Journ. Micr. Sci., 46: 225–262, 2 pls.

Prell, H. 1910. Beiträge zur Kenntnis der Lebensweise einiger Pantopoden. Bergens Mus. Aarbok. Naturvidenskabaleg Raekke, No. 10, 30 pp., 12 figs.

Prentiss, C. W. 1901. The otocyst of decapod Crustacea: its structure, development, and functions. Bull. Mus. Comp. Zool., 36: 165–251, 10 pls.

Pryor, M. G. M. 1940. On the hardening of the ootheca of Blatta orientalis. Proc. R. Soc. London, Ser. B, No. 852, vol. 128: 378–393.

Pumphrey, R. J., and Rawdon-Smith, A. F. 1936. Hearing in insects: the nature of the response of certain receptors to auditory stimuli. Proc. R. Soc. London, Ser. B, No. 820, vol. 121: 18–27, 2 pls.

Purcell, W. F. 1909. Development and origin of the respiratory organs in Araneae. Quart. Journ. Micr. Sci., 54: 1–110, 7 text figs., 7 pls.

Purcell, W. F. 1910. The phylogeny of the tracheae in Araneae. Quart. Journ. Micr. Sci., 54: 519–564, 1 pl.

Qadri, M. A. H. 1940. On the development of the genitalia and their ducts of orthopteroid insects. Trans. R. Ent. Soc. London, 90: 121–175.

Raw, F. 1925. The development of Leptoplastus salteri and other trilobites. Quart. Journ. Geol. Soc., 81: 223–324, 4 pls.

Raymond, P. E. 1920. The appendages, anatomy, and relationships of trilobites. Mem. Connecticut Acad. Arts Sci., 7, 169 pp., 11 pls.

Reese, A. M. 1944. The anatomy of the venom glands in the black widow spider, Latrodectus mactans. Trans. American Micro. Soc., 63: 171–174, 1 pl.

Rehn, A. G. 1945. Man's uninvited fellow traveller—the cockroach. Science Monthly, 61: 265–276, 11 figs.

Reinecke, G. 1910. Beiträge zur Kenntnis von Polyxenus. Jenaische Zeitschr. Naturwiss., 46 (N.F. 39): 845–896, 21 text figs., 5 pls.

Reitzenstein, W. von. 1905. Untersuchungen über die Entwicklung der Stirnaugen von Periplaneta orientalis und Cloeon. Zool. Jahrb., Anat., 21: 161–180, 8 text figs., 2 pls.

Rémy, P. 1931. Un nouveau type de pauropode: Decapauropus cuenoti, nov. gen., nov. spec. Arch. Zool. Exp. Gén., 71: 67–83, 12 figs.

Richards, A. G., and Korda, Frances H. 1947. Electron micrographs of centipede setae and microtrichia. Ent. News, 58: 141–145, 3 figs.

Richards, A. G., and Korda, Frances H. 1950. Studies on arthropod cuticle. IV. An electron microscope survey of the intima of arthropod tracheae. Ann. Ent. Soc. America, 43: 49–71, 3 pls.

Ripper, W. 1931. Versuch einer Kritik der Homologiefrage der Arthropodentracheen. Zeitschr. wiss. Zool., 138: 303–369, 18 figs.

Robinson, L. E., and Davidson, J. 1913–1914. The anatomy of Argas persicus (Oken 1818). Parasitology, 6: 20–48, 217–256, 382–424, pls. 1–5, 14–17, 25–28.

Sargent, W. D. 1937. The internal thoracic skeleton of the dragonflies (Odonata; Anisoptera). Ann. Ent. Soc. America, 30: 81–95, 2 pls.

Sargent, W. D. 1951. The flight of the dragonfly. Biol. Rev., City Coll. of New York, 13: 8–10, 1 fig.

Sayce, O. A. 1908. On Koonunga cursor, a remarkable new type of malacostracous crustaceans. Trans. Linn. Soc. London, Zool., 2d Ser., 11: 1–16, 2 pls.

Scheuring, L. 1914. Die Augen der Arachnoideen. II. (Phalangida und Araneida). Zool. Jahrb., Anat., 37: 369–464, 16 text figs., 4 pls.

Schimkewitsch, W. 1895. Über Bau und Entwicklung des Endosternits der Arachniden. Zool. Jahrb., Anat., 8: 191–216, 2 pls.

Schlottke, E. 1933. Darm und Verdauung bei Pantopoden. Zeitschr. mikro-anatom. Forschung, 32: 633–658, 14 text figs.

Schlottke, E. 1934. Histologische Beobachtungen über die intrazellulare Verdauung bei Dendrocoelum lacteum (Müll) und Euscorpius carpathicus (L.). Sitzungsb. u. Abhandl. naturf. Gesell. Rostock, Ser. 3, vol. 4: 76–86, 3 figs.

Schmidt, W. 1915. Die Muskulatur von Astacus fluviatilis (Potomobius astacus L.). Zeitschr. wiss. Zool., 113: 165–251, 26 figs.

Schulze, P. 1941. Das Geruchsorgan der Zecken. Zeitschr. Morph. Ökol. Tiere, 37: 491–564, 90 figs.

Seifert, B. 1932. Anatomie und Biologie des Diplopoden Strongylosoma pallipes Oliv. Zeitschr. Morph. Ökol. Tiere, 25: 362–507, 86 figs.

Silvestri, F. 1902. Ordo Pauropoda. Acari Myriopoda et Scorpiones hucusque in Italia reperta, 85 pp., 56 figs. Portici.

Silvestri, F. 1903. Classis Diplopoda, vol. 1, Anatome. Acari Myriopoda et Scorpiones hucusque in Italia reperta, 272 pp., 346 text figs., 4 pls. Portici.

Silvestri, F. 1950. Segmentazione del capo dei Colobognati (Diplopodi). Proc. Eighth Internat. Congr. Ent. Stockholm, pp. 571–576, 5 figs.

Smith, C. N., Cole, M. M., and Gouck, H. K. 1946. Biology and control of the American dog tick. U.S. Dep. Agr., Tech. Bull. 905, 74 pp., 37 figs.

Smith, G. 1909. On the Anaspidacea, living and fossil. Quart. Journ. Micr. Sci., 53: 489–578, 62 text figs., 2 pls.

Snodgrass, R. E. 1935. Principles of Insect Morphology, 667 pp., 319 figs. New York and London.

Snodgrass, R. E. 1948. The feeding organs of Arachnida, including mites and ticks. Smithsonian Misc. Coll., 110, No. 10, 93 pp., 29 figs.

Snodgrass, R. E. 1951. Comparative studies on the head of mandibulate arthropods, 181 pp., 37 figs. Ithaca, N.Y.

Starling, J. H. 1944. Ecological studies of the Pauropoda of the Duke Forest. Ecological Monographs, 14: 291–310, 21 figs.

Stiles, C. W. 1910. The taxonomic value of the microscopic structure of the stigmal plates in the tick genus Dermacentor. Hygienic Lab. Bull. No. 62, 67 pp., 43 pls.

Störmer, L. 1933. Are the trilobites related to the arachnids? American Journ. Sci., 26: 147–157, 2 figs.

Störmer, L. 1934. Merostomata from the Downtonian sandstone of Ringerike, Norway. Skrift. Utg. Norske. Videnskaps-Akad., Oslo. Mat.-Naturv., 1933, No. 10, 125 pp., 39 text figs., 12 pls.

Störmer, L. 1936. Eurypteriden aus dem rheinischen Unterdevon. Abhandl. preussischen geol. Landesanstalt, n.f. 175: 1–74, 10 text figs., 12 pls.

Störmer, L. 1939. Studies on trilobite morphology. Part I. The thoracic appendages and their phylogenetic significance. Norsk Geol. Tidssk., 19: 143–273, 35 text figs., 12 pls.

Störmer, L. 1942. Studies on trilobite morphology. Part II. The larval development, the segmentation and the sutures, and their bearing on trilobite classification. Norsk Geol. Tidssk., 21: 49–164, 2 pls., 19 text figs.

Störmer, L. 1944. On the relationships and phylogeny of fossil and recent Arachnomorpha. Skrift. Utg. Norske Videnskaps-Akad., Oslo. Mat.-Naturv., 1944, No. 5, 158 pp., 30 figs.

Tiegs, O. W. 1940. The embryology and affinities of the Symphyla, based on a study of Hanseniella agilis. Quart. Journ. Micr. Sci., 82: 1–225, 41 text figs., 9 pls.

Tiegs, O. W. 1947. The development and affinities of the Pauropoda, based on a study of Pauropus silvaticus. Quart. Journ. Micr. Sci., 88: 165–336, 29 text figs., 11 pls.

Totze, R. 1933. Beiträge zur Sinnesphysiologie der Zecken. Zeitschr. vergl. Physiol., 19: 110–161, 38 figs.

True, G. H. 1932. Studies on the anatomy of the Pajaroella tick, Ornitho-

dorus coriaceus Koch. I. The alimentary canal. Univ. California Pubs.
Ent., 6: 21—48, 17 text figs., 3 pls.

Verhoeff, K. W. 1906. Vergleichend-morphologische Studie über coxo-
pleuralen Korperteile der Chilopoden. Nova Acta. Acad. Leop. Carol.,
86, No. 2, 153 pp., 44 figs.

Verhoeff, K. W. 1926—1931. Diplopoda. In Bronns Klassen und Ordnungen des Tierreichs, 5, Abt. 2, Buch 2, Lieferungen, 1—13.

Verhoeff, K. W. 1933. Symphyla. In Bronns Klassen und Ordnungen des
Tierreichs, 5, Abt. 2, Buch 3, Lief. 1, pp. 1—120, 62 figs.

Versluys, J., and Demoll, R. 1920. Die Verwandtschaft der Merostomata
mit den Arachnida und den anderen Abteilungen der Arthropoda. Proc.
K. Akad. Wetens. Amsterdam, 23: 739—765, 6 figs.

Vitzthum, H. G. 1931. Acari—Milben. In Kükenthal and Krumbach,
Handbuch der Zoologie, 3, second half: 1—160, 161 figs.

Vitzthum, H. G. 1940—1943. Acarina. In Bronns Klassen und Ordnungen
des Tierreichs, 5: 1—1011, 498 figs.

Voges, E. 1916. Myriapodenstudien. Zeitschr. wiss. Zool., 116: 75—
135, pls. 3—5.

Walcott, C. D. 1908. Mount Stephen rocks and fossils. Canadian Alpine
Journ., 1, No. 2: 232—248, 4 pls.

Walcott, C. D. 1910. Olenellus and other genera of the Mesonacidae.
Smithsonian Misc. Coll., 53, No. 6: 231—422, 22 pls.

Walcott, C. D. 1916. Cambrian geology and paleontology III. No. 5,
Cambrian trilobites. Smithsonian Misc. Coll., 64, No. 5: 303—456, 23
pls.

Warburg, Elsa. 1925. The trilobites of the Leptaena limestone in Dalarne.
Bull. Geol. Inst. Univ. Upsala, 17: 1—450, 11 pls.

Weiss, S. 1923. Untersuchungen über die Lunge und die Atmung der
Spinnen. Zool. Jahrb., allg. Zool., 39: 535—545, 2 figs.

Wernitzsch, W. 1910. Beiträge zur Kenntnis von Craspedosoma simile
und des Tracheensystems der Diplopoden. Jenaische Zeitschr. Naturw.,
46 (N.F. 39): 225—284, 12 text figs., 2 pls.

Widmann, E. 1908. Über den feineren Bau der Augen einiger Spinnen.
Zeitschr. wiss. Zool., 90: 258—312, 4 text figs., 3 pls.

Willem, V. 1892. Les ocelles de Lithobius et de Polyxenus. Ann. Soc. R.
Malacologique Belgique, 27: 69—71, 1 fig.

Williams, S. R. 1907. Habits and structure of Scutigerella immaculata
Newport. Proc. Boston Soc. Nat. Hist., 33: 461—485.

Wirén, E. 1918. Zur Morphologie und Phylogenie der Pantopoden.
Zool. Beidrag Uppsala, 6: 41—181, 40 text figs., 8 pls.

With, C. J. 1904. The Notostigmata, a new suborder of Acari. Videnskab.
Medd. Naturhist. Foren. Kjöbenhavn, 56: 137—192, 3 pls.

REFERENCES

Yeager, J. F., and Hendrickson, G. O. 1934. Circulation of blood in wings and wing pads of the cockroach, Periplaneta americana Linn. Ann. Ent. Soc. America, 27: 257–272, 1 fig.

Ziegler, H. E. 1907. Die Tracheen bei Julus. Zool. Anz., 31: 776–782, 3 figs.

SUBJECT INDEX

[Numbers in boldface type refer to text figures.]

353

SUBJECT INDEX

[Numbers in boldface type refer to text figures.]

353

AUTHOR INDEX

[Numbers in boldface type refer to text figures.]